VLSI and Modern Signal Processing

PRENTICE-HALL INFORMATION AND SYSTEM SCIENCES SERIES

Thomas Kailath, *Editor*

VLSI and Modern Signal Processing

S.Y. KUNG, H.J. WHITEHOUSE AND T. KAILATH

Editors

PRENTICE-HALL, INC., *Englewood Cliffs, New Jersey 07632*

Library of Congress Cataloging in Publication Data
Main entry under title:

VLSI and modern signal processing.

    Includes bibliographical references and index.
    1. Signal processing—Digital techniques—Addresses,
essays, lectures.   2. Integrated circuits—Very large
scale integration–Addresses, essays, lectures.
I. Kung, S. Y. (Sun Yuan) II. Whitehouse, H. J.
(Harper J.) III. Kailath, Thomas.
TK5102.5.V18 1985   621.38′043   84-11512
ISBN 0-13-942699-X

Editorial/production supervision: Virginia Huebner
Cover design: Photo Plus Art
Manufacturing buyer: Anthony Caruso

Printed in the United States of America

10 9 8 7 6 5 4 3 2 1

ISBN 0-13-942699-X 01

Prentice-Hall International, Inc., *London*
Prentice-Hall of Australia Pty. Limited, *Sydney*
Editora Prentice-Hall do Brasil, Ltda., *Rio de Janeiro*
Prentice-Hall Canada Inc., *Toronto*
Prentice-Hall of India Private Limited, *New Delhi*
Prentice-Hall of Japan, Inc., *Tokyo*
Prentice-Hall of Southeast Asia Pte. Ltd., *Singapore*
Whitehall Books Limited, *Wellington, New Zealand*

Contents

PART II

CONCURRENT ARRAY PROCESSORS: ARCHITECTURE AND LANGUAGES

PART III

APPLICATION OF CONCURRENT ARRAY PROCESSORS

Preface

The rapidly growing interaction between VLSI technology and modern signal processing has a natural basis. The ever-increasing demands of speed and performance in modern signal processing clearly point to the need for tremendous computation capability. On the other hand, low-cost, high-density, fast VLSI devices make such super-computing increasingly practical, in terms of volume, speed, and cost. As a result, VLSI microelectronics technology has aroused many innovative perspectives in signal processing, precipitating a new era of VLSI signal processing. This prospective trend is now becoming a major focus of attention for government, industry, and academia. In just the last five years, there has been a dramatic worldwide growth in research and development efforts on mapping VLSI into various signal processing applications.

In striving for a useful treatment of VLSI signal processing, it is crucial to have a fundamental understanding of the computational and architectural needs of modern signal processing as well as of the potential and the constraints of VLSI devices and CAD technology. The future trend of VLSI signal processing research is bound to be highly cross-disciplinary, involving close interactions between VLSI, computer engineering, and signal processing. In line with this new trend, this book puts together several selected articles addressing three major themes:

Part I: Modern Signal Processing (MSP) in VLSI

Part II: VLSI-oriented Concurrent Array Processors (CAPs)

Part III: Application of Concurrent Array Processors

Part I describes some of the new theory and new algorithms of modern signal processing, taking VLSI's potential impact into account. Fueled by the potential of the abundant computing hardware in VLSI, many leading researchers are now enthusiastically advocating raising the level of sophistication in signal processing methods. The new trend is to enhance, in both hardware and software, the currently prevailing computations (such as FFT or convolution) with more sophisticated linear algebraic or nonlinear adaptive methods. Forthcoming VLSI computing hardware packages are likely to include modern techniques such as Kalman filtering, adaptive transversal or lattice filtering, SVD-based spectrum analysis, two-dimensional processing, and so on. This will definitely boost the ultimate goal of achieving optimal performance and high-speed (real-time) signal processing. The areas of linear prediction and modeling, high-resolution spectral analysis, and phased-sensor processing arrays represent a major portion of advanced research in modern signal processing. Part I includes several articles focusing on high-performance processing techniques recently developed in the literature.

Part II deals primarily with novel VLSI architecture designs for concurrent array processors (CAPs). The practicality of the modern signal processing methods discussed in Part I depends critically on the compatibility of the state-of-the-art computing hardware. General-purpose super-computers, very often incurring unnecessary system overhead, are not suitable for handling computations in real time. Fortunately, most processing methods mentioned above fall into the classes of FFT, digital filter, or matrix-type algorithms. With the exception of the FFT, most of the algorithms above possess some useful common properties, such as regularity, recursiveness, and localized data transactions. These properties may be efficiently utilized so as to maximize the potential of VLSI technology and to circumvent its limitations. This provides the theoretical footing for the design of locally interconnected CAPs, such as systolic, wavefront, or engagement processors. New issues of interest are the reconfigurability, fault-tolerant design, and partitioning of CAPs. The advance of VLSI architectures also challenges programming language theory to describe precisely as well as to take advantage of the concurrency hidden in computational problems. These fundamental issues constitute the major theme of Part II.

Part III addresses the application of concurrent array processor concepts to signal processing. In the past, major work has been done on mapping various signal processing applications into specific VLSI architectures, and vice versa. The insight gained from such experiences will certainly greatly enhance the understanding of VLSI's impact on signal processing. Nevertheless, with a better theoretical understanding, and with closer interaction among the different types of applications, a much greater impact may be expected. To this end, Part III includes articles concerning implementation of processor chips and FFT-type signal processors and various applications of concurrent array processors. Important application areas emphasized in this part are image processing, image understanding, and pattern recognition.

This book is an outgrowth of the USC Workshop "VLSI and Modern Signal Processing," held in Los Angeles, November 1–3, 1982, cosponsored by the University of Southern California and the Office of Naval Research. Throughout the entire process of organizing the workshop and putting together these final manuscripts, we have been fortunate to have had the enthusiastic support of colleagues, reviewers, and authors. We wish to express our thanks to all of them. Especially, we would like to thank Ms. Linda B. Varilla of USC for her dedication and efficiency in assisting with this editorial work.

S. Y. Kung, H. J. Whitehouse and T. Kailath

List of Contributors

H. M. AHMED
Codex Corporation
20 Cabot Boulevard
Mansfield, MA 02048

K. S. ARUN
University of Southern California
Department of Electrical Engineering—Systems
University Park
Los Angeles, CA 90089

D. V. BHASKAR RAO
University of Southern California
Department of Electrical Engineering—Systems
University Park
Los Angeles, CA 90089

G. BIENVENU
Thomson-CSF
Division des Activités Sous-Marines
B.P. 53
06801 Cagnes-sur-Mer CEDEX
France

K. BROMLEY
Naval Ocean Systems Center
San Diego, CA 92152

J. J. CHANG
Department of Electrical Engineering
University of Southern California
Los Angeles, CA 90007

C. CHEN
Computer Science Department
California Institute of Technology
Pasadena, CA 91125

A. B. CREMERS
Computer Science Department
University of Dortmund
P.O. Box 50 05 00

D-4600 Dortmund 50
Federal Republic of Germany

G. CULLER
CHI Systems, Inc.
Goleta, CA 93117

E. DEPRETTERE
Department of Electrical Engineering
Delft University of Technology
Mekelweg 4
2628 CD Delft
The Netherlands

P. DEWILDE
Department of Electrical Engineering
Delft University of Technology
Mekelweg 4
2628 CD Delft
The Netherlands

A. L. FISHER
Department of Computer Science
Carnegie-Mellon University
Pittsburgh, PA 15213

K.-S. FU
School of Electrical Engineering
Purdue University
West Lafayette, IN 47907

B. K. GILBERT
Mayo Foundation
Rochester, MN 55900

E. GREENWOOD
Government Electronics Group
Motorola, Inc.
Scottsdale, AZ 85200

G. HALLNOR
TRW Defense Systems Group
7600 Colshire Drive
McLean, VA 22102

D. HARRISON
Government Electronics Group
Motorola, Inc.
Scottsdale, AZ 85200

D. HELLER
Physics and Computer Science Department
Shell Development Company
Bellaire Research Center
P.O. Box 481
Houston, TX 77001

T. N. HIBBARD
Jet Propulsion Laboratory
California Institute of Technology
4800 Oak Grove Drive
Pasadena, CA 91103

K. HWANG
School of Electrical Engineering
Purdue University
West Lafayette, IN 47907

T. KAILATH
Information Systems Laboratory
Department of Electrical Engineering
Stanford University
Stanford, CA 94305

T. M. KINTER
Mayo Foundation
Rochester, MN 55900

G. E. KOPEC
Fairchild Laboratory for Artificial Intelligence
  Research
Palo Alto, CA 94304

L. M. KRUEGER
Mayo Foundation
Rochester, MN 55900

P. J. KUEKES
ESL, Inc.
Sunnyvale, CA 94086

R. H. KUHN
Department of EECS
Northwestern University
Evanston, IL 60201

H. T. KUNG
Department of Computer Science
Carnegie-Mellon University
Pittsburgh, PA 15213

S.-Y. KUNG
University of Southern California

Department of Electrical Engineering—Systems
University Park
Los Angeles, CA 90089

C. A. MEAD
Computer Science Department
California Institute of Technology
Pasadena, CA 91125

H. MERMOZ
Ingénieur Général des Télécommunications
582, Chemin de la Calade
83140 Six Fours Les Plages
France

J. G. NASH
Hughes Research Laboratories
Malibu, CA 90265

B. A. NAUSED
Mayo Foundation
Rochester, MN 55900

R. NOUTA
Department of Electrical Engineering
Delft University of Technology
Mekelweg 4
2628 CD Delft
The Netherlands

G. R. NUDD
Hughes Research Laboratories
Malibu, CA 90265

N. L. OWSLEY
Naval Underwater Systems Center
New London, CT 06320

B. N. PARLETT
Department of Mathematics
Computer Science Division of the Department of
  Electrical Engineering and Computer Science
University of California at Berkeley
Berkeley, CA 94720

I. S. REED
Department of Electrical Engineering
University of Southern California
Los Angeles, CA 90007

A. ROSENFELD
Computer Vision Laboratory
Computer Science Center
University of Maryland
College Park, MD 20742

R. SCHREIBER
Department of Computer Science

Stanford University
Stanford, CA 94305

D. J. Schwab
Mayo Foundation
Rochester, MN 55900

H. M. Shao
Department of Electrical Engineering
University of Southern California
Los Angeles, CA 90007

L. Snyder
Purdue University
West Lafayette, IN 47906

J. M. Speiser
Naval Ocean Systems Center
San Diego, CA 92152

E. E. Swartzlander, Jr.
TRW Defense Systems Group
One Space Park
Redondo Beach, CA 90278

R. H. Travassos
Systolic Systems, Inc.
1408 Petal Way
San Jose, CA 95129

T. K. Truong
Communication System Research
Jet Propulsion Laboratory

California Institute of Technology
Pasadena, CA 91103

L. Uhr
Department of Computer Sciences
University of Wisconsin
Madison, WI 53706

W. Van Nurden
Mayo Foundation
Rochester, MN 55900

B. W. Wah
School of Electrical Engineering
Purdue University
West Lafayette, IN 47907

H. J. Whitehouse
Naval Ocean Systems Center
San Diego, CA 92152

R. Wood
CHI Systems, Inc.
Goleta, CA 93117

C.-S. Yeh
Department of Electrical Engineering
University of Southern California
Los Angeles, CA 90007

R. Zucca
Rockwell International Microelectronics
Research and Development Center
Thousand Oaks, CA 91300

Part I

SIGNAL PROCESSING: THEORY AND ALGORITHMS

Modern signal processing is characterized by a growing interplay among several fields: signal analysis, system theory, statistical methods, and numerical analysis. The very large scale integration (VLSI) revolution is forcing closer cooperation between the integrated circuits and computer science disciplines. The bringing together of VLSI and modern signal processing makes for an even more exciting blend of all these disciplines. In each of them, it forces us to look again at the traditional solutions, which were "optimum" in their own settings.

This is the main theme of Chapter 1, by Kailath, which gives several illustrations of the fruitful interplay that can exist between mathematical solutions and currently promising implementation technologies. An example that is considered is that of the popular Levinson algorithm used in the linear predictive coding (LPC) technique for speech analysis and synthesis. The Levinson algorithm exploits certain stationarity assumptions to reduce the computations for an nth-order synthesis model from $O(n^3)$ to $O(n^2)$. The number of computations will be proportional to the calculation time if we use a single processor. With the advent of VLSI it is natural to ask if parallel processing can be used to speed up the computation. It turns out that because the Levinson algorithm requires the formation of inner products of n-dimensional vectors, even with n processors operating in parallel the computation time can only be reduced from $O(n^2)$ to $O(n \log n)$. However, by using a lesser known (but older) alternative to the Levinson algorithm, the inner products can be avoided and the computation time with n processors reduced to $O(n)$. Moreover, the structure is quite amenable to VLSI implementation, and in fact a prototype VLSI chip has been designed (see Chapter 17).

This alternative, the Schur algorithm, has many other interesting properties, including a close relationship to transmission-line theory. In the VLSI era, when memory costs are declining steeply, transmission-line models of linear systems turn out to be useful alternatives to the currently more popular, at least in control and system theory, minimal (memory unit) state-space models. Some of these points are described in Chapter 1 and the references therein.

A surprising feature of modern signal processing is the wide range of mathematical ideas that arise in apparently straightforward problems. However, a basic core consists of linear algebra and linear operator theory. Specific tasks that need to be performed in real time in modern signal processing systems include matrix multiplication for covariance estimation, solution of linear equations for adaptive processing, and eigenstructure computation for high-resolution direction finding and adaptive beamforming. For each input data vector of N samples, these tasks typically require of the order of N^3 or N^4 elementary operations (multiplications and additions). To perform these computations at a sufficient rate to keep up with the input data flow is difficult even for only modest bandwidth applications. Chapter 2, by Whitehouse, Speiser, and Bromley, starts by quantifying these issues and then discusses concurrent array processor architectures that offer a promising solution to these problems.

However, there is an important caution that must be observed in plunging ahead to invent elegant new parallel computing structures for a host of linear algebraic calculations. This is a lesson learned somewhat painfully in digital filter theory over the last two decades: that close attention must be paid to such issues as the propagation of round-off errors in even simple numerical calculations, and the occurrence of overflow oscillations and limit-cycle phenomena in digital filters. It has been found [see S. Y. Kung, Proc. IEEE, (1984)] that several proposed parallel systolic arrays for matrix-vector and matrix-matrix multiplication, recurrence evaluation, GCD computation, and so on, are rearrangements of direct-form realizations of digital filter theory. The main consequence of the rearrangement is "pipelineability," which permits higher data-throughput rates. However, it is well known in digital filter theory that direct-form realizations have poor numerical properties and therefore are generally avoided in favor of cascade realizations. Cascades of orthogonal sections are particularly desirable because they completely avoid limit-cycle and overflow oscillation problems. It has been shown that pipelineable versions of cascade orthogonal filters can be obtained and may be quite effective in practice. Chapter 15 and a paper of Rao and Kailath [46] describe the computation and analysis of such structures.

Numerical issues of other kinds enter strongly into the problem of eigenvalue and eigenvector computation, especially for large matrices. Chapter 3, by Kung *et. al.*, Chapter 4, by Owsley, and Chapter 5, by Bienvenu, describe the evolution and present status of high-resolution spectral estimation and direction-finding techniques. These new direction-finding and closely related adaptive beamforming methods have been shown to be very effective. It is striking that they rely heavily on the computation of specific (minimum) eigenvalues and eigenvectors.

The eigenvalue problem has been of intense interest to numerical analysts since at least 1947, when eigenvalues began to be computed on desktop calculators. In Chapter 6, which concludes this section, Parlett makes some very cogent remarks on many aspects of this problem, including the provocative (at this time) issue of how effective parallel computation can be for extracting useful information about eigenvalues and singular values.

THOMAS KAILATH

1

Signal Processing in the VLSI Era

Thomas Kailath

Stanford University
Stanford, California

1.1 INTRODUCTION

Nearly two decades ago, the reintroduction by Cooley and Tukey of the fast Fourier transform (FFT) and the growing expertise in digital-circuit technology led to a revolution in signal processing. Among other things, this revolution was marked by the growth of the new subject called digital filtering and digital signal processing (see, e.g., the well-known textbook of Oppenheim and Schafer, Prentice-Hall, 1973).

Just over a decade ago, the pioneering work of Atal and his colleagues at the Bell Laboratories, and of Itakura and Saito in Japan, reintroduced time-domain ideas into the field of speech analysis and synthesis (see, e.g., [45]). Combined with the growth of integrated-circuit technology, this development is heralding another revolution in signal processing, as witnessed for example by the Texas Instruments' Speak-and-Spell chip and the mounting interest in parallel (or highly concurrent) algorithms and architectures (see, e.g., the special January 1982 issue of *IEEE Computer Magazine* on High-Speed Parallel Computing).

A major characteristic of the new direction is the emphasis on formulating optimization criteria for signal processing problems. Without tying ourselves beforehand to specific implementations, the aim is to set up a mathematical model with a specific performance criterion, satisfaction of which will specify the signal processing operations that need to be carried out. Of course, there are always limitations in setting up realistic but not overly complicated mathematical models and tractable optimization criteria. However, these issues have already been met and encouragingly resolved in the fields of statistical communications, control, and system identification.

The special constraints of signal processing problems often make it desirable and feasible to impose linearity as a major desideratum, and therefore a substantial part of current signal processing theory and practice deals ultimately with the solution of sets of linear equations. This may sound disappointing, because such problems have been studied for so long that there would seem to be very little left to say about them. But the point is that in most interesting physical problems, direct numerical solution of the equations is of little value; our interest is really in the special structure of the equations and in ways of matching current technological tools to these structures so as to reduce the computational burden of determining the solutions and the technological burden of implementing them. Thus at one time, formulation in terms of FFTs and array processor computer configurations had a dramatic impact on traditional notions of how to compute empirical spectral estimates. Similarly, state-space models in control theory brought about a revolution in the 1960s in our ability to handle large data-processing problems, by introducing the notion that it was enough to keep track of the evolution of a fixed finite-dimensional Markovian state vector. Nowadays the cheap availability of memory devices makes it less pressing to summarize past information in a fixed-dimension state vector; it can now often be easier to allow a growing memory, allowed to be as large as the data itself. We give some illustrations in this chapter. So also the emerging family of floating-point arithmetic chips may soon be changing some of our notions of what constitutes good algorithms for specific problems. Many tricks used to circumvent finite precision effects become irrelevant when computation in double-precision form becomes routinely available.

In this chapter we attempt to illustrate some of the interplay that is emerging between signal processing algorithms and integrated-circuit technologies. In Section 1.2 we discuss some signal processing problems associated with stationary stochastic models, or equivalently, problems leading to linear equations with Toeplitz coefficient matrices. This special structure was one of the first to be assumed in statistical signal processing, in the work of Wiener and Levinson in the 1940s and 1950s. However, technological constraints on growing memory implementations led, as mentioned above, to an emphasis in the 1960s and 1950s on fixed-memory (state-space) Kalman filter types of solutions (see, e.g., [26]). The Levinson algorithm continued to be used in geophysical (seismic) signal processing problems, where large off-line computing resources were feasible; in the late 1960s it was rediscovered by the speech synthesis community (Atal, Schroeder, Itakura, Saito, and others). We shall elaborate on their results in Section 1.2 by showing how the Toeplitz structure turns out to have very nice consequences from the viewpoint of current technologies—VLSI and perhaps also optical and wave technologies.

Despite its many nice consequences, there are often situations where the assumption of stationarity is unsatisfactory. We illustrate this in Section 1.3 and explain how some relatively new theories of the displacement structure of processes and matrices can allow us to use additional memory to effectively restore the benefits of stationarity to several nonstationary signal processing problems.

Needless to say, for a variety of reasons our presentation will be sketchy and without proofs, which can be found in the references. Our aim is to illustrate some aspects of the continuing interplay between signal processing theory and signal processing technology. This volume contains several other examples of this theme.

1.2 SOME ALGORITHMS FOR TOEPLITZ MATRICES

The by-now well-known linear predictive coding (LPC) methods for speech analysis are based on the hypothesis that speech waveforms can be modeled as the output of a linear time-invariant filter-driven white noise. This is predicated on the fact, first clearly expressed by Norbert Wiener, that information-bearing processes are inherently random. For speech analysis, the air issuing from the lungs can be thought of as white noise which is modulated by the vocal system (vocal codes, vocal tract, nose, and lips) to produce the speech waveform.

The linear time-invariant filter can be modeled in many ways, but for a variety of reasons early attention was focused on using "all-pole" models, or equivalently on modeling speech as a stationary autoregressive discrete-time random process:

$$y_t + A_{N,1} y_{t-1} + \cdots + A_{N,N} y_{t-N} = e_{N,t}$$

where $\{e_{N,t}\}$ is a zero-mean white-noise process. The modeling problem is to choose the order N, the coefficients $\{A_{N,i}\}$, and the noise variance, $R_{e,N}$ say, so as best to fit the observed speech signal $\{y_t, t \geq 0\}$.

The standard procedure is to form the "sample covariance" estimate of the second-order statistic,

$$R_l = E[y_t y_{t+l}]$$

of the stationary process $\{y_t, t \geq 0\}$, and to notice that the coefficients $\{A_{N,i}\}$ can be obtained by solving the Yule–Walker equations

$$[A_{N,N} \quad \cdots \quad A_{N,1} \quad I] \begin{bmatrix} R_0 & R_1 & \cdots & R_N \\ R_1 & R_0 & & \vdots \\ \vdots & & \ddots & R_1 \\ R_N & \cdots & R_1 & R_0 \end{bmatrix} = [0 \quad \cdots \quad 0 \quad I] \qquad (1.1)$$

We shall examine the assumptions underlying this formulation in more detail in Section 1.3, and in fact we shall find some room for disagreement with it. However, one reason the formulation above is popular is that the special constant-along-diagonals (Toeplitz) nature of the coefficient matrix in equation (1.1) lends itself to a convenient fast recursive solution algorithm—the Levinson algorithm:†

$$\begin{bmatrix} A_{N+1}(z) \\ B_{N+1}(z) \end{bmatrix} = \begin{bmatrix} z & -k_{N+1} \\ -zk_{N+1}z & 1 \end{bmatrix} \begin{bmatrix} A_N(z) \\ B_N(z) \end{bmatrix} \qquad \begin{bmatrix} A_0(z) \\ B_0(z) \end{bmatrix} = \begin{bmatrix} 1 \\ 1 \end{bmatrix} \qquad (1.2)$$

---

†Apparently first given by Levinson in 1947 [37]; several other names can be associated with this algorithm, especially Durbin [18]. We may note that the recursions in (1.2) are those of the classical Szegö [57] orthogonal polynomials on the unit circle.

where

$$A_N(z) = A_{N,N} + A_{N,N-1}z + \cdots + A_{N,1}z^{N-1} + z^N$$

$$B_N(z) = \text{the "reverse polynomial" of } A_N(z)$$

$$= A_{N,N}z^N + A_{N,N-1}z^{N-1} + \cdots A_{N,1}z + 1$$

and

$$k_{N+1} = \frac{A_{N,N}R_1 + A_{N,N-1}R_2 + \cdots + A_{N,1}R_N + R_{N+1}}{R_N^e} \tag{1.3a}$$

$$R_{N+1}^e = R_N^e(1 - k_{N+1}^2) \qquad R_0^e = R_0 \tag{1.3b}$$

The point is that we successively build up the solutions $\{A_{N,i}, N = 0, 1, \ldots\}$, with the major computational burden being that of forming the "inner product" for k_{N+1}, which requires N multiplications and N additions. Therefore, computing $\{k_1, \ldots, k_N\}$ will require of the order of

$$1 + 2 + \cdots + N = \frac{N(N+1)}{2}$$

or $O(N^2)$ elementary computations, which is an order of magnitude less than the $O(N^3)$ computations required to solve an arbitrary set of linear equations, that is, one without the special Toeplitz structure of (1.1).

The Levinson algorithm can be used to produce a family of autoregressive models of increasing order, and we have to decide in some way on the appropriate order. There are various tests (e.g., Akaike's information criterion) and other considerations (e.g., practical design limits on integrated-circuit implementations) involved in this choice, which we shall not elaborate here. What we do wish to emphasize is that Levinson's algorithm has a special structure that allows us some flexibility in making a decision on the order.

The traditional way of implementing a filter to compute $e_{N,t}$ is via a transversal (tapped-delay-line) filter with coefficients $\{A_{N,0}, \ldots, A_{N,N}\}$ (see Figure 1.1); we should remark, to avoid confusion, that the transfer function of the filter that computes $e_{N,t}$ is not $A_N(z)$ but rather

$$A_{N,N} + A_{N,N-1}z^{-1} + \cdots + z^{-N} = z^{-N}A_N(z)$$

However, if we were uncertain about our choice of order and wished to compute, say, $e_{N+2,t}$, we would have not only a longer transversal filter but we would have to reset all the tap-gain coefficients from $\{A_{N,0}, \ldots, A_{N,N}\}$ to $\{A_{N+2,0}, \ldots, A_{N+2,N+2}\}$. There are many examples in which it is desirable to compute the filter response over a range of values of N before deciding on a fixed order, and this is not convenient with transversal filter implementations.

Now the Levinson recursion (1.2) shows that the $\{k_i, i = 1, \ldots, N\}$, together with the fact that $A_0(z) = B_0(z) = 1$, provide an alternative parametrization of the filter: knowing these values allows us to construct $A_N(z)$, and hence the coefficients

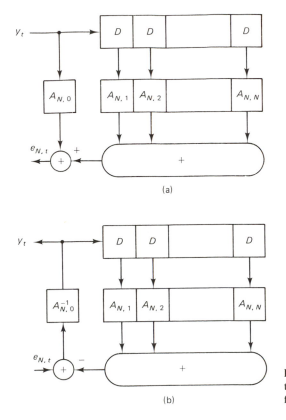

Figure 1.1 Tapped-delay-line implementation: (a) whitening filter; (b) modeling filter.

$\{A_{N,i}\}$, from (1.2). If we wish to go to a different order, say $N + 2$, in this parameterization we need only to add two more coefficients $\{k_{N+1}, k_{N+2}\}$, without having to change any of the earlier values. This invariance property of the $\{k_i\}$ parametrization can be exploited by using a different implementation of the filter in terms of the $\{k_i, 1 \le i \le N\}$ rather than the $\{A_{N,i}\}$. Examination of (1.2) suggests that we can build the filter as a cascade of "lattice sections," as in Figure 1.2(a), or in a certain *normalized* [see (1.5) below] form, as in Figure 1.2(b).

It is easy to calculate the inverse of the lattice filters in Figure 1.2 by using signal flow graph rules. The result of doing this for Figure 1.2(b) is shown in Figure 1.3 and provides the "modeling" filter. It has the form of a discrete transmission line and helps explain why the $\{k_i\}$ are often called *reflection coefficients*.

This physical interpretation suggests that we must have

$$|k_i| \le 1 \qquad (1.4)$$

a fact that also follows from (1.3b) (since the variances $\{R_N^e\}$ must be nonnegative). There is a corresponding constraint on $A_N(z)$: the roots of the polynomial $z^N A(z)$ must lie within the unit circle, but this condition does not translate easily into

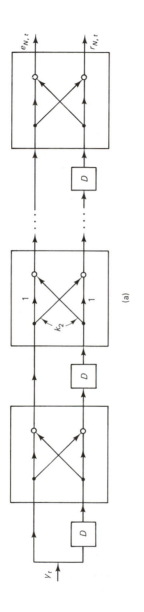

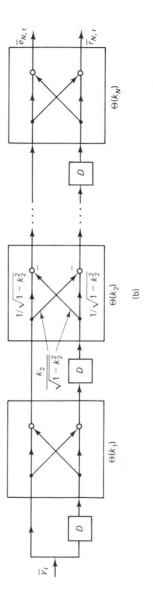

Figure 1.2 Lattice filter implementations of whitening filters; (a) unnormalized form; (b) normalized form.

10

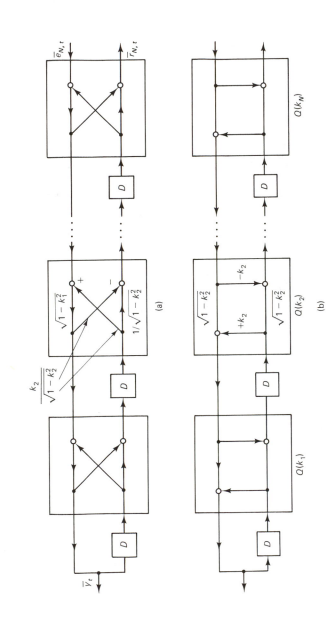

Figure 1.3 Modeling filters: (a) feedback lattice form—normalized; (b) transmission line form.

11

bounds on the coefficients $\{A_{N,i}\}$. For this and other reasons the numerical properties of the (normalized) lattice filter tend to be better than those of the transversal filter. Therefore, it may not be surprising that Texas Instruments chose to use this structure when building their very successful speech synthesis chip [60]; the modular structure, local interconnections, and rhythmic data flow all make for convenient VLSI implementation [27,28,30,39].

1.2.1 CORDIC Implementations

Each section in Figure 1.2(a) requires two multiplications and two additions; the multiplications tend to be more expensive and therefore rearrangements have been devised that use scaling to manage with only one multiplication (see, e.g., [38, Sec. 5.5]). For numerical reasons, one often goes to a section with four multipliers, as in Figure 1.2(b). Somewhat ironically, it turns out that this section can, in fact, be implemented without using any explicit multiplications at all.

To introduce this, note first that by exploiting the condition $|k| < 1$, a typical lattice section can be written in a normalized form, as in fact was used in Figure 1.2(b):

$$\boldsymbol{\Theta} = \frac{1}{\sqrt{1 - k^2}} \begin{bmatrix} 1 & -k \\ -k & 1 \end{bmatrix} \tag{1.5a}$$

The reason for the name is the property

$$\boldsymbol{\Theta} J \boldsymbol{\Theta}^* = J \qquad J = \begin{bmatrix} 1 & 0 \\ 0 & -1 \end{bmatrix} \tag{1.5b}$$

where the * denotes (conjugate) transpose. This property implies that the matrix $\boldsymbol{\Theta}$ preserves the "length" of any vector $u^* = [u_1^* \quad u_2^*]$ that it operates on, provided that length is measured in the J metric, that is,

$$\|\boldsymbol{\Theta}u\|^2 = \|u\|^2 \qquad \|u\|^2 = \|u_1\|^2 - |u_2|^2$$

The matrix $\boldsymbol{\Theta}$ is often called *J-orthogonal*, and its action described as a *J-rotation*, or *a hyperbolic rotation*, in contrast to the usual circular rotation.

Now, as pointed out by Lee [31], J. Volder had as early as 1959 [58] proposed a simple scheme for implementing rotations (hyperbolic, circular, and linear) by using what he called "coordinate digital computers" or CORDICs. These schemes use a circuit built only of registers, shifters, and adders to iteratively compute functions such as $\sqrt{x^2 + y^2}$, $\tan^{-1}(y/x)$, $\sqrt{x^2 - y^2}$, $\tanh^{-1}(y/x)$, $x \cos \alpha - y \sin \alpha$, and so on; they have been employed for this purpose in several pocket calculators, for example, the HP-35 (see [11,54,59]).

In recent years, CORDICs have been applied to a variety of one- and two-dimensional VLSI computing structures of the systolic and wavefront array types (see, e.g., [2,46,47], as well as Chapters 15 and 16 of this volume. Despain [14,15] was one of the first to draw attention to CORDICs for signal processing of the FFT type, and more recently Haviland and Tuszynski [20] have described a VLSI CORDIC chip for arithmetic calculations. We should note that there are several factors (speed, accuracy, design overhead, etc.) that have to be considered before settling on an actual implementation via CORDICs or via some other schemes (e.g., array multipliers, perhaps coupled with stored trigonometric tables, etc.).

Our point here is mainly that recognition of the special mathematical structure of the lattice section has suggested a new macro building block for signal processing applications.

1.2.2 Parallel Computation and the Schur Algorithm

A natural question with VLSI technology is to explore the question of how much the determination of the reflection coefficients can be speeded up by using parallel computation—say with N processors working together. The hope is that if a processor takes one unit of time for each elementary computation, we would hope to use N processors to obtain the answer in time $O(N)$ compared to $O(N^2)$ with a single serial processor.

However, it is not hard to see that a parallel implementation of the Levinson algorithm will take time $O(N \log N)$ because of the inner product operation needed to form the $\{k_i\}$; N processors can carry out the N multiplications needed to form k_{N+1} in parallel in one time unit, but the additions required will take at least $\log N$ steps.

This seems disappointing, but there is hope. It turns out that there is a mathematically equivalent algorithm, traceable to Schur [55], that avoids the inner product step in forming the $\{k_i\}$ and thus will allow implementation in time $O(N)$ with N processors.

The Schur algorithm was originally presented in 1917 as a test to see if a power series was analytic and bounded in the unit disk. In our problem it reduces to a simple three-step procedure (see [17,35]) for the computation of the $\{k_i\}$ associated with a covariance sequence $\{R_0, R_1, \ldots, R_N\}$, $R_0 = 1$.

1. Start with a generator matrix (a superscript * denotes a matrix transpose)

$$G_0^* := \begin{bmatrix} R_0 & R_1 & \cdots & R_N \\ 0 & -R_1 & \cdots & -R_N \end{bmatrix}$$

and shift the first column down one row to get

$$\tilde{G}_1^* = \begin{bmatrix} 0 & R_0 & \cdots & R_{N-1} \\ 0 & -R_1 & \cdots & -R_N \end{bmatrix}$$

2. Compute k_1 as the ratio of the (2, 2) and (2, 1) entries of $\tilde{G}_1$.

3. Form a matrix

$$\Theta(k_1) = \frac{1}{\sqrt{1 - k_1^2}} \begin{bmatrix} 1 & -k_1 \\ -k_1 & 1 \end{bmatrix}$$

This page

and apply it to $\tilde{G}_1$ to obtain a new Schur-reduced generator of the form

$$\tilde{G}_1^* = \Theta(k_1)\tilde{G}_1^* = \begin{bmatrix} 0 & x & x & \cdots & x \\ 0 & 0 & x & \cdots & x \end{bmatrix}$$

where the x denotes elements whose exact value is not relevant at the moment; Θ is known as a J-rotation matrix because

$$\Theta(k)J\Theta^*(k) = J \qquad J = \begin{bmatrix} 1 & 0 \\ 0 & -1 \end{bmatrix}$$

Its role is to rotate in the J metric (i.e., to hyperbolate) the second row of $\tilde{G}$ to lie along the first coordinate direction.

Now we can repeat steps 1, 2, and 3 to obtain k_2 and a new reduced generator matrix G_2. And so on.

It can be shown that the $\{k_i\}$ computed in the Schur algorithm are exactly the same quantities as defined in the Levinson algorithm. However, note that the Schur algorithm requires only sets of 2×1 row vector by 2×2 matrix multiplications, which can be carried out in parallel at each stage. Therefore, we need only $O(N)$ time units to carry out the Schur algorithm with N processors, compared to $O(N \log N)$ for the Levinson algorithm. In fact, a parallel and pipelined lattice VLSI computing structure based on the Schur algorithm has already been designed and built (see [29] and Chapter 17 in this book).

1.2.3 Cholesky Factorization

The Schur algorithm is also closely connected with the Cholesky factorization of the Toeplitz covariance matrix R in (1.1). Such factorizations are important in many calculations involving the corresponding stochastic process. The Cholesky factor, say C, is the unique lower triangular matrix with positive diagonal entries such that

$$R = CC^*$$

C is immediately determined by the Schur algorithm:

$$i\text{th column of } C = \text{first column of } G_i \qquad (1.6)$$

where the $\{G_i\}$ are the reduced generator matrices obtained in the Schur algorithm. Thus we have a fast algorithm, using $O(N^2)$ computations, for Cholesky factorization of a Toeplitz matrix, compared to $O(N^3)$ in general.

We should mention here that the Schur procedure for fast Cholesky factorization of Toeplitz matrices was, and continues to be, rediscovered in different contexts. We may mention Bareiss [3], Morf [40], and Rissanen [50]; Robinson [51] essentially rediscovered the Schur algorithm in solving an inverse scattering problem for a layered-earth model. The Schur algorithm was perhaps first explicitly identified in engineering by P. Dewilde in the early 1970s for the problem of cascade synthesis of impedance functions of passive systems. Its applications to stochastic estimation were first described in Dewilde et al. [17] and explored further by Dewilde and Dym [16], Lev-Ari [33,34], and Lev-Ari and Kailath [35]; applications in digital filter theory have been made by Deprettere and Dewilde [13] (see also Chapter 15 of this volume and [46,47]). The new structures obtained in these papers seem to be excellent candidates for VLSI implementation.

1.2.4 Transmission-Line Interpretations and Inverse Scattering Problems

A good indication of a natural connection between the Schur algorithm and potential VLSI or other implementation methods can be obtained from the graphical representations of the Schur algorithm shown in Figure 1.4. Then the result above on the relation between C and the $\{G_i\}$ means that the values at the input of the ith column of the Cholesky factor C. Consequently, we can also say that the values at the inputs of the delay elements at any time l are the entries of the lth row of the Cholesky factor C. (Incidentally, this interpretation removes much of the mystique sometimes associated with the fast Cholesky-by-column and fast Cholesky-by-row algorithms.)

This transmission-line interpretation suggests that the computations could be carried out by exciting structures as in Figure 1.4 with light or other electromagnetic waves, depending on the medium chosen for implementation (e.g., fiber optics, acoustic waves in solids, etc.).

In fact, we might mention that starting with the graphical representations in Figure 1.4 and using some fundamental wave propagation and energy conservation concepts, we can obtain simple proofs of all the foregoing results and in the process develop intimate and useful connections between transmission-line theory, Cholesky and inverse Cholesky factorization, inverse scattering theory, and the Schur and dual Schur algorithms (see [6]). This transmission-line picture was also helpful in obtaining certain orthogonal cascade implementations of ARMA (pole-and-zero) filters with excellent numerical properties [46,47].

On the other hand, we can approach the results from quite another direction by noting that Cholesky factorization is actually a way of testing for positive definiteness of a matrix; if we now ask what matrix structures lend themselves to easy testing in this way, we shall be led to a class of matrices (with "displacement" structure) of which the Toeplitz family is only one special case [35]. By this route, all the previously mentioned results on parallel computation and transmission-line interpretations and implementations can be carried over to a much larger class of problems, as we review briefly in the next section.

1.2.5 Doubling Algorithms

A final interesting result that may be noted here is that the Schur algorithm as depicted can be used to obtain what is perhaps the simplest version of a "doubling" (or "divide and conquer") algorithm for the computation of the $\{k_i\}$ and of the

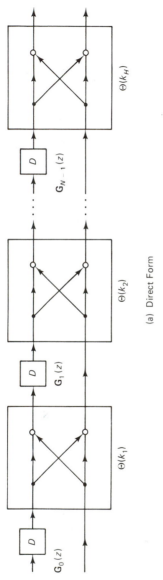

(a) Direct Form

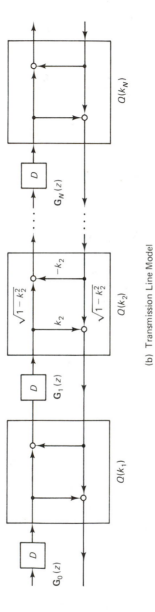

(b) Transmission Line Model

Figure 1.4 Representations of the Schur algorithm: (a) direct form; (b) transmission-line model.

16

solution of the Yule–Walker equations. By using fast convolution algorithms to combine the impulse responses of pairs of adjacent Θ sections in Figure 1.4, the amount of computation can be reduced to $12N \log^2 N$ (see [42,43]); the crossover point with the direct Schur algorithm for the solution of the Toeplitz equations is $N = 128$. Such $O(N \log^2 N)$ results were first obtained by Brent et al. [5], Morf [41], and Bitmead and Anderson [4]; however, the actual algorithms were somewhat complicated, and, for example, Sexton et al. [56] calculated that for the method of Bitmead and Anderson, the coefficient of the $N \log^2 N$ terms was 7000! The algorithm based on the Schur procedure not only has a lower coefficient but also lends itself better to potential VLSI implementation.

1.3 EXTENSIONS TO NEAR-TOEPLITZ MATRICES; DISPLACEMENT STRUCTURE

In Section 1.2 we started out by stating that a common technique for fitting autoregressive models to time-series data was via solution of the Yule–Walker linear equations with a Toeplitz coefficient matrix. We mentioned at the time that this was not necessarily the only, or the best, way of solving this problem, and we shall explain why now.

Briefly stated, the problem is the following: Given a set of observations $\{y_0, \ldots, y_T\}$, we wish to choose a set of coefficients $\{a_1, \ldots, a_n\}, n < T$, such that

$$\sum_t e_{n,t}^2 = \text{minimum} \tag{1.7}$$

where $e_{n,t}$ is the error (residual in trying to predict y_t from n past values)

$$e_{n,t} = y_t + a_1 y_{t-1} + \cdots + a_n y_{t-n} \tag{1.8}$$

The problem arises when $0 \leq t < n$, because then we do not have all n previous values. One choice is only to compute residuals for $t \geq n$, but for a number of reasons (some of which will become clear presently), we often accommodate the case $t < n$ by assuming that the missing data $\{y_{-n}, y_{-n+1}, \ldots, y_{-1}\}$ are zero. One reason is the hope that for $T \gg n$, the arbitrariness of this initial assumption will not affect the overall solution very much. This is not unreasonable, but sometimes a potentially more drastic assumption is made: namely, residuals are computed for $T < t \leq T + n$ by assuming that the "future" data $\{y_{T+1}, \ldots, y_{T+n}\}$ are also zero. We shall see the reason for this presently, but first let us display the set of residuals we have defined in matrix form as

$$e = Ya \tag{1.9}$$

where

$$e^T = [e_{n,0} \quad e_{n,1} \quad \cdots \quad e_{n,T+n}]$$

$$a = [1 \quad a_1 \quad \cdots \quad a_n]$$

The least-squares solution, $\hat{a}$, that minimizes $\| e \|^2$ can now be computed as the

$$Y = \begin{bmatrix} y_0 & & & \\ y_1 & y_0 & & \\ \vdots & \ddots & & \\ \hline y_n & \cdots & & y_0 \\ y_t & \cdots & & y_{t-n} \\ y_T & \cdots & & y_{T-n} \\ \hline & & \ddots & \\ & & & y_T \end{bmatrix}$$

solution of the "normal" equations

$$[Y^T Y]a = [R_n^e \quad 0 \quad \cdots \quad 0]^T \tag{1.10}$$

where $R_n^e = \min \| e \|^2$. We can now state one reason for extending the residuals $\{e_{n,t}\}$ to both sides of the given sample $\{y_0, \ldots, y_T\}$: Only in this case will the coefficient matrix $Y^T Y$ be Toeplitz, and in particular be the sample covariance matrix as used in the Yule–Walker equations of Section 1.2.

However, one may very well argue that one should make no assumptions at all about unavailable data and therefore only use the residuals $\{e_{n,n}, \ldots, e_{n,T}\}$; or perhaps, allow only assumptions on past missing data and thus use $\{e_{n,0}, \ldots, e_{n,T}\}$. In either case, we shall no longer have $Y^T Y$ as a Toeplitz matrix, and therefore it would seem that the nice results of Section 1.2 [on reduction of complexity to $O(N^2)$ or $O(N \log^2 N)$ with serial processors, and on lattice filter implementations and transmission-line interpretations] would no longer apply.

This would be disappointing except that closer examination shows that the situation can be saved in large measure by introducing the concept of displacement structure (Kailath et al. [23–25]), which we later discovered defined the natural family of covariance matrix structures that allow fast computation of the Cholesky factorization (see Lev-Ari and Kailath [35]).

We shall not describe here the elegant and comprehensive theory that has been developed. Detailed presentations can be found in the paper just cited and in Lev-Ari, Kailath, and Cioffi [36], and the Ph.D. dissertations of Lev-Ari [33], Morgan [42], Cioffi [8], and the earlier ones of Lee [31], Porat [44], and Delosme [12].

Nevertheless, it will be useful to give some of the flavor of the results. Let us define a "lower-shift" matrix

$$Z = \text{matrix with 1's on the first subdiagonal and 0's elsewhere}$$

With Z, we can introduce

$$R - ZRZ^T := \textit{displacement of } R \tag{1.11}$$

The reason for the name is that, except on the boundaries, the (i, j) element of $R - ZRZ^T$ is $[R_{i,j} - R_{i-1,j-1}]$. When R is Toeplitz, only the first row and column of $R - ZRZ^T$ will be nonzero. We shall define the *displacement rank* of R as

$$\text{rank } \{R - ZRZ^T\} = \alpha \tag{1.12a}$$

and the *displacement inertia* of R as

$$\text{inertia } \{R - ZRZ^T\} = \{1, -J\} \tag{1.12b}$$

where J is a diagonal matrix with entries $+1$ or -1. In the Toeplitz case, $\alpha = 2$ and $J = 1$. The matrices $Y^T Y$ arising in the normal equations at the beginning of this section have $\{\alpha \equiv 1, J = 0\}$ when we use the residuals $\{e_{n,0}, \ldots, e_{n,T}\}$, which defines the "prewindowed" case, and $\{\alpha = 3, J = \text{diag } \{1, 1\}\}$ when we use only the residuals $\{e_{n,n}, \ldots, e_{n,T}\}$, which defines the "covariance" case.

1.3.1 Generalized Time-Invariant Lattice Sections

For matrices with displacement rank α and displacement inertia $\{1, -J\}$, we can again have cascade implementations of general lattice sections, but each section will be defined by an $(\alpha - 1)$-dimensional row vector K (rather than by a scalar, see Figure 1.5) such that

$$1 - KJK^* \geq 0 \tag{1.13}$$

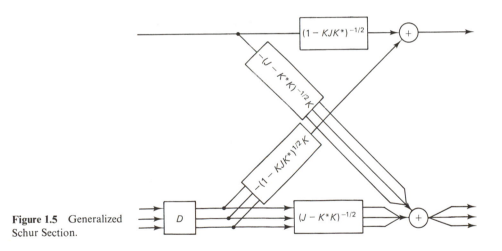

Figure 1.5 Generalized Schur Section.

The $\{K_i\}$ can be found by a generalization of the Schur algorithm. We see that the major hardware cost is that of having $\alpha - 1$ delay units per section rather than 1 as in the Toeplitz case; however, this may not be a steep price to pay with current technology, especially since we continue to have the advantages of regular structures with local interconnections.

1.3.2 Time-Variant Lattice Sections

An alternative lattice implementation can be obtained with a traditional two-line lattice but with time-variant reflection coefficients. In fact, it can be shown [23] that such a lattice structure can be used for *any* nonstationary process. However, by using the displacement structure, simple updating formulas can be obtained for the reflection coefficients, using only $O(\alpha)$ elementary computations per step. For N computations and N reflection coefficients, this will require $O(N^2\alpha)$ computations compared to $O(N^3)$ in the general case.

The time-update formulas can be compactly stated. The reflection coefficient of the nth section obeys the relation

$$k_{n+1,t} \equiv (1 - \eta_{n,t}\eta_{n,t}^T)^{1/2} k_{n+1,t-1}(1 - \mu_{n,t-1}\mu_{n,t-1}^T)^{1/2} + \eta_{n,t}\mu_{n,t-1}^T \qquad (1.14)$$

where $\{\eta, \mu\}$ are α-dimensional row vectors obeying the recursions

$$\eta_{n+1,t} = F\{\eta_{n,t}, k_{n+1,t}, \mu_{n,t-1}\} \qquad (1.14a)$$

$$\mu_{n+1,t} = F\{\mu_{n,t-1}, k_{n+1,t}^T, \eta_{n,t}^T\} \qquad (1.14b)$$

and the function $F(\cdot)$ is defined as

$$F\{A, B, C\} \equiv (1 - BB^T)^{-1/2}(A - BC^T)(1 - CC^T)^{-T/2}$$

These recursions can be depicted as in Figure 1.6.

In special cases, these general results simplify further (see [36]). Thus in the "prewindowed least-squares" formulation, $\eta_{n,t}$ and $\mu_{n,t}$ are apart from a scale factor, equal to the variance-normalized forward residual $\bar{e}_{n,t}$, and the corresponding delayed "backward" residual $\bar{r}_{n,t-1}$, which are the signals flowing on the lines of the lattice of Figure 1.6.

In this case, the lattice filter itself can be used to update the reflection coefficients. Some calculation will yield the result

$$k_{n+1,t} = \left(1 - \frac{|\bar{e}_{n,t}|^2}{\phi_{n,t-1}}\right)^{1/2} k_{n+1,t-1} \left(1 - \frac{|\bar{r}_{n,t-1}|^2}{\phi_{n,t-1}}\right)^{1/2} + \frac{\bar{e}_{n,t}\bar{r}_{n,t-1}}{\phi_{n,t-1}} \qquad (1.15)$$

where $\phi_{n,t-1}$ is a quantity between zero and unity. This quantity has some useful statistical interpretations (e.g., as a log-likelihood ratio) that have been exploited in special applications (see, e.g., [32]). We mention this here in part because the fore-

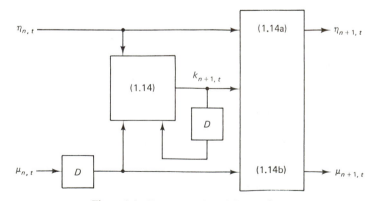

Figure 1.6   Representation of time updates.

going updating formula for $k_{n+1,t}$ with $\phi_{n,t-1} \equiv 1$ has been derived previously by several investigators as an *approximate* adaptive lattice filter known variously as an Itakura–Saito filter or a gradient lattice filter. It is a surprising, pleasing, and important fact that we have such a nice form as (1.15) for the *exact* solution, exact in the sense that it minimizes the sum of the squared residuals at each t; the importance arises from the fact that exact least-squares solutions are known to have the important properties of consistency, unbiasedness, rapid convergence, and so on. Therefore, such algorithms and close variants, such as fixed-order transversal filter forms, have been applied to problems of signal tracking, channel equalization, echo cancellation, adaptive line enhancement, and others (see, e.g., [8,9,10,21,48,49,53]). Special-purpose VLSI chips have already been developed for at least one of these applications (see, e.g., [19]), and others are no doubt under way.

As an example of another application, we might mention here a systolic array architecture for measurement update calculations in the Kalman filter (see [22]); this architecture is based on a special square-root array method for state-space estimation.

Another area in which the ideas similar to these presented in this chapter are being further developed is computed tomography (see, e.g., the thesis of Buonocore [7]). The speech, communications, and image-processing areas abound with other potential applications: voice modeling and recognition, voice-mail chips, CODECs, vocoders, transmultiplexers, residue encoding, and many others. Other chapters in this book discuss sonar and radar applications in more detail.

1.4 CONCLUDING REMARKS

We have attempted to illustrate some of the many ways in which the mathematical development of signal processing algorithms and the new possibilities of VLSI technology can fruitfully interact with each other, sometimes suggesting new physical implementations and sometimes stimulating new theoretical developments. However, only the surface has been scratched in such work and there is plenty of opportunity for stimulating and rewarding exploration of signal processing and system theory techniques and applications in the VLSI era; a recent report (Rao and Kailath [47] shows, among other things, the close relationships between the systolic arrays of Kung and Leiserson (Mead and Conway [39]) and the direct (or controller and observer) forms of digital filtering and system theory.

ACKNOWLEDGMENTS

This work was supported in part by the Air Force Office of Scientific Research, Air Force Systems Command under Contract AFOSR83-0228, the U.S. Army Research Office, under Contract DAAG29-83-K-0028, and by the Defense Advanced Research Projects Agency, under Contract MDA903-79-C-0680.

REFERENCES

[1] H. M. Ahmed, "Signal Processing Algorithms and Architectures," Ph.D. dissertation, Department of Electrical Engineering, Stanford University, Stanford, Calif., 1982.

[2] H. M. Ahmed, J. M. Delosme, and M. Morf, "Highly Concurrent Computing Structures for Digital Signal Processing and Matrix Arithmetic," *IEEE Comput. Mag.*, (Feb. 1982).

[3] E. H. Bareiss, "Numerical Solution of Linear Equations with Toeplitz and Vector Toeplitz Matrices, *Numer. Math., 13:* 404–424 (1969).

[4] R. R. Bitmead, and B. D. O. Anderson, "Asymptotically Fast Solution of Toeplitz and Related Systems of Linear Equations," *J. Linear Algebra Appl., 34*:103–116 (Dec. 1980).

[5] R. P. Brent, F. G. Gustavson, and D. Y. Y. Yun, "Fast Solution of Toeplitz Systems of Equations and Computation of Pade Approximants," *J. Algorithms, 1*(3):259–295 (1980).

[6] T. Kailath, A. Bruckstein and D. Morgan, "Fast Matrix Factorizations via Discrete Transmission Lines," submitted for publication, 1984.

[7] M. Buonocore, "Fast Minimum Variance Estimators for Limited Angle Computed Tomography Image Reconstruction," Ph.D. dissertation, Department of Electrical Engineering, Stanford University, Stanford, Calif., 1981.

[8] J. Cioffi, "Fast Transversal Filters for Communications Applications," Ph.D. dissertation, Department of Electrical Engineering, Stanford University, Stanford, Calif., 1984.

[9] J. Cioffi, and T. Kailath, "An Efficient, Exact Least-Squares Fractionally Spaced Equalizer Using Intersymbol Interpolation," *IEEE Journal on Selected Areas in Communications, Special Issue on Voiceband Data Transmission*, (May 1984).

[10] J. Cioffi, and T. Kailath, "Fast Recursive Least Squares Transversal Filter Algorithms for Adaptive Filtering," *IEEE Trans. Acoust. Speech Signal Process., ASSP-32* (1984).

[11] D. Cochran, "Algorithms and Accuracy in the HP-35," *Hewlett-Packard J.*, 1972, pp. 10–11.

[12] J. M. Delosme, "Algorithms and Architectures for Finite Shift-Rank Processes," Ph.D. dissertation, Department of Electrical Engineering, Stanford University, Stanford, Calif., Sept. 1982.

[13] E. Deprettere and P. Dewilde, "Orthogonal Cascade Realization of Real Multiport Digital Filters," *Circuit Theory Appl., 8*:245–272 (1980).

[14] A. M. Despain, "Fourier Transform Computers Using CORDIC Iterations," *IEEE Trans. Comput., C-23*:993–1001 (1974).

[15] A. M. Despain, "Very Fast Fourier Transform Algorithms for Hardware Implementation," *IEEE Trans. Comput., C-28*(5):333–341 (1979).

[16] P. Dewilde and H. Dym, "Schur Recursions, Error Formulas, and Convergence of Rational Estimators for Stationary Stochastic Processes," *IEEE Trans. Inf. Theory, IT-27*(4):446–461 (1981).

[17] P. Dewilde, A. C. Vieira, and T. Kailath, "On a Generalized Szegö–Levinson Realization Algorithm for Optimal Linear Predictors Based on a Network Synthesis Approach," *IEEE Trans. Circuits Syst., CAS-25*(9):663–675 (1978).

[18] J. Durbin, "The Fitting of Time-Series Models," *Rev. Int. Inst. Stat., 28*:233–244 (1960).

[19] D. L. Duttweiler, "A Twelve-Channel Digital Echo Canceler," *IEEE Trans. Commun., COM-28*:647–653 (1978).

[20] G. L. Haviland and A. Tuszynski, "A CORDIC Arithmetic Processor Chip," *IEEE Trans. Comput.*, C-29(2):68–79 (1980).

[21] W. S. Hodgkiss and J. A. Presley, "Adaptive Tracking of Multiple Sinusoids Whose Power Levels Are Widely Separated," *IEEE Trans. Acoust. Speech Signal Process.*, ASSP-29(3):710–721 (1981).

[22] J. M. Jover and T. Kailath, "A Parallel Architecture for Kalman Filter Measurement Update," *Proc. IFAC Congress*, Budapest, Hungary, July 1984.

[23] T. Kailath, "Time-Variant and Time-Invariant Lattice Filters for Nonstationary Processes," *Proc. Fast Algorithms for Linear Dynamical Systems*, Aussois, France, Sept. 21–25, 1981, pp. 417–464. Reprinted as *Outils et modèles mathématiques pour l'Automatique, l'Analyse de Systèmes et le Traitement du Signal*, Vol. 2, I. D. Landau, ed., CNRS, France, 1982, pp. 417–464.

[24] T. Kailath, S. Y. Kung, and M. Morf, "Displacement Ranks of Matrices and Linear Equations," *J. Math. Anal. Appl.*, 68(2):395–407 (1979).

[25] T. Kailath, S. Y. Kung, and M. Morf, "Displacement Ranks of a Matrix," *Bull. Am. Math. Soc.*, 1(5):769–773 (1979).

[26] T. Kailath, *Lectures on Wiener and Kalman Filtering*, Springer-Verlag, New York, 1981.

[27] H. T. Kung, "Let's Design Algorithms for VLSI Systems," *Proc. First Caltech VLSI Symp.*, 1979, pp. 65–90.

[28] H. T. Kung, "Why Systolic Arrays?" *IEEE Computer*, 15(1):37–46 (1982).

[29] S. Y. Kung and H. Hu, "A Highly Concurrent Algorithm and Pipelined Architecture for Solving Toeplitz Systems," *IEEE Trans. Acoust. Speech Signal Process.*, ASSP-31(1):66–75 (1983).

[30] S. Y. Kung, K. S. Arun, R. J. Gal-Ezer, and D. V. Bhaskar Rao, "Wavefront Array Processor: Language, Architecture and Applications," *IEEE Trans. Comput.*, C-31:1054–1066 (1982).

[31] D. T. Lee, "Canonical Ladder Form Realizations and Fast Algorithms," Ph.D. dissertation, Department of Electrical Engineering, Stanford University, Stanford, Calif., 1980.

[32] D. T. Lee and M. Morf, "A Novel Innovations Based Approach to Pitch Detection," *Proc. 1980 IEEE Int. Conf. Acoust. Speech Signal Process.*, Denver, Colo., Apr. 9–11, 1980, pp. 40–44.

[33] H. Lev-Ari, "Parameterization and Modeling of Nonstationary Processes," Ph.D. dissertation, Stanford University, Stanford, Calif., 1983.

[34] H. Lev-Ari, "Modular Architectures for Adaptive Multichannel Lattice Algorithms," *IEEE Int. Conf. Acoust. Speech Signal Process.*, Boston, Apr. 1983, pp. 455–458.

[35] H. Lev-Ari and T. Kailath, "Lattice Filter Parametrization and Modeling of Nonstationary Processes," *IEEE Trans. Inf. Theory*, IT-30, (Jan. 1984).

[36] H. Lev-Ari, T. Kailath, and J. Cioffi, "Least-Squares Adaptive Lattice and Transversal Filter for Nonstationary Processes," *IEEE Trans. Inform. Theory*, IT-30, (Mar. 1984).

[37] N. Levinson, "The Wiener RMS (Root-Mean-Square) Error Criterion in Filter Design and Prediction," *J. Math. Phys.*, 25(4):261–278 (1947).

[38] J. Markel and A. H. Gray, Jr., *Linear Prediction of Speech*, Springer-Verlag, New York, 1976.

[39] C. Mead and L. Conway, *Introduction to VLSI Systems*, Addison-Wesley, Reading, Mass., 1980.

[40] M. Morf, "Fast Algorithms for Multivariable Systems," Ph.D. dissertation, Department of Electrical Engineering, Stanford University, Stanford, Calif., 1974.

[41] M. Morf, "Doubling Algorithms for Toeplitz and Related Equations," *Proc. IEEE Int. Conf. Acoust. Speech Signal Processing*, Denver, Colo., Apr. 9–11, 1980, pp. 954–959.

[42] D. R. Morgan, "Orthogonal Triangularization Methods for Fast Estimation Algorithms," Ph.D. dissertation, Department of Electrical Engineering, Stanford University, Stanford, Calif., Dec. 1984.

[43] B. Musicus, "Levinson and Fast Cholesky Algorithms for Toeplitz and Almost Toeplitz Matrices," MIT Internal Report, submitted for publication, 1981.

[44] B. Porat, "Contributions to the Theory and Applications of Lattice Filters," Ph.D. dissertation, Dept. of Electrical Engineering, Stanford University, Stanford, Calif., 1982.

[45] L. R. Rabiner and R. W. Schafer, *Digital Processing of Speech Signals*, Prentice-Hall, Englewood Cliffs, N. J., 1978.

[46] S. K. Rao and T. Kailath, "Orthogonal Digital Filters for VLSI Implementation," *IEEE Trans. Circuits Systems, CAS-31*, (1984).

[47] S. K. Rao and T. Kailath, "Digital Filtering in VLSI," Technical Report ISL, Stanford Univ., Stanford, Calif., Jan. 1984.

[48] V. U. Reddy, B. Egardt, and T. Kailath," Optimized Lattice-Form Adaptive Line Enhancer for a Sinusoidal Signal in Broadband Noise," *IEEE Trans. Circuits Syst., CAS-28*:542–550 (June 1981).

[49] V. U. Reddy, T. J. Shan, and T. Kailath, "Application of Modified Least-Squares Algorithms to Adaptive Echo Cancellation," *Proc. 1983 Int. Conf. Acoust. Speech Signal Process.*, Boston, 1983, pp. 53–56.

[50] J. Rissanen, "Algorithms for Triangular Decomposition of Block Hankel and Toeplitz Matrices," *Math. Comp., 27*:147–154 (1973).

[51] E. A. Robinson, "Dynamic Predictive Deconvolution," *Geophys. Prospect., 23*:779–797 (1975).

[52] E. A. Robinson, "Spectral Approach to Geophysical Inversion by Lorentz, Fourier and Radon Transforms," *Proc. IEEE, 70*:1039–1054 (1982).

[53] E. Satorius and J. Pack, "Application of Least-Squares Lattice Algorithms to Adaptive Equalization," *IEEE Trans. Commun., COM-29*:136–142 (1981).

[54] C. W. Schelin, "Calculator Function Approximation," *Am. Math. Monthly, 90*:317–325 (1983).

[55] I. Schur, "Über Potenzreihen, die im Innern des Einheitskreises beschrankt sind," *J. Reine Angew. Math., 147*:205–232 (1917).

[56] H. Sexton, M. Shensa, and J. Speiser, "Remarks on a Displacement-Rank Inversion Method for Toeplitz Systems," *Linear Algebra Appl., 45*:127–130 (1982).

[57] G. Szegö, *Orthogonal Polynomials*, 4th ed., Colloquium Publications, 23, American Mathematical Society, Providence, R. I., 1975. Originally published in 1939.

[58] J. E. Volder, "The CORDIC Trigonometric Computing Technique," *IRE Trans. Electron. Comput., EC-8*(3):330–334 (1959).

[59] J. S. Walther, "A Unified Algorithm for Elementary Functions," *Proc. 1971 Spring Joint Comput. Conf.*, 1971, pp. 379–385.

[60] R. Wiggins and L. Brantingham, "Three Chip System Synthesizes Human Speech," *Electronics, 51*(18):109–116 (1978).

2

Signal Processing Applications of Concurrent Array Processor Technology

H. J. WHITEHOUSE, J. M. SPEISER, AND K. BROMLEY

Naval Ocean Systems Center
San Diego, California

2.1 INTRODUCTION

It has previously been shown that the major computational requirements for many important real-time signal processing tasks can be reduced to a common set of basic matrix operations [1–3]. These include matrix-vector multiplication, matrix-matrix multiplication and addition, matrix inversion, solution of linear systems, least-squares approximate solution of linear systems, eigensystem solution, generalized eigensystems solution, and singular-value decomposition of matrices. Extensive research in numerical linear algebra has resulted in two sets of numerically stable, well-documented software routines for performing these operations on single-processor sequential-operation computers: One set, called LINPACK [4], covers the solution of linear systems and least-squares problems. The other, called EISPACK [5], covers eigensystem problems. These algorithms are extremely computation intensive, requiring of the order of N^3 or N^4 multiplications for each set of N input data samples. By way of comparison, today's workhorse signal processing algorithm, the FFT, requires on the order of only $N \log_2 N$ multiplications and yet its real-time computation taxes today's single-chip APUs at even audio bandwidths. Clearly, orders-of-magnitude increases in computation rate are required for real-time implementation of these advanced algorithms.

Despite the tremendous growth in digital integrated-circuit technology over the last decade, one cannot simply look to further advances in device fabrication to satisfy this need. Presently, state-of-the-art VLSI chips are fabricated with a minimum feature size of 1 to 2 μm. The Department of Defense's VHSIC phase II program aims to reduce this to 0.5 μm by about 1987, thereby increasing the functional throughput rate (i.e., clock speed times gate density) per chip from

5×10^{11} Hz gates/cm^2 to 10^{13} Hz gates/cm^2. It will probably require at least three additional years of effort to improve the yield and wafer throughput to the point where this technology is commercially viable. Thus we can expect little more than a 20-fold improvement in functional throughput rate of VLSI chips by 1990. Furthermore, most current projections indicate that further reductions in feature size below 0.5 µm will come only at the expense of greatly increasing effort. The authors conclude that, barring any presently unforeseen breakthroughs in signal processor implementation technologies, the orders-of-magnitude throughput gains necessary for the real-time computation of the LINPACK/EISPACK algorithms must come from *architectural* advances—the efficient utilization of *parallelism* in computation.

The most straightforward approach to parallel signal processing architectures is simply to connect a number of CPUs to a common bus. Indeed, most of today's commercial microprocessor board sets boast this "multiprocessing" capability. However, performance improves linearly with the number of processors only up to the point that bus-contention problems become the limitation. Minsky's famous conjecture is that, for a broad range of algorithms, the conflict between N processors for access to shared resources along the common bus limits the performance improvement to $\log_2 N$. Modern "supercomputer" designers have utilized a number of parallel-processing stratagems to improve on this state of affairs and are achieving performance improvements commensurate with Amdahl's law: namely, $N/\log_2 N$. In 1978, however, H. T. Kung presented his pioneering work on systolic-array architectures [6], which yield a perfectly efficient improvement factor of N. Figure 2.1 graphically depicts the speed-up achieved through parallelism for the three cases of Minsky's conjecture, Amdahl's law, and systolic arrays.

Parallel processing architectures have been surveyed [1–3] and it was concluded that the systolic and wavefront architectures [6,7] provide the most promising combination of characteristics for utilizing VLSI/VHSIC technology for real-time signal processing: modular parallelism with throughput directly proportional

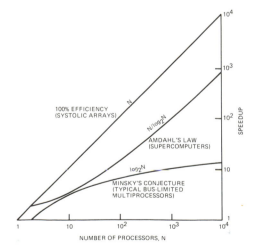

Figure 2.1 Representative speedup by advanced architectures with N processors working concurrently.

to the number of cells, simple control, synchronous data flow, local interconnects, and sufficient versatility for implementing the matrix operations needed for signal processing.

Previously reported systolic and wavefront architectures [6,7] include linear array configurations for matrix-vector multiplication and solution of linear equations with triangular coefficient matrices, and hexagonal configurations for matrix multiplication-addition and L-U decomposition of matrices. These architectures are particularly attractive when the matrices have only a few bands occupied about the main diagonal.

This chapter discusses new architectures for the multiplication of dense matrices, configurations for partitioned matrix operations, and applications to cross-ambiguity function calculation. The reader interested in hardware implementations of systolic and wavefront architectures is referred to [8] for a 200-MOP one-dimensional array, [9] for a two-dimensional test bed, Chapter 8 of this volume for the design of a custom LSI systolic chip, Chapter 22 for the design of two custom VLSI systolic chips, and Chapter 17 for the design of a VLSI chip for the Toeplitz system solver [10]. The reader interested in systolic architectures for more advanced algorithms than those discussed here is referred to [11] for matrix triangularization, [12] for singular-value decomposition, [13,14] for eigensystem computation, and [10] for Toeplitz system solution.

2.2 LEAST-SQUARES PROBLEMS

Three types of least-squares problems will be considered: the deterministic problem, the stochastic problem with known second moments, and the stochastic problem with estimated second moments. It will be shown that the stochastic problem with estimated second moments is computationally identical to a deterministic least-squares problem, and thus a baseline method for real-time solution in many signal processing applications may be implemented by forming and solving the normal equations via systolic matrix multiplication, matrix-vector multiplication, and matrix inversion. Alternative methods for the least-squares problem are known [19] and a systolic implementation is discussed in [11].

The deterministic least-squares problem is the selection of a vector $\mathbf{x}$ to minimize ε_1^2 in equation (2.1) where A is a known matrix and $\mathbf{y}$ is a known vector.

$$\varepsilon_1^2 = \| A\mathbf{x} - \mathbf{y} \|^2 = \mathbf{x}^T A^T A \mathbf{x} - 2\mathbf{y}^T A \mathbf{x} + \| \mathbf{y} \|^2 \tag{2.1}$$

The superscipt T denotes the transpose of a matrix or vector.

The stochastic least-squares problem is the selection of a deterministic weight vector $\mathbf{w}$ to minimize the expected squared error ε_2^2 in (1.2), where $\mathbf{z}$ is a random vector and d is a random variable. The scalar random variable $\mathbf{w}^T \mathbf{z}$ is a linear combination of the elements of the random vector $\mathbf{z}$, and is used as a linear estimator of the random variable d.

$$\varepsilon_2^2 = E(\mathbf{w}^T\mathbf{z} - d)^2 = \mathbf{w}_z^T \mathbf{R} \mathbf{w} - 2\mathbf{R}_{dz}^T \mathbf{w} + E(d^2) \tag{2.2}$$

TABLE 2.1 DESIRED IDENTIFICATIONS
FOR COMPUTATIONAL EQUIVALENCE
OF LEAST-SQUARES PROBLEMS

Deterministic	Stochastic with estimated second moments
$(A)_{ns}$	$N^{-0.5}z(n, s)$
$y(n)$	$N^{-0.5}d(n)$
$x(s)$	$w(s)$
$A^T A$	$\hat{R}_z$
$y^T A$	$\hat{R}_{dz}^T$
$\| y \|^2$	$E(d^2)$

Statistical expectation is denoted by the operator E, and R_z denotes the second moment matrix of the random vector z: $R_z = Ezz^T$. Similarly, the vector R_{dz} is defined as $R_{dz} = E(dz)$.

In order to compare the two problems, it will be assumed that A is $N \times S$, x is $S \times 1$, y is $N \times 1$, z is $S \times 1$, and w is $S \times 1$. For the stochastic least-squares problem with estimated second moments, it will be assumed that N samples are observed of the random vector z and the random variable d, say $z(n, s)$ and $d(n)$ for $n = 1, \ldots, N$ and $s = 1, \ldots, S$.

The standard unbiased estimators of the second moments are shown in (2.3) to (2.5). The corresponding error term for the estimation of d by $w^T z$ is shown in (2.6).

$$(\hat{R}_z)_{s_1, s_2} = \frac{1}{N} \sum_{n=1}^{N} z(n, s_1)z(n, s_2) \tag{2.3}$$

$$(\hat{R}_{dz})_{s_1} = \frac{1}{N} \sum_{n=1}^{N} d(n)z(n, s_1)$$

$$= \frac{1}{N} \sum_{n=1}^{N} z^T(s_1, n)d(n) \tag{2.4}$$

$$(Ed^2) = \frac{1}{N} \sum_{n=1}^{N} d^2(n) \tag{2.5}$$

$$\varepsilon_3^2 = w^T \hat{R}_z w - 2\hat{R}_{dz}^T w + (Ed^2) \tag{2.6}$$

The correspondence shown in Table 2.1 may then be used to show the computational equivalence of the stochastic least-squares problem with estimated second moments and the corresponding deterministic least-squares problem. It is therefore only necessary to consider computational methods for the deterministic least-squares problem.

2.2.1 Sonar Applications of Least-Squares Solutions

Least-squares solutions have three important applications for improving sonar performance: noise cancellation, interference cancellation, and maximum entropy spectrum analysis. For the noise cancellation problem, it is assumed that the hydrophone output is a linear combination of the ambient acoustic field with local noise sources, such as one's own ship's machinery noises. Auxiliary transducers may be coupled closely to the machinery to provide almost pure noise sources. As a further refinement of this technique, inputs to the canceler can include delayed copies of each noise reference. Linear combinations of the delayed copies can then approximate the filtered noise received by the array element after propagation through the ship's hull.

The interference cancellation problem is similar to the noise cancellation problem except that the inputs to the adaptive combiner are the outputs of preformed conventional beams. To form one beam with interference cancellation, one conventional beam is steered in the desired-look direction, and S "null beams" are generated. The null beams have zero sensitivity in the desired-look directions, and are used to provide interference references. In general, as many linearly independent null beams will be required as interfering sources to be canceled. If the beam outputs are combined with fixed weight for the beam steered in the desired-look direction, minimizing the total power output minimizes the interference contribution to the output. If the interference is uncorrelated with the desired signal, the signal contribution to the output power remains constant. Under certain conditions, however, this assumption can be violated. Under conditions of multipath propagation, the signal may be received on a null beam as well as on the beam in the desired-look direction. In this case, adjusting the weights for minimum total power output can reduce the signal contribution as well as the noise.

Noise and interference cancellation have been typically implemented via gradient descent adaptive transversal filters [15]—the Widrow LMS algorithm. Such implementations are designed to provide adaptations with a reduced multiplication rate and hence relatively simple hardware. Unfortunately, the adaptation rate is reduced if there is a large spread in the eigenvalue distribution of the data covariance matrix [16]. Unfortunately, a strong interfering source will result in a large eigenvalue spread and may result in a convergence time which is greater than the stationarity time of the problem. In such a context, faster convergence can be obtained via direct inversion of the sample covariance matrix to solve the normal equations [17], making more statistically efficient use of the available data at any given time.

Recent methods of spectrum analysis provide improved resolution by utilizing a parametric model of the signal. In the maximum-entropy method [18], the signal is modeled at the output of an all-pole filter driven by white noise. The inverse of such a filter is a transversal filter that converts the signal to white noise and whose tap weights are calculated by solving the linear one-step prediction problem for the

TABLE 2.2 CORRESPONDENCES FOR SONAR APPLICATIONS
OF LEAST-SQUARES SOLUTIONS

Problem	z	d
Noise cancellation	$z_1, \ldots, z_S$ (noise references)	Signal + noise
Interference cancellation via adaptive combiner with fixed beams	$b_1, \ldots, b_S$ (outputs of preformed null beams)	b_0 (output of preformed beam steered in desired-look direction)
Maximum-entropy spectrum analysis	$x_1, \ldots, x_S$ (time samples of signal)	x_{S+1}

signal. It is well known that for a stationary process, the prediction error of the least-squares linear predictor is a white-noise process. Then the estimated spectral density function is proportional to the reciprocal of the magnitude squared of the transfer function of the prediction error filter. This spectrum estimation technique, which provides improved resolution when the model is applicable and the signal-to-noise ratio is sufficiently high, has also been applied to beamforming [18].

The least-squares problems corresponding to the noise cancellation, interference cancellation, and maximum-entropy spectrum analysis are summarized in Table 2.2.

2.3 THE ENGAGEMENT PROCESSOR

A systolic architecture may be viewed as the implementation of a set of recurrence relations by a set of identical (or mostly identical) computational cells. For the computation of a matrix product $C = AB$, each element c_{pq} of C is the inner product of a row of A with a column of B, as shown in (2.1).

$$c_{pq} = \sum_{s=1}^{N} a_{ps} b_{sq} \tag{2.7}$$

It may also be viewed as the solution of a set of recurrences as shown in (2.8), where $c_{pq}^{(N)}$ will be the desired result.

$$c_{pq}^{(0)} = 0 \tag{2.8a}$$

$$c_{pq}^{(s)} = c_{pq}^{(s-1)} + a_{ps} b_{sq} \qquad \text{for } s = 1, \ldots, N \tag{2.8b}$$

Such summations may be evaluated via recurrences having intermediate terms moving through the structure [4,5], or by accumulating partial sums in place.

An important systolic cell for linear algebra operations is the inner product step processor [6] shown in Figure 2.2 in both its hexagonal and orthogonal forms. The latter form will be used exclusively in the remainder of this chapter. By routing the C_{out} port around as a feedback path into C_{in}, a cell configuration particularly well suited for use in inner product evaluation via (2.8a) and (2.8b) results. If it is assumed that C_{in} is initially set equal to zero at time $n = 0$, then at time $n = N$, C_{out} is the desired inner product. It will be noted that the value of the sum in (2.7) remains unchanged under a permutation of the summation index. Two special permutations will be utilized later in this chapter: a reversal and a cyclic permutation.

The engagement processor, shown in Figure 2.3, performs a matrix-matrix multiplication or multiplication-addition using an array of cells similar to those shown in Figure 2.2, using an in-place accumulation of partial sums. The rows of matrix A and the columns of matrix B are supplied to the processor via an interface memory. A unit delay is provided between the addressing of successive rows or columns, so that the memory emulates a set of delay lines providing a skew of one arithmetic cycle time between the introduction of successive rows or columns. Each cell of the engagement processor is configured as shown in Figure 2.2, so that it computes the sum of an inner product and the initial value of C_{out}. To compute a matrix product, c_{11} is initialized to zero at time zero, c_{21} and c_{12} are initialized to zero by time 1, c_{13}, c_{22}, and c_{31} are initialized to zero by time 2, and so forth, with the kth antidiagonal initialized to zero by time $k - 1$. At time $k + N$, all of the inner products corresponding to the positions on the kth antidiagonal will be accumulated in place. If the elements are not initially set equal to zero, the sums are offset by the initial values, or a matrix multiplication-addition is performed: $C_{final} = C_{initial} + A * B$. The accumulation of the last term is completed at time $T = (N - 1) + (N - 1) + N = 3N - 2$. That is, the engagement processor performs matrix multiplication or multiplication-addition for full $N \times N$ matrices in time $3N - 2$ using N^2 cells, thus providing about a fourfold improvement in efficiency over the original hexagonal systolic matrix multiplier configuration when full matrices must be multiplied. Since signal processing applications frequently require

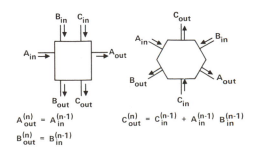

Figure 2.2 Geometries for the inner product step processor.

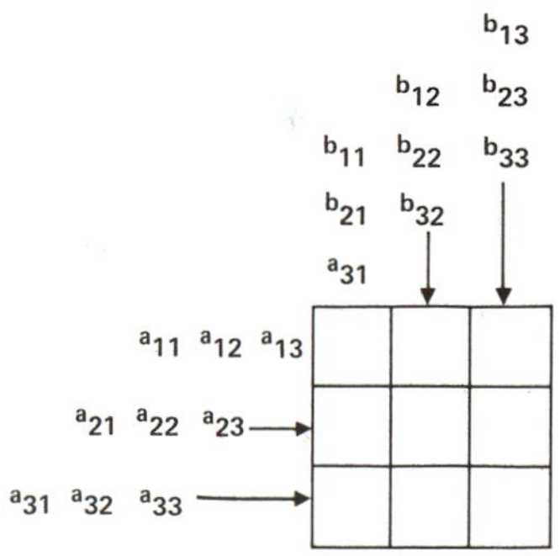

Figure 2.3 Multiplication of full matrices by an engagement processor.

the multiplication of full matrices, the improvement is believed to be significant. The engagement processor calculates the matrix product of (2.1) using the recursions of (2.9) and (2.10), where $c_{jk}^{(n)}$ denotes the contents of cell (j, k) at time n.

$$c_{jk}^{[(j-1)+(k-1)]} = 0 \tag{2.9}$$

$$c_{j,k}^{(n+1)} = c_{j,k}^{(n)} + a_{j,(N-n)+(j-1)+(k-1)} * b_{(N-n)+(j-1)+(k-1),k} \tag{2.10}$$

At time $n = (j-1) + (k-1) + N$, the (j, k) cell then contains the (j, k) element of the matrix product.

If the efficiency is defined as shown in (2.11), it will be seen that in the nonpipelined mode, the engagement processor has an efficiency of $N/(3N - 2)$, which is greater than $\frac{1}{3}$.

$$\text{efficiency} = \frac{\text{number of multiplication-additions performed}}{\text{number of cells} \times \text{total time}} \tag{2.11}$$

An engagement processor may also be used to perform the multiplication of non-square matrices. If A is $J \times N$ and B is $N \times K$, the cellular array size needed is $J \times K$, and the last computation is completed at time $N + (J - 1) + (K - 1) = N + J + K - 2$. In this case, the efficiency is $(NJK)/JK(N + J + K - 2) = N/(N + J + K = 2)$, which is high as long as N is not small compared to $J + K$. It will be shown later in this chapter that frequently it is possible to pipeline matrix multiplications using an engagement processor to provide nearly 100% efficiency. Next, several special cases of matrix-matrix multiplication will be considered using an engagement processor or an augmented engagement processor.

2.3.1. Outer Product of Vectors Using an Engagement Processor

The outer product of two vectors is an important degenerate case of matrix-matrix multiplication. It is defined in vector and component notations in (2.12), and is illustrated in (2.13). The outer product of two vectors is needed in implementing Householder transformations [19] and in ambiguity function calculation.

$$C = ab^T \tag{2.12a}$$

$$c_{jk} = a_j b_k^* \tag{2.12b}$$

$$\begin{bmatrix} a_1 \\ a_2 \\ a_3 \end{bmatrix} [b_1^* \quad b_2^* \quad b_3^*] = \begin{bmatrix} a_1 b_1^* & a_1 b_2^* & a_1 b_3^* \\ a_2 b_1^* & a_2 b_2^* & a_2 b_3^* \\ a_3 b_1^* & a_3 b_2^* & a_3 b_3^* \end{bmatrix} \tag{2.13}$$

The computation of an outer product by the engagement processor is shown in Figure 2.4. The required computation time is $2N - 1$.

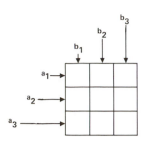

Figure 2.4   Outer product computation using an engagement processor.

2.3.2 Multiplication of a Hankel Matrix by an Arbitrary Matrix

The multiplication of a Hankel matrix by an arbitrary matrix is illustrated in (2.14). It is useful for performing multichannel cross-correlation or as part of the calculation of a cross-ambiguity function.

$$AH = \begin{bmatrix} a_{11} & a_{12} & a_{13} & b_{-2} & b_{-1} & b_0 \\ a_{21} & a_{22} & a_{23} & b_{-1} & b_0 & b_1 \\ a_{31} & a_{32} & a_{33} & b_0 & b_1 & b_2 \end{bmatrix} \qquad (2.14)$$

Since a Hankel matrix is constant along antidiagonals, it is completely defined by its values on its leftmost column and lowest row, and hence an $N \times N$ Hankel matrix may be stored as a vector of length $2N - 1$. In order to use such compressed storage without increasing the time required to load an engagement processor, a single element is used to load a row of the engagement processor via a "bus expander." The use of a bus expander with an engagement processor to perform the multiplication of a Hankel matrix by an arbitrary matrix is shown in Figure 2.5.

Since the long vector is converted to an equivalent matrix input by the bus expander, the computation time required for this configuration is exactly the same as that of the general matrix-matrix multiplication by the configuration of Figure 2.3 (i.e., $3N - 2$).

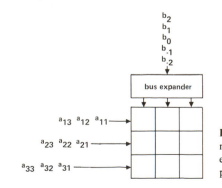

Figure 2.5   Multiplication of a Hankel matrix by an arbitrary matrix using an engagement processor with a bus expander.

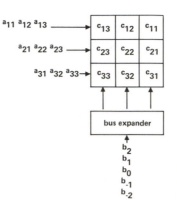

Figure 2.6 Multiplication of a Toeplitz matrix by an arbitrary matrix using an engagement processor with a bus expander.

2.3.3 Multiplication of a Toeplitz Matrix by an Arbitrary Matrix

A Toeplitz matrix is constant along diagonals, so that the multiplication of a Toeplitz matrix by an arbitrary matrix is very similar to the multiplication of a Hankel matrix by an arbitrary matrix, but corresponds to multichannel convolution rather than correlation. It is illustrated in (2.15), and an implementation using an engagement processor with a bus expander is shown in Figure 2.6. The required computation time is $3N - 2$.

$$AT = \begin{bmatrix} a_{11} & a_{12} & a_{13} \\ a_{21} & a_{22} & a_{23} \\ a_{31} & a_{32} & a_{33} \end{bmatrix} \begin{bmatrix} b_0 & b_{-1} & b_{-2} \\ b_1 & b_0 & b_{-1} \\ b_2 & b_1 & b_0 \end{bmatrix} \tag{2.15}$$

2.3.4 Skewed Outer Product Formation

The skewed outer product computation, shown in (2.16), is useful for cross-ambiguity function calculation

$$c_{jk} = a_j b_{j-k} \tag{2.16}$$

It may be implemented as a special case of outer product formation, using an engagement processor with a bus expander as shown in Figure 2.7. The required computation time is $2N - 1$.

2.3.5 Skewed Vector Triple-Product Calculation

The skewed triple product of three vectors is defined by (2.17). Like the skewed outer product, the skewed triple product is also useful for cross-ambiguity function calculation. The diagonal portion $P(s, s)$ of the skewed triple product is called a "triple-product convolution" and is useful for beamforming applications [20].

$$P(u, s) = \sum_{k=0}^{N-1} a_k b_{k-u} c_{k+s} \tag{2.17}$$

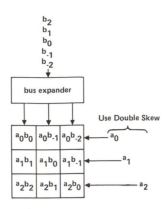

Figure 2.7 Skewed outer product formation using an engagement processor with a bus expander.

It may be computed as a skewed outer product of a with b, followed by a Hankel matrix postmultiplication by c, with a total computation time of $5N - 3$.

2.3.6 Two-Dimensional Discrete Fourier Transforms Using a Matrix Processor

An $N \times N$ matrix multiplier can perform the discrete Fourier transform (DFT) of N vectors of length N as a single matrix multiplication. If F is the DFT matrix defined by (2.18), then FX is a matrix whose columns are the DFTs of the columns of X, and XF is the matrix whose rows are the DFTs of the rows of X. Similarly, FXF is the two-dimensional DFT of X.

$$F_{pq} = e^{-i2\pi pq/N} \tag{2.18}$$

2.3.7 Cross-Ambiguity Calculation Using an Engagement Processor with a Bus Expander

The cross-ambiguity function defined by (2.19) is a generalized cross-correlation, providing a search through relative frequency offset as well as delay [21]. Table 2.3 shows how the cross-ambiguity calculation is implemented by current optical processors, and may be used to reduce this calculation to a combination of the matrix operations described in this chapter.

$$A_{12}(\tau, f) = \int g_1(t) g_2^*(t + \tau) e^{-i2\pi ft} \, dt \tag{2.19}$$

The two-dimensional Fourier transform method requires a matrix-vector multiplication to form the one-dimensional Fourier transform of g_2, an outer product, and two-dimensional Fourier transform. The τ-slice method corresponds to forming a skewed outer product and then performing a matrix multiplication. The time-integrating correlator configuration corresponds to forming a vector pointwise

TABLE 2.3 CROSS-AMBIGUITY CALCULATION METHODS

Two-Dimensional Fourier transform

$$\int [g_1(t)g_2^*(u)e^{-i2\pi ut}]e^{-i2\pi(ft+u\tau)} \, dt \, du$$

τ Slice

$$\int [g_1(t)g_2^*(t+\tau)]e^{-i2\pi ft} \, dt$$

Time Integrating

$$e^{-i\pi f^2} \int [g_1(t)e^{-i\pi t^2}][g_2^*(t+\tau)]e^{i\pi(t-f)^2} \, dt$$

Space Integrating

$$\int [g_1(t)e^{-i2\pi ft}][g_2^*(t+\tau)] \, dt$$

product, a skewed outer product, and then premultiplying a Toeplitz matrix. The space-integrating correlator configuration corresponds to scaling the columns of a matrix, followed by postmultiplication by a Hankel matrix.

The required computation times may be substantially reduced by using more than one bus expander, and by configuring memory to avoid the need for row or column serial transfers of intermediate matrices to the interface memories. Also, partitioning of the matrix operations will be required when the number of delay/frequency cells to be computed exceeds the size of the processor array.

2.3.8 Partitioned Matrix Multiplication

An $N \times N$ processor array will frequently be required to perform the multiplication of matrices larger than $N \times N$. Although the algebra for such operations is elementary, using the partitioned matrix multiplication identity of (2.20), it is desirable to provide a speed improvement by providing parallel storage and recall of the intermediate matrix products.

$$
\begin{matrix} N & N \end{matrix} \quad\quad \begin{matrix} N & N \end{matrix} \quad\quad \begin{matrix} N & N \end{matrix}
$$
$$
\begin{matrix} N \\ N \end{matrix} \left(\begin{array}{c|c} A & B \\ \hline C & D \end{array}\right) \begin{matrix} N \\ N \end{matrix} \left(\begin{array}{c|c} E & F \\ \hline G & H \end{array}\right) = \begin{matrix} N \\ N \end{matrix} \left(\begin{array}{c|c} AE+BG & AF+BH \\ \hline CE+DG & CF+DH \end{array}\right) \quad (2.20)
$$

For this reason it is desirable to augment the engagement processor of Figure 2.3 by a "vertical memory" with addressing orthogonal to the plane of the array, so that an entire $N \times N$ matrix can be stored and subsequently reloaded in one memory access time. With such a vertical memory, the storage and retrieval times needed to implement the partitioned matrix multiplication of (2.20) are negligible, and the $2N \times 2N$ matrix multiplication time is approximately eight times the $N \times N$ matrix multiplication or multiplication-addition time. The partitioning may of course be iterated to handle still larger matrices, limited only by the space available in the vertical memory. Although techniques for performing a $2N \times 2N$ matrix multiplication using fewer than eight $N \times N$ matrix multiplications are known, they are less numerically stable, and appear to also require greatly increased memory space to store the intermediate computational terms.

2.3.9 Matrix Multiplication in Time N Using an
Engagement Processor with N Bus Expanders

It was mentioned earlier that an inner product is unchanged by a cyclic permutation of the products to be summed. This idea will now be used to provide a matrix multiplication using an $N \times N$ enhanced engagement processor in such a way that all of the cells are active all of the time—a result ordinarily achieved only when pipelining is possible. If the summation index of (2.7) is interpreted as an integer modulo N, then the value of the sum is unchanged if the summation index is cyclically shifted by j, as shown in (2.21), where $C = BA$:

$$c_{jk} = \sum_{s=0}^{N-1} b_{js} a_{sk} = \sum_{n=0}^{N-1} b_{j,\,j+n} a_{j+n,\,k} \tag{2.21}$$

The implementation of this technique proceeds as follows.

1. Parallel load $A^{(0)} = A$ from local vertical memory.
2. Parallel load $B^{(0)}$ with $B^{(0)}_{j,k} = b_{jj}$ via N bus expanders across the rows of the engagement processor.
3. After each multiplication-addition cycle, A moves upward, that is, $A^{(n+1)}_{jk} = A^{(n)}_{j+1,k}$.
4. Parallel load $B^{(n)}$ with $B^{(n)}_{jk} = b_{j,\,j+n}$.

The address control would typically be via counters modulo N to address each row of B. It will be noted that the jth row of $B^{(n)}$ is thus loaded with the nth extended diagonal of B. This is accomplished by providing a cyclic offset in the addressing registers used to load the rows of B.

2.3.10 Matrix Inversion

A hexagonal systolic array may be used to perform the L-U factorization of a matrix into a product of a lower-triangular matrix and an upper-triangular matrix [6], essentially the first half of Gaussian elimination. Such an array uses inner product step processors together with one exceptional cell on the boundary which is required to perform division. It has been observed [3,7] that if diagonal interconnections are added to a rectangular array, as shown in Figure 2.8, it may also function as a hexagonal array. It has also been shown [6] that a linear systolic array with one boundary processor capable of performing division may be used to per-

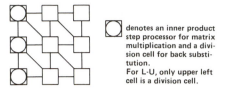

denotes an inner product step processor for matrix multiplication and a division cell for back substitution.
For L-U, only upper left cell is a division cell.

Figure 2.8   Combined rectangular/hexagonal array.

form the solution of a linear system of equations with a triangular coefficient matrix, that is, the "back-substitution" half of Gaussian elimination. Therefore, an $N \times N$ rectangular array with diagonal interconnects and N boundary processors, as shown in Figure 2.8, may invert $N \times N$ matrices in time $O(N)$ by using it first as a hexagonal array and then as N linear arrays.

2.3.11 Partitioned Matrix Inversion Using an Engagement Processor

The requirements for inversion of a partitioned matrix using an engagement processor will be similar to those for partitioned matrix multiplication, but two additional factors must be considered: (1) Do the required inverses of submatrices exist? (2) Does the amount of memory required increase as a result of the partitioned algorithm?

Let the $2N \times 2N$ matrix R be partitioned into $N \times N$ submatrices as follows:

$$R = \begin{pmatrix} A & B \\ C & D \end{pmatrix} \tag{2.22}$$

Then the inverse is given in (2.23), assuming that the submatrices A and D are both invertible [22]. In many signal processing applications, R will be symmetric and nonnegative definite. In this case, invertibility of R implies the invertibility of A and D.

$$R^{-1} = \begin{pmatrix} A^{-1} + A^{-1}B\varDelta^{-1}CA^{-1} & -A^{-1}B\varDelta^{-1} \\ -\varDelta^{-1}CA^{-1} & \varDelta^{-1} \end{pmatrix} \tag{2.23}$$

with $\varDelta = D - CA^{-1}B$. If the computation sequence of Figure 2.9 is used, it will readily be seen that an $N \times N$ engagement processor with a vertical memory can compute the inverse of a $2N \times 2N$ matrix in approximately eight times the amount of time required to invert an $N \times N$ matrix, with no increase in memory required beyond that needed to store the partitioned $2N \times 2N$ matrix.

Storage contents after successive submatrix operations:

A	$\tilde{R} = A^{-1}$	$\tilde{R}$	$\tilde{R}$	$\tilde{R}$	$\tilde{R}$	$\tilde{R} + MT$	$\tilde{R} + MT$
B	B	$M = \tilde{R}B$	M	M	M	M	$S - M\varDelta^{-1}$
C	C	C	C	$Q = C\tilde{R}$	$T = \varDelta^{-1}Q$	T	T
D	D	$\varDelta = D - CM$	$\varDelta^{-1}$	$\varDelta^{-1}$	$\varDelta^{-1}$	$\varDelta^{-1}$	$\varDelta^{-1}$

Operation count: 2 inversions, 6 matrix multiplications or multiplication-additions.

Figure 2.9   Storage requirements and operation count for partitioned matrix inversion.

2.3.12 Modular DFT Using an Engagement Processor

An engagement processor using an $N \times N$ array of cells can be used to provide the equivalent of a radix N FFT with N^2 parallelism. This will be illustrated for a length N^2 DFT as defined by

$$G_k = \sum_{n=0}^{N^2-1} g_n e^{-i2\pi nk/N^2} \qquad \text{for } k = 0, 1, \ldots, N^2 - 1 \qquad (2.24)$$

The "lexical scan" indexing of (2.25) may be used to interpret a long one-dimensional DFT as a partial DFT, followed by a pointwise matrix multiplication, followed by a partial DFT on the orthogonal direction [23] as shown in (2.26).

$$\left.\begin{array}{l} n = n_1 N + n_2 \\ k = k_1 + k_2 N \end{array}\right\} \quad \text{for } n_1, n_2, k_1, k_2 = 0, 1, \ldots, N - 1 \qquad \begin{array}{l} (2.25a) \\ (2.25b) \end{array}$$

$$G_{k_1} + k_2 N = \sum_{n_2=0}^{N-1} \sum_{n_1=0}^{N-1} e^{-i2\pi k_2 n_2 /N} e^{-i2\pi k_1 n_2 /N^2} e^{-i2\pi n_1 k_1 /N} g_{n_1 N + n_2} \qquad (2.26)$$

If these operations are performed using a complex $N \times N$ engagement processor with vertical memory and N bus expanders, approximately $2N$ arithmetic cycle times are required for a length N^2 DFT, or with $M = N^2$, the computation time is about $2M^{0.5}$ using M cells.

Since multiplication by the "twiddle factors" of $e^{-i2\pi k_1 n_2/N^2}$ can be performed in parallel in one arithmetic cycle time, it consumes a negligible amount of time compared to the matrix multiplications. Thus, eliminating the twiddle factors would appear to complicate the data accessing while providing only a minor speed improvement.

2.4 CONCLUDING REMARKS

It has been shown that a new systolic architecture, the engagement processor, provides efficient multiplication of dense matrices. Two types of enhancements to engagement processors were described: a memory with three-dimensional organization for storing matrix intermediate computation terms, and a bus expander for loading a row or column of the engagement processor in parallel. Enhanced engagement processors were shown to perform efficiently partitioned matrix multiplication and inversion, one- and two-dimensional discrete Fourier transforms, and cross-ambiguity function calculation.

REFERENCES

[1] J. M. Speiser, and H. J. Whitehouse, "Architectures for Real-Time Matrix Operations," *Proc. 1980 Government Microcircuits Appl. Conf.*, Houston, Nov. 1980, pp. 19–21.

[2] J. M. Speiser, H. J. Whitehouse, and K. Bromley, "Signal Processing Applications for

Systolic Arrays," *14th Asilomar Conf. Circuits, Syst. Comput.*, Pacific Grove, Calif., Nov. 17–19, 1980, IEEE Catalog 80CH1625-3, pp. 100–104.

[3] S. Y. Kung, "VLSI Array Processors for Signal Processing," *MIT Conf. Adv. Res. VLSI*, Cambridge, Mass., Jan. 1980.

[4] J. J. Dongarra, et al., *LINPACK User's Guide*, SIAM, Philadelphia, 1979.

[5] B. S. Garbow, et al., *Matrix Eigensystem Routines—EISPACK Guide Extensions*, Springer-Verlag, New York, 1977.

[6] H. T. Kung, "Systolic Arrays for VLSI," in I. S. Duff and G. W. Stewart, eds., *Sparse Matrix Proceedings*, 1978, SIAM, Philadelphia, 1979.

[7] S. Y. Kung, et al., "Wavefront Array Processor: Language, Architecture, and Application," *IEEE Trans. Comput.*, Special Issue on Parallel and Distributed Computers, *C-31*(11): 1054–1066 (Nov. 1982).

[8] G. A. Frank, J. Blackmer, and P. Kuekes, "A 200 MOPS Systolic Processor," *Proc. SPIE Symp.*, Vol. 298: *Real-Time Signal Processing IV*, San Diego, Aug. 1981.

[9] J. Symanski, et al., "A Systolic-Array Processor Implementation," *Proc. SPIE Symp.*, Vol. 298: *Real-Time Signal Processing IV*, San Diego, Aug. 1981.

[10] S. Y. Kung, and Y. H. Hu, "A Highly Concurrent Algorithm and Pipelined Architectures for Solving Toeplitz Systems," *IEEE Trans. Acoust. Speech Signal Process.*, *31*(1):66–76 (Feb. 1983).

[11] W. M. Gentleman, and H. T. Kung, "Matrix Triangularization by Systolic Arrays," *Proc. SPIE Symp.*, Vol. 298: *Real-Time Signal Processing IV*, San Diego, Aug. 1981.

[12] A. M. Finn, F. T. Luk, and C. Pottle, "Systolic-Array Computation of the Singular Value Decomposition," *Proc. SPIE Symp.*, Vol. 341: *Real-Time Signal Processing V*, Arlington, Va., May 1982.

[13] R. P. Brent and F. T. Luk, "A Systolic Architecture for Almost Linear-Time Solution of the Symmetric Eigenvalue Problem," Cornell University Tech. Rep. TR82-525, Aug. 1982.

[14] S. Y. Kung and R. J. Gal-Ezer, "Eigenvalue, Singular Value and Least Squares Solvers via the Wavefront Array Processors," in L. Synder et al., eds., *Algorithmically Specialized Computer Organizations*, Academic Press, New York, 1983.

[15] B. Widrow, et al., "Adaptive Noise Cancelling: Principles and Applications," *Proc. IEEE*, *63*:1692–1716 (Dec. 1975).

[16] L. J. Griffiths, "A Continuously-Adaptive Filter Implemented as a Lattice Structure," *1977 IEEE Int. Conf. Acoust. Speech Signal Process.*, IEEE Catalog 77CH1197-3, pp. 683–686.

[17] R. A. Monzingo and T. W. Miller, *Introduction to Adaptive Arrays*, Wiley, New York, 1980, Chap. 6: "Direct Inversion of the Sample Covariance Matrix."

[18] S. Haykin, ed., *Nonlinear Methods of Spectral Analysis*, Springer-Verlag, New York, 1979.

[19] C. L. Lawson and R. J. Hanson, *Solving Least Squares Problems*, Prentice-Hall, Englewood Cliffs, N.J., 1974.

[20] J. M. Speiser, H. J. Whitehouse, and N. J. Berg, "Signal Processing Architectures Using Convolutional Technology," *Proc. SPIE Symp.*, Vol. 154: *Real-Time Signal Processing*, 1978, pp. 66–80.

[21] A. W. Rihaczek, *Principles of High-Resolution Radar*, McGraw-Hill, New York, 1969.

[22] E. Bodewig, *Matrix Calculus*, 2nd ed., North-Holland/Interscience, New York, 1959.

[23] H. J. Whitehouse and J. M. Speiser, "Linear Signal Processing Architectures," in *Aspects of Signal Processing, with Emphasis on Underwater Acoustics*, Part 2, D. Reidel, Dordrecht, The Netherlands, 1977.

3

Spectral Estimation: From Conventional Methods to High-Resolution Modeling Methods

S. Y. KUNG, D. V. BHASKAR RAO, AND K. S. ARUN

University of Southern California
Los Angeles, California

3.1 INTRODUCTION

Spectral analysis forms the basis of a major part of signal processing, typically for distinguishing and tracking signals of interest, and for extracting information from the relevant data. Given a finite number of noisy measurements of a discrete-time stochastic process, or its first few covariance lags, the classical spectral estimation problem was that of estimating the shape of its continuous power spectrum. For some modern applications of signal processing, such as radar, sonar, and phased arrays, the spectrum of interest is a line spectrum, and the modern spectral estimation problem is that of estimating the locations of these spectral lines. As a result, the notions of bias and variance that once referred to estimates of spectral shape now refer to estimates of frequency. Another measure of performance in the classical problem was frequency resolution that defined how finely a spectrum could be examined [1]. In the modern context, resolution is the ability to distinguish and identify spectral lines that are closely spaced in frequency [2]. In many modern applications, the spectral estimates have to be based on short data records (in radar, for example, only a few data samples are available in each radar pulse), and yet low-bias, low-variance, high-resolution estimates are desired.

The frequency resolution in conventional Fourier transform methods is roughly equal to the reciprocal of the data record length. So some additional constraints (or prior information) have to be incorporated to improve the resolution capability. To this end, in modern methods the data are modeled as the output of a linear system driven by white noise. When the model is appropriate, these methods will lead to enhanced performance. The price paid for this improvement is usually in the form of increased computational complexity compared to conventional methods that utilize FFT. The advent of VLSI, however, has reduced the cost of computation hardware, providing a major impetus toward the utilization and development of computationally more sophisticated spectral estimation methods for improved performance.

In this chapter we first discuss the conventional methods for power spectrum estimation. Then in the later sections we focus our attention on the modern methods, with special emphasis on those modeling methods that are suited for resolving narrowband signals. A more detailed discussion on this subject can be found in [3,4].

3.2 CONVENTIONAL METHODS [5]

Suppose that the process $x(n)$ is a zero-mean, wide-sense-stationary discrete-time, real stochastic process; then the quantity of interest is the power spectrum, which represents the distribution of power over frequencies, and is defined as [6]

$$P(\omega) = \sum_{n=-\infty}^{\infty} r_{xx}(n)e^{-j\omega n} \tag{3.1}$$

Here, $r_{xx}(m)$ is the autocorrelation sequence, defined as

$$r_{xx}(m) = E[x(n)x(m+n)]$$

where $E[\cdot]$ denotes the expectation operator. The indirect approach to power spectrum estimation due to Blackman and Tukey is obtained by first estimating the autocorrelation over a finite segment and then by taking the Fourier transform, inherently assuming that the unavailable covariance lags are zero.

$$P_{\text{BT}}(\omega) = \sum_{n=-M+1}^{M-1} \hat{r}_{xx}(n)e^{(-j\omega n)} \tag{3.2}$$

Very often the autocorrelation function is weighted by a window function (to smooth the transition to zero) before the Fourier transformation.

$$P_{\text{BT}}^{M}(\omega) = \sum_{n=-M+1}^{M-1} \hat{r}_{xx}(n)W(n)e^{(-j\omega n)} \tag{3.3}$$

The definition of the power spectrum may use an alternative route. Using the classical Wiener–Khinchin theorem [5], the power spectrum of an ergodic process is also given by

$$P(\omega) = \lim_{N\to\infty} E \frac{1}{N}\left|\sum_{n=0}^{N-1} x(n)e^{-jn\omega}\right|^2 \tag{3.4}$$

This leads to a more direct computational method. In practice we have a finite record length N. So the direct method uses the above definition with a finite N and drops the ensemble averaging. The power spectrum estimate based on the direct approach is called the periodogram, a term first used by Schuster [7] and it is obtained by taking the magnitude squared of the Fourier transform of the finite-length data.

$$P_{\text{PER}}(\omega) = \frac{1}{N} \left| \sum_{n=0}^{N-1} x_n e^{(-j\omega n)} \right|^2 \tag{3.5}$$

The periodogram estimate is the same as the indirect estimate of (3.2) when the following biased estimate $\hat{r}(m)$ is used for the covariance in the Blackman-Tukey approach:

$$\hat{r}(m) = \frac{1}{N} \sum_{n=0}^{N-m-1} x(n)x(n+m)$$

These approaches have become popular due to the computational efficiency of the fast Fourier transform (FFT) algorithm. However, the zero extension of data/covariance outside the observation interval is unnatural, and limits the resolution to roughly the reciprocal of the data length. Moreover, the abrupt transition to zero causes large side lobes (Gibbs phenomenon) in the spectrum estimate. A proper choice of window functions can improve the statistical stability of the estimate and alleviate sidelobe leakage, but at the same time resolution is further compromised. For details regarding the trade-off between resolution and sidelobe leakage in the choice of window functions, we refer the reader to [5,8,9].

It can be shown that both the direct and indirect estimates of power spectrum are not consistent; that is, the variance does not approach zero as the record length increases [5]. Bartlett and Medhi [10] and Welch [11] independently suggested segmental averaging as a means of achieving some sort of pseudo-ensemble averaging, thereby improving the asymptotic statistical properties of the spectral estimate. Here the data are split into segments (possibly overlapping), the power spectrum is computed for each segment, and the overall power spectrum is obtained by averaging over all segments. However, here resolution is compromised even further and since the data records are often short in most applications of our interest, the segmental averaging technique is not suitable.

In summary, all these conventional spectral estimation methods take advantage of the computationally efficient FFT algorithm, but do not provide sufficient resolution. The frequency resolution is bounded by a fundamental uncertainty type limit, equal to the reciprocal of the covariance length used in the discrete Fourier transform. Hence the only means for achieving high resolution, given a short segment of the covariance sequence, is by extending the covariance outside the given segment based on some prior knowledge. To this end, modern methods model the data as the output of a linear rational system driven by input white noise, and in effect, provide a covariance extrapolation outside the observation interval.

3.3 MODEL-BASED METHODS

In many applications, the underlying physical environment generating the signal can be modeled well by a linear rational system of low order. In speech, for instance, it is known that a good model for the speech-generating mechanism is an all-pole linear system driven by white noise. In model-based methods, such prior information is exploited to improve the frequency resolution.

Conventional methods assume that the covariance is zero outside the given segment, but in practice, especially when dealing with narrowband signals, the covariance tends to be repetitive, and a zero extension is most inappropriate. Model-based methods extend the covariance sequence outside the given segment via certain recurrence relations determined by the model parameters. These parameters, estimated from the available covariance segment, completely specify the infinite covariance extension and the corresponding power spectrum. Hence, the modern spectral estimation problem is basically a parameter estimation problem and such parametric spectral estimates achieve very high resolution when the model is appropriate.

In general, a linear, rational system has an input–output relationship described by a linear difference equation:

$$y(t) = \sum_{i=1}^{p} a_i\, y(t-1) + \sum_{i=1}^{q} b_i\, v(t-i) + v(t) \tag{3.6}$$

where the input is $\{v(t)\}$ and the output is $\{y(t)\}$. This model is known as an autoregressive moving average model (**ARMA**).

The transfer function $H(z)$ of this system is

$$H(z) = \frac{B(z)}{A(z)}$$

where

$$B(z) = 1 + \sum_{i=1}^{q} b_i\, z^{-i} \quad \text{and} \quad A(z) = 1 - \sum_{i=1}^{p} a_i\, z^{-i}$$

When the input $\{v(t)\}$ is white noise of variance σ^2, the power spectrum of $y(t)$ is

$$P(\omega) = S(z)\Big|_{z=e^{j\omega}}$$

where

$$S(z) = \sigma^2 H(z)H(z^{-1}) = \sigma^2\, \frac{B(z)B(z^{-1})}{A(z)A(z^{-1})} \tag{3.7}$$

$S(z)$ is also the two sided z transform of the doubly infinite covariance of $y(t)$,

$$S(z) = \sum_{k=-\infty}^{\infty} r(k)z^{-k}$$

If $h(n)$ is the impulse response of the system, then $H(z^{-1})$ must be the z transform of the time-reversed sequence $h(-n)$. Therefore, (3.7) indicates that when the system $H(z)$ is driven by the sequence $\sigma^2 h(-n)$, the output sequence will be the covariance $r(m)$. This implies

$$r(m) = \sum_{i=1}^{p} a_i r(m-i) + \sigma^2 \sum_{i=0}^{q} b_i h(-m+i). \qquad (3.8)$$

However, since the impulse response $h(n)$ is a causal sequence, we have $h(-m+i) = 0$ for all $m > i$. As a result, the above equation simplifies to

$$r(m) = \sum_{i=1}^{p} a_i r(m-i) \qquad \text{for all } m > q. \qquad (3.9)$$

Thus linear rational modeling implies that the covariance sequence can be extended to infinity from only p lags, by means of a recurrence relation, and so ARMA models can vastly improve resolution.

An interesting special case of the general rational model (3.6) is when $A(z)$ is constrained to be equal to 1. Then the model is normally called a moving average model (MA). The covariance of such a model is finite in length, and so, MA models cannot improve the frequency resolution beyond Fourier transform methods.

3.3.1 Autoregressive Models

A more popular model used for spectral estimation is the special case of (3.6) when $B(z)$ is constrained to be equal to 1. Then

$$y(t) = \sum_{i=1}^{p} a_i y(t-i) + v(t) \qquad (3.10)$$

Here output $y(t)$ is generated as a linear regression of its past values, and hence such a model is known as an autoregressive (AR) model. [In retrospect, the general model (3.6) is a composite of AR and MA models, and so is appropriately called an autoregressive moving average (ARMA) model.]

In AR models, $B(z) = 1$, so that

$$P(\omega) = \sigma^2 \left. \frac{1}{A(z)A(z^{-1})} \right|_{z = e^{j\omega}}$$

and the recurrence relation (3.9) holds for all $k > 0$.

$$r(k) = \sum_{i-1}^{p} a_i r(k-i) \qquad k > 0$$

For $k = 0$, recurrence relation (3.9) specializes to

$$r(0) = \sum_{i=1}^{p} a_i r(-i) + \sigma^2.$$

Given M exact covariance lags $\{r(k)\}$, where $M \geq p$, the parameters $\{a_i\}$ of the AR(p) model can be estimated from the recurrence equations above for the first p values of k. They are known as the Yule–Walker equations or normal equations and are sometimes also referred to as the discrete-time Wiener–Hopf equations. In matrix form,

$$
\begin{bmatrix}
r(0) & r(1) & r(2) & \cdots & r(p) \\
r(1) & r(0) & r(1) & \cdots & r(p-1) \\
r(2) & r(1) & r(0) & \cdots & r(p-2) \\
\vdots & \vdots & & & \vdots \\
r(p) & r(p-1) & r(p-2) & \cdots & r(0)
\end{bmatrix}
\begin{pmatrix}
1 \\
-a_1 \\
-a_2 \\
\vdots \\
-a_p
\end{pmatrix}
=
\begin{bmatrix}
\sigma^2 \\
0 \\
0 \\
\vdots \\
0
\end{bmatrix}
\tag{3.11}
$$

Thus AR parameter estimation involves the solution of a symmetric positive-definite Toeplitz system, which can be computed very efficiently using the now well-known Levinson algorithm [32,34,36,37,38,48].

There are two justifications for the AR modeling approach. The first is in terms of minimum mean-square inverse filtering or the best linear prediction filter. More precisely, since an AR model is all-pole, the inverse filter is all-zero and has the same structure as the prediction error filter. Furthermore, the coefficients of the linear prediction filter that minimize the prediction error also satisfy the normal equations. Hence AR modeling is equivalent to the linear prediction approach.

The second justification is that under the Gaussian assumption the infinite covariance extension to an M lag segment, provided by an AR(M) model, maximizes the entropy of the corresponding time series. In other words, among all possible extensions, the time series corresponding to the AR extension is the "whitest" and has the "flattest" spectrum. In fact, the AR method is equivalent to the now popular maximum-entropy method [12, 13].

3.3.2 Maximum-Entropy Method (MEM) [12–14]

The maximum-entropy method (MEM) was first developed by Burg [12], who argued that the covariance extrapolation should be made so that the data characterized by the extended autocorrelation sequence has maximum entropy. A heuristic explanation for the choice of this criterion is that it maximizes the randomness or whiteness of the process. The entropy rate is defined as

$$
H = \int_{-\pi}^{+\pi} \ln\,[P(\omega)]\,d\omega
$$

where ln refers to the natural logarithm, and

$$
P(\omega) = \sum_{n=-\infty}^{+\infty} r(n)e^{-j\omega n}
$$

To maximize H, we shall take the derivative of H with respect to $\{r(i), |i| = N + 1,$

$N + 2, \ldots\}$. This leads to

$$\int_{-\pi}^{\pi} \frac{e^{-j\omega m}}{P(\omega)}\, d\omega = 0 \qquad \text{for } |m| = N + 1, N + 2, \ldots$$

This implies that $P^{-1}(\omega)$ has a finite Fourier expansion, that is,

$$P^{-1}(\omega) = \sum_{m=-p}^{p} T(m)e^{-j\omega m} \tag{3.12}$$

By the spectral factorization theorem, since $P^{-1}(\omega)$ is nonnegative, we know that

$$\sum_{m=-p}^{p} T(m)e^{-jm\omega} = \frac{1}{K} \sum_{m=0}^{p} a(m)e^{j\omega m} \sum_{m=0}^{p} a(m)e^{-j\omega m} \tag{3.13}$$

for some $\{a(m)\}$ [where $a(0) = 1$ and K is the adjusting constant].

In other words, combining (3.12) and (3.13), we have

$$P(\omega) = \frac{K}{a(z)a(z^{-1})\big|_{z=e^{j\omega}}}$$

This proves that the MEM yields an AR solution. The actual computation of the $\{a(m)\}$ coefficients involves the solution of equation (3.11).

It can be shown [3] that the entropy rate is also given by

$$H = \lim_{N \to \infty} \tfrac{1}{2} \ln\, [\det \mathbf{R}_N] = \tfrac{1}{2} \ln\, [P_\infty] \tag{3.14}$$

where $\mathbf{R}_N$ is the $N \times N$ covariance matrix of the process and P_∞ is the prediction error variance in predicting the process from the infinite past. Now from (3.14) a process with maximum entropy is one with the maximum value of P_∞; in other words, it is least predictable. So the MEM method produces the *flattest spectrum* that is consistent with the given correlation lags.

The simplest example is when the only constraint on the covariance data is $r(0)$ (the signal power). The process with maximum entropy is a white-noise process with variance $r(0)$, since it yields the maximum prediction error $P_\infty\, [= r(0)]$.

When the spectrum is constrained to match $r(0)$, $r(1)$, $\ldots$, $r(p)$, it is equivalent to specifying the prediction errors $P_1, P_2, \ldots, P_p$, where P_k denotes the prediction error variance in predicting the process from the k immediate past values. Since the prediction error can only decrease (and at worst be unchanged) when more of the past is available for the purposes of prediction, hence P_{k+1} must be always less than (or equal to) P_k. Consequently, for a prespecified $(P_1, P_2, \ldots, P_p)$, the spectrum with the largest P_∞ is the one that keeps P_k at the constant level of P_p for all $k > p$. The AR(p) spectrum provides such a constant extension to the prediction error sequence, and consequently, it is also the MEM spectrum or the least predictable ("whitest") spectrum consistent with the given first p covariance lags. A theoretical justification for the MEM criterion was provided by Shore [53,54] who showed that for Gaussian processes, the MEM estimate *minimizes* the so-called relative entropy or the

Kullback-Leibler mutual information [55] for discrimination between the probability density corresponding to the estimate, and the probability density of a white process.

3.3.3 Pisarenko's Method |15,16|

The above interpretation of MEM as a method that produces the flattest possible spectrum estimate, raises the question of its applicability to spectral line estimation. For instance, maximizing entropy will not be appropriate if it is known a priori that the signal consists only of noise-free sinusoids, which can be predicted with zero error ($P_\infty = 0$), and has minimum entropy. In this case, our aim should be to produce a line spectrum, not the flattest possible spectrum. Moreover, the correlation matching constraint imposed by MEM is not justifiable when it is known beforehand that uncorrelated additive noise is present, as the AR model generated will match the wrong correlation lags.

The observed covariance matrix $\boldsymbol{R}$ is the sum of the signal covariance matrix $\boldsymbol{R}_x$ and the noise covariance matrix $\boldsymbol{R}_n$ (i.e., $\boldsymbol{R} = \boldsymbol{R}_x + \boldsymbol{R}_n$). Intuition tells us that it is desirable to reduce the noise contribution to $\boldsymbol{R}$ before fitting a model to the covariance matrix. For simplicity we assume that the noise is white (i.e., $\boldsymbol{R}_n = \sigma^2\boldsymbol{I}$). Here an AR model fitted to the modified covariance ($\boldsymbol{R} - \sigma^2\boldsymbol{I}$) results in enhanced resolution. Often σ^2 is unknown, and hence an estimate $\hat{\sigma}^2$ has to be used. The larger the estimate $\hat{\sigma}^2$, the sharper is the spectral estimate that matches ($\boldsymbol{R} - \hat{\sigma}^2\boldsymbol{I}$), and also the entropy is reduced (as the prediction error power P_∞ decreases). The largest value of $\hat{\sigma}^2$ that can be subtracted [so that ($\boldsymbol{R} - \hat{\sigma}^2\boldsymbol{I}$) remains semipositive definite] is the minimum eigenvalue of $\boldsymbol{R}$. For the limiting case, when $\hat{\sigma}^2$ is equal to the minimum eigenvalue of $\boldsymbol{R}$, the matrix ($\boldsymbol{R} - \hat{\sigma}^2\boldsymbol{I}$) becomes singular (yet nonnegative definite), $P_k = 0$ for all $k > p$, and the entropy is minimum. For this limiting case of minimum entropy, the prediction filter polynomial turns out to be the null vector of ($\boldsymbol{R} - \hat{\sigma}^2\boldsymbol{I}$), or equivalently, the minimum eigenvector of $\boldsymbol{R}$.

The approach above was originally suggested by Pisarenko [16] for the problem of retrieving sinusoids in white noise. Let us now take a closer look at the sinusoids plus noise model. Suppose that we have p complex exponentials with amplitudes $\{q_i, i = 1, 2, \ldots, p\}$, at frequencies $\{\omega_i, i = 1, 2, \ldots, p\}$, corrupted by uncorrelated additive white noise. Then the $(p + 1) \times (p + 1)$ Toeplitz matrix formed from exact autocorrelations should ideally have the following form:

$$\boldsymbol{R} = \boldsymbol{R}_x + \boldsymbol{R}_n$$

$$= \boldsymbol{F}\boldsymbol{A}\boldsymbol{F}^* + \sigma^2\boldsymbol{I}$$

where * denotes complex conjugate transpose,

$$\boldsymbol{F} = \begin{bmatrix} 1 & 1 & \cdots & 1 \\ e^{-j\omega_1} & e^{-j\omega_2} & & e^{-j\omega_p} \\ \vdots & \vdots & & \vdots \\ e^{-j\omega_1 p} & e^{-j\omega_2 p} & & e^{-j\omega_p p} \end{bmatrix}$$

and

$$A = \text{diag}\,(q_i^2,\, i = 1, 2, \ldots, p)$$

Now we have the following observations.

1. Note that FAF^* has only rank p; therefore, σ^2 has to be an eigenvalue of the $(p + 1) \times (p + 1)$ matrix R.
2. In fact, σ^2 has to be the minimum eigenvalue, so that $[R - \sigma^2 I]\ (=R_x)$ remains semipositive definite.
3. Finally, by an important Caratheodory theorem [35], the roots associated with the minimum eigenvector **a** will be equal to $\{e^{j\omega_i},\, i = 1, \ldots, p\}$, $\omega_i \neq \omega_j$ when $i \neq j$.

All these results combined lead to Pisarenko's method as summarized below.

1. Compute the minimum eigenvalue of matrix R.
2. Compute the corresponding eigenvector **a**.
3. Solve for $a(z) = 0$ to obtain the locations of the spectral lines.
4. The power in each of the sinusoidal components can be obtained by solving

$$\tilde{F} \cdot \begin{bmatrix} q_1^2 \\ \vdots \\ q_p^2 \end{bmatrix} = \begin{bmatrix} r(0) - \lambda_{\min} \\ r(1) \\ \vdots \\ r(p - 1) \end{bmatrix}$$

where $\tilde{F}$ is obtained from matrix F by dropping the last row.

The minimum eigenvalue of R can be computed efficiently by a modified Rayleigh quotient iteration scheme that exploits the Toeplitz structure of the matrix and furthermore, is VLSI implementable [32,36].

When the number of sinusoids is not known a priori, we could start with a large $(N \times N)$ Toeplitz matrix, and examine its eigenvalue distribution. Ideally, the minimum eigenvalue will have multiplicity $(N - p)$, and p can be estimated from the eigenvalue distribution. In most practical situations, however, a cluster of eigenvalues are detected in the neighborhood of the minimum eigenvalue rather than multiple minimum eigenvalues. Then a direct utilization of Pisarenko's method is not appropriate. One should resort to methods such as MUSIC [29] (see Section 3.3.4) or a Toeplitz approximation method (see Section 3.4).

Apart from the AR model and Pisarenko's sinusoids-plus-noise model, ARMA models have also been used for high-resolution spectral estimates. The interested reader is referred to [19–21,24,33,39–44,49–51].

3.3.4 Application to Phased-Array Processing

The foregoing methods used in modern spectrum estimation can be immediately applied to beamforming and phased-array processing applications in underwater passive sonar systems. One of the main functions of a phased-array processing system is determining the number of sources present in the medium, and their characteristic parameters, such as direction and intensity [26–28], from the signals received by N sensors.

The similarity between spectral estimation and angle estimation becomes clear by noting the duality that exists between the problem of spectral-line estimation from a time series and angle estimation from a spatial data sequence. Here, θ, representing the source angle, is the dual of frequency and the spatial samples provided by the sensors take the role of the time-samples [52].

As a result, the following substitutions and interpretational changes are necessary before the above-mentioned spectral estimation techniques can be used. For simplicity, it is assumed that a uniformly spaced linear array of sensors are used to measure the data. In this stationary environment, the temporal correlation matrix $\mathbf{R}$ has to be replaced by the spatial correlation matrix $\mathbf{S}$ (i.e., the elements of $\mathbf{S}$ represent correlation of data between sensors). Second, ω, which denotes frequency, now is related to the spatial angular direction θ. The phasing vector $\mathbf{a}(\theta)$ will be given by

$$\mathbf{a}(\theta) = [1, e^{j\omega}, e^{j2\omega}, \ldots, e^{j(N-1)\omega}]^T$$

where ω is now given by $\omega = 2(d/\lambda) \sin \theta$, d being the spacing between sensors and λ the wavelength of the incoming signal. So instead of the power spectrum, the spatial directivity spectrum $P(\theta)$ is computed and the location of peaks in $P(\theta)$ leads to estimates of the source directions.

The conventional beamforming technique uses the following formulation [26, 29], where * denotes complex conjugate transpose:

$$P_{\mathrm{BF}}(\theta) = \mathbf{a}^*(\theta)\mathbf{S}\mathbf{a}(\theta) \tag{3.15}$$

As $\mathbf{a}(\theta)$ is a Fourier coefficient vector, (3.15) reduces to the classical transform-based method for spectrum estimation as follows:

$$P_{\mathrm{BF}}(\theta) = \mathbf{a}^*(\theta)\mathbf{S}\mathbf{a}(\theta)$$

$$= \sum_{n=-(N-1)}^{N-1} W(n)S(n)e^{-j\omega n}$$

This is exactly the Fourier transform of the correlation function $S(n)$ with Bartlett (triangular) window, $W(n)$ [see (3.3)].

It is important to note that a similar formulation can be naturally extended to the randomly spaced sensor array situation. In that case, the phasing vector $\mathbf{a}(\theta)$ will be dependent on the geometry of the sensor array; and also quite understandably, the spatial correlation matrix will not necessarily observe the Toeplitz structure, i.e., the correlation between ith and jth sensors will in general differ from that between the $(i + k)$th and $(j + k)$th sensors. Except for this observation, the conventional beamforming spectrum is still defined as in (3.15).

A popular method originally developed for array processing by Capon [17,18] results in a directivity spectrum [29] given by

$$P_C(\theta) = \frac{1}{\mathbf{a}^*(\theta)\mathbf{S}^{-1}\mathbf{a}(\theta)} \tag{3.16}$$

Here, the basic idea is to find a p-weight beam former $\mathbf{c}$ that has unity gain along the direction θ, and at the same time has minimum output power $\mathbf{c}^*\mathbf{S}\mathbf{c}$, so that the contribution from other directions is minimal. Optimization under the unity gain constraint $\mathbf{c}^*\mathbf{a}(\theta) = 1$ leads to the minimum power of $(\mathbf{a}^*(\theta)\mathbf{S}^{-1}\mathbf{a}(\theta))^{-1}$. At the minimum, the output is mainly due to the power from the direction θ, and this leads to the estimate of $P(\theta)$ given by (3.16).

For the stationary case, the maximum entropy principle can also be applied, and it leads to the following directivity spectrum [29]:

$$P_{ME}(\theta) = \frac{1}{\mathbf{a}^*(\theta)\mathbf{c}\mathbf{c}^*\mathbf{a}(\theta)}$$

where $\mathbf{c}$ is the solution to the Weiner–Hopf equations using $\mathbf{S}$ and is also the first column of $\mathbf{S}^{-1}$ (to within a scaling factor).

Pisarenko's method can also be applied to the problem. By the procedure outlined in Section 3.3.3, the minimum eigenvector of matrix $\mathbf{S}$ can be used to estimate the source angles. However, a more popular approach used in this area is the one based on eigenvalue–eigenvector decomposition of the correlation matrix $\mathbf{S}$, as proposed by Owsley [30], Bienvenu [27,28], Schmidt [29], and Johnson and DeGraaf [31]. They are all similar but differ somewhat in the actual use of the eigenvectors. As an example, the MUSIC (Multiple Signal Classification) method by Schmidt [29] uses the following formulation:

$$P_{MUSIC}(\theta) = \frac{1}{\mathbf{a}^*(\theta)E_N E_N^*\mathbf{a}(\theta)}$$

where E_N is a matrix formed from the eigenvectors associated with the minimum $(N - M)$ eigenvalues of the correlation matrix $\mathbf{S}$, where M is the number of targets. A closely related method was independently developed by Bienvenu (see [27,28]). Theoretically, $(N - M)$ minimum eigenvalues must be identical, but in practice, they are only closely clustered. So, unlike Pisarenko's method, where only one minimum eigenvector is used, here all the eigenvectors corresponding to the smaller eigenvalues are used. Owsley developed another method, based on the principal eigenvectors. In general, compared to Pisarenko, these methods have been found to have better performance and are less sensitive to perturbations. The next two chapters by Owsley and Bienvenu deal with these methods in greater detail.

3.4 SVD-BASED SIGNAL MODELING

Among the methods discussed in the preceding section, the Fourier transform-based methods, although very popular and well established, suffer from severe bias and poor resolution; while the maximum entropy method yields considerably better resolution, and so does Pisarenko's method. However, for high-resolution spectral line estimation, the eigenvalue decomposition based approach (such as the MUSIC method [29] suggested for phased array processing) appears to be more appropriate.

In the presence of colored noise, covariance errors, and in a finite-precision environment, a cluster of eigenvalues will arise in the neighborhood of the minimum eigenvalue, as opposed to multiple minimum eigenvalues, and the simple Pisarenko approach will fail. The sensitivity exhibited by Pisarenko's method to covariance perturbations and finite-precision inaccuracies [47] appears to be rather severe, and more robust schemes are called for.

An SVD-based signal modeling method that has been fast attracting a lot of attention is the method proposed by Tufts and Kumaresan [22]. Instead of obtaining the prediction polynomial **a** as the minimum eigenvector of R as in Pisarenko's method, they compute **a** from the principal eigenvectors corresponding to the p largest eigenvalues. The sinusoid frequencies are then obtained from the roots of **a**.

One can avoid the numerically undesirable step of polynomial rooting, by using state-space parameters for the sinusoids-plus-noise model. This provides a flexibility in the choice of coordinates, and simulation examples [56] lead us to believe this could result in better numerical properties.

The sinusoids-plus-noise problem admits a very special state-space representation. For example, a purely sinusoidal data record $y(k)$ can be described by

$$\mathbf{x}_{k+1} = F\mathbf{x}_k$$

$$y_k = h\mathbf{x}_k$$

where $\mathbf{x}(k)$ or x_k, the state vector, is an $n \times 1$ vector process, F and h are constant matrices of sizes $n \times n$ and $1 \times n$, respectively, and the eigenvalues of F are required to lie on the unit circle. The data record is completely specified by the state-space triple $(F, \mathbf{x}_0, h)$, but the triple $(F, \mathbf{x}_0, h)$ itself is unique only up to a similarity transformation. An interesting representation for the model is when F is diagonal. Then

$$F = \text{diag}\,(e^{j\omega_1} \quad e^{j\omega_2} \quad \cdots \quad e^{j\omega_p})$$

and

$$h^{(i)} \cdot \mathbf{x}_0^{(i)} = \rho_i \cdot e^{j\phi_i}$$

where ω_i, ρ_i, and ϕ_i are the frequency, amplitude, and phase, respectively, of the ith complex exponential, and the ith element of a vector is denoted by the superscript (*i*). It can be easily shown that

$$r(m) = hPF^{*m}h^* = hF^m Ph^* \qquad m \geq 0$$

where the state variance P satisfies $P = FPF'$.

An examination of the Toeplitz covariance matrix R indicates that it is factorizable as follows:

$$
R = \begin{bmatrix}
r(0) & r(1) & \cdots & \cdots & r(n) \\
r(1) & r(0) & & & \\
& & \ddots & & \\
& & & r(0) & r(1) \\
r(n) & & & r(1) & r(0)
\end{bmatrix}
$$

$$
= \underbrace{\begin{pmatrix} h \\ hF \\ \vdots \\ hF^n \end{pmatrix}}_{\Delta\theta} \underbrace{[Ph^*, F^{-1}Ph^*, \ldots, F^{-n}Ph^*]}_{\mathscr{R}}
$$

This factorization into the observability matrix θ and the matrix $\mathscr{R}$ indicates that the matrix R is singular. Furthermore, it is of rank p if the frequencies are distinct (i.e., $\omega_i \neq \omega_j$, $i \neq j$ and if $\rho_i \neq 0$ for all i).

In the Toeplitz approximation method [23,46], the first step is enforcing the p-rank property on an estimated covariance matrix R which will generally be full rank. This is done via singular value decomposition and an approximate factorization of R into θ and $\mathscr{R}$ is obtained as explained below.

Let the SVD of R be

$$R = U\Sigma^2 V = [U_1 \quad U_2]\begin{bmatrix} \Sigma_1^2 & 0 \\ 0 & \Sigma_2^2 \end{bmatrix}\begin{bmatrix} V_1 \\ V_2 \end{bmatrix}$$

where Σ_1 is $p \times p$ and Σ_2 is $(N - p) \times (N - p)$. The observability matrix is obtained from the principal singular vectors U_1 and the principal singular values Σ_1^2. In the presence of white noise, though the singular vectors are unchanged, the singular values are affected. In fact all the singular values are increased by the noise variance, and so the smallest singular value has to be subtracted to compensate for this effect, i.e.

$$\hat{\Sigma}_1^2 = \Sigma_1^2 - \sigma_N^2 I,$$

where σ_N^2 is the smallest singular value of R. Hence, the observability-type matrix is given by

$$\theta = U_1 \cdot \hat{\Sigma}_1$$

while

$$R = \hat{\Sigma}_1 \cdot V_I.$$

Such an approximation of $(\boldsymbol{R} - \sigma_N^2 \boldsymbol{I})$ by $\boldsymbol{\theta} \cdot \mathcal{R}$ is optimal in the spectral norm [25,46].

The second step involves determining the model parameters. Ideally, $\boldsymbol{\theta}$ would have the exact observability matrix structure, and a $p \times p$ solution $\boldsymbol{F}$ to the matrix equation

$$\boldsymbol{\theta} F = \begin{pmatrix} \boldsymbol{hF} \\ \boldsymbol{hF}^2 \\ \boldsymbol{hF}^3 \\ \vdots \end{pmatrix} = \boldsymbol{\theta}\uparrow$$

would exist. However, because of the approximation, no exact solution exists and we have to resort to a least-squares solution that minimizes $\| \boldsymbol{\theta} F - \boldsymbol{\theta}\uparrow \|_{\mathrm{E}}$, where subscript E denotes the Euclidean norm and $\boldsymbol{\theta}\uparrow$ is obtained by shifting $\boldsymbol{\theta}$ one row upwards. This leads to least-squares estimates of the state-space parameters, as

$$\boldsymbol{F} = \boldsymbol{\theta}^\dagger \cdot \boldsymbol{\theta}\uparrow,$$

$$\boldsymbol{h} = \text{the first row of } \boldsymbol{\theta},$$

$$\boldsymbol{Ph}^* = \text{the first column of } \mathcal{R},$$

where the superscript ($\dagger$) stands for pseudoinverse. The eigenvalues of $\boldsymbol{F}$ give the frequencies of the sinusoids.

Starting from data one could either estimate the covariance lags using the unbiased estimator and construct a Toeplitz covariance matrix $\bar{\boldsymbol{R}}$ or simply use as in [47,22],

$$\hat{\bar{R}} = (\boldsymbol{DD'})/2(L - N + 1)$$

where

$$\boldsymbol{D} = \begin{bmatrix} y(1) & y(2) & \cdots & y(L - N + 1) & y(L) & y(L - 1) & \cdots & y(N) \\ y(2) & y(3) & \cdots & y(L - N + 2) & y(L - 1) & y(L - 2) & \cdots & y(N - 1) \\ y(3) & y(4) & \cdots & y(L - N + 3) & y(L - 2) & y(L - 3) & \cdots & y(N - 2) \\ \vdots & \vdots & \cdots & \vdots & \vdots & \vdots & \cdots & \vdots \\ y(N) & y(N + 1) & \cdots & y(L) & y(L - N + 1) & y(L - N) & \cdots & y(1) \end{bmatrix}$$

We now provide a simulation example from [56], to demonstrate the power of the principal components approach. The problem considered is the retrieval of a single sinusoid (effectively, the problem of resolving two closely spaced complex exponentials) in additive white noise, from 25 data samples. The Toeplitz approximation method (TAM) applied on 11×11 sized $\bar{\boldsymbol{R}}$ and 13×13 sized $\hat{\boldsymbol{R}}$ were both tested on the harmonic process corrupted by two hundred different pseudonoise sequences. The results in terms of the mean[1], standard deviation, root mean square (rms) error of the frequency estimates, and the failure rate (failure to resolve the spectral lines) are tabulated below. Other simulation parameters are the rank of the

---

[1] The mean and st. dev. of the estimates are computed only for the resolved cases, while the rms error takes the unresolved samples also into consideration.

approximant (model order) and the signal to noise ratio. Although, the rank can be determined by examining the singular values, for convenience, the rank of the approximant is predetermined to be 2 in our simulation study. The signal to noise ratio (SNR) is defined as the ratio of the power in each exponential to the variance of the noise.

TABLE 1

Frequencies to be resolved	(0.97, 1.03)	(0.98, 1.02)	(0.99, 1.01)
SNR	0 dB	10 dB	20 dB
TAM (on $\bar{R}$)			
Miss ratio	7/200	2/200	100/200
Rms error	0.020154	0.003738	0.008007
Mean	0.952043	0.979278	0.995135
St. dev.	0.008113	0.003090	0.001359
TAM (on $\hat{R}$)			
Miss ratio	72/200	9/200	5/200
Rms error	0.023608	0.007115	0.003163
Mean	0.959822	0.980397	0.990353
St. dev.	0.016156	0.005831	0.002752

Source: Ref. [56].

The results indicate that TAM on $\bar{R}$ (the unbiased estimate) is very suitable for low SNR situations, while TAM on the covariance estimate $\hat{R}$, performs well for high resolution problems. For instance, note that for the low SNR (0 db) case, TAM on $\bar{R}$ performs better, with 7 failures in 200 trials. For the high SNR (20 db) case, TAM on $\hat{R}$ performs better, with 5 failures in the 200 trials.

3.5 CONCLUDING REMARKS

The invention of the fast Fourier transform algorithm nearly two decades ago provided the first major impetus to the advancement of digital signal processing. In the past two decades, FFT has been a major influence in directing the growth of signal processing theory. In spectral estimation, for instance, transform-based methods have traditionally been the dominant approach, mainly due to the computational efficiency of FFT, the existence of an extensive FFT software library, and the commercial availability of high-speed FFT processors. But today, VLSI offers very inexpensive computing power as well as enables the use of tremendous parallelism in modern signal processing systems. This will definitely encourage many computationally intensive algorithms, such as the signal modeling methods discussed in this chapter. We have seen that when the model used is realistic, the improvement in performance over the conventional FFT spectral estimate can be quite dramatic, especially for short data records. More promisingly, computations for the modern signal processing methods are often reducible to basic matrix operations such as eigenvalue and singular value decomposition, correlation matrix computation, and matrix inversion, which have very high potential for VLSI parallel computing. More importantly, SVD offers attractive numerical stability [45], and opens up a promising area for future developments.

REFERENCES

[1] A. V. Oppenheim and R. W. Schafer, *Digital Signal Processing*, Prentice-Hall, Englewood Cliffs, N.J., 1975.

[2] O. L. Frost, "Power Spectrum Estimation," *Proc. 1976 NATO Advanced Study Institute Signal Process. Emphasis Underwater Acoust.*, Portovener, Italy, Aug. 30–Sept. 11, 1976.

[3] S. S. Haykin, ed., *Nonlinear Methods of Spectral Analysis*, Springer-Verlag, New York, 1979.

[4] S. M. Kay and S. L. Marple, Jr., "Spectrum Analysis—A Modern Perspective," *Proc. IEEE, 69*(11):1380–1418 (Nov. 1981).

[5] G. M. Jenkins and D. G. Watts, *Spectral Analysis and Its Applications*, Holden-Day, San Francisco, 1966.

[6] A. Papoulis, *Probability, Random Variables, and Stochastic Processes*, McGraw-Hill, New York, 1965.

[7] A. Schuster, "On the Investigation of Hidden Periodicities with Application to a Supposed 26 Day Period of Meteorological Phenomena," *Terrest. Magn., 3*:13–41 (Mar. 1898).

[8] F. J. Harris, "On the Use of Windows for Harmonic Analysis with the Discrete Fourier Transform," *Proc. IEEE, 66*:51–83 (Jan. 1978).

[9] A. H. Nuttall, "Some Windows with Very Good Sidelobe Behavior," *IEEE Trans. Acoust. Speech Signal Process., ASSP-29*:84–89 (Feb. 1981).

[10] M. S. Bartlett and J. Medhi, "On the Efficiency of Procedures for Smoothing Periodograms from Time Series with Continuous Spectra," *Biometrika, 42*:143–150 (1955).

[11] P. D. Welch. "The Use of Fast Fourier Transform for the Estimation of Power Spectra: A Method Based on Time Averaging over Short Modified Periodograms," *IEEE Trans. Audio Electroacoust., AU-15*:70–73 (June 1967).

[12] J. P. Burg, "Maximum Entropy Spectral Analysis," Ph.D. dissertation, Stanford University, Stanford, Calif., 1975.

[13] T. J. Ulrych and T. N. Bishop, "Maximum Entropy Spectral Analysis and Autoregressive Decomposition," *Rev. Geophys. Space Phys., 13*:183–200 (Feb. 1975).

[14] A. van den Bos, "Alternative Interpretation of Maximum Entropy Spectral Analysis," *IEEE Trans. Inf. Theory, IT-17*:493–494 (July 1971).

[15] V. F. Pisarenko, "On the Estimation of Spectra by Means of Nonlinear Functions of the Covariance Matrix," *Geophys. J. R. Astron. Soc., 28*:511–531 (1970).

[16] V. F. Pisarenko, "The Retrieval of Harmonics from a Covariance Function," *Geophys. J. R. Astron. Soc., 33*:347–366 (1973).

[17] J. Capon, "High-Resolution Frequency-Wavenumber Spectrum Analysis," *Proc. IEEE*, *57*:1408–1418 (Aug. 1969).

[18] R. T. Lacoss, "Data Adaptive Spectral Analysis Method," *Geophysics*, *36*:661–675 (Aug. 1971).

[19] D. Graupe, D. J. Krause, and J. B. Moore, "Identification of Autoregressive Moving Average Parameters of Time Series," *IEEE Trans. Autom. Control*, *AC-20*:104–107 (Feb. 1975).

[20] G. E. P. Box and G. M. Jenkins, *Time Series Analysis: Forecasting and Control*, Holden-Day, San Francisco, 1970.

[21] J. A. Cadzow, "Spectral Estimation: An Overdetermined Rational Model Equation Approach," *Proc. IEEE*, *70*(9):907–939 (Sept. 1982).

[22] D. W. Tufts and R. Kumaresen, "Estimation of Frequencies of Multiple Sinusoids: Making Linear Prediction Perform like Maximum Likelihood," *Proc. IEEE*, *70*(9):975–989 (Sept. 1982).

[23] S. Y. Kung, "A Toeplitz Approximation Method and Some Applications," *Int. Symp. Math. Theory Networks Syst.*, Santa Monica, Calif., Aug. 5–7, 1981.

[24] S. Y. Kung and K. S. Arun, "A Novel Hankel Approximation Method for ARMA Pole Zero Estimation from Noisy Covariance Data," *Topical Meet. Signal Recovery*, Optical Society of America, Jan. 1983.

[25] S. Y. Kung, "A New Identification and Model Reduction Algorithm via Singular Value Decomposition," *12th Asilomar Conf. Circuits, Syst. Comput.*, Pacific Grove, Calif., Nov. 1978.

[26] A. B. Baggeroer, "Sonar Signal Processing," in A. V. Oppenheim, ed., *Applications of Digital Signal Processing*, Prentice-Hall, Englewood Cliffs, N.J., 1978, pp. 331–437.

[27] G. Bienvenu and L. Kopp, "Adaptivity to Background Noise Spatial Coherence for High Resolution Passive Methods," *Proc. IEEE ICASSP*, Denver, Colo., 1980, pp. 307–310.

[28] G. Bienvenu and L. Kopp, "Source Power Estimation Method Associated with High Resolution Bearing Estimation," *Proc. IEEE ICASSP*, Atlanta, Ga., 1981, pp. 153–156.

[29] R. Schmidt, "Multiple Emitter Location and Signal Parameter Estimation," *Proc. RADC Spectral Estimation Workshop*, Rome, N.Y., 1979, pp. 243–258.

[30] N. L. Owsley, "Modal Decomposition of Data Adaptive Spectral Estimates," *Yale Univ. Workshop Appl. Adaptive Syst. Theory*, New Haven, Conn., 1981.

[31] D. H. Johnson and S. R. DeGraaf, "Improving the Resolution of Bearing in Passive Sonar Arrays by Eigenvalue Analysis," *IEEE Trans. Acoust. Speech Signal Process.*, *ASSP-30*(4):638–647 (Aug. 1982).

[32] S. Y. Kung and Y. H. Hu, "A Highly Concurrent Algorithm and Pipelined Architecture for Solving Toeplitz Systems," *IEEE Trans. Acoust. Speech Signal Process.*, *ASSP-31*:No. 1, pp. 66–76 (Feb. 1983).

[33] A. A. Beex and L. L. Scharf, "Covariance Sequence Approximation for Parametric Spectrum Modeling," *IEEE Trans. Acoust. Speech Signal Process.*, *ASSP-29*(5):1042–1052 (Oct. 1981).

[34] T. Kailath, "A View of Three Decades of Linear Filtering Theory," *IEEE Trans. Inf. Theory*, *IT-20*(2):146–179 (Mar. 1974).

[35] G. Szego, "Ein Grenzwertsatz über die Toeplitzschen Determinanten einer reellen positiven Funktion," *Math. Ann.*, *76*:490–503 (1915).

[36] S. Y. Kung and Y. H. Hu, "Highly Concurrent Toeplitz System Solver for High Resolution Spectral Estimation," *Proc. ICASSP*, Boston, Apr. 1983, pp. 1422–1425.

[37] N. Levinson, "The Wiener rms (root mean square) Error Criterion in Filter Design and Prediction," *J. Math. Phys.*, vol. 25, pp. 261–278, Jan. 1947.

[38] N. Levinson, "A heuristic exposition of Wiener's mathematical theory of prediction and filtering," *J. Math. Phys.*, vol. 26, pp. 110–119, July 1947.

[39] H. P. Zeiger and A. J. McEwen, "Approximate Linear Realizations of Given Dimension via Ho's Algorithm," *IEEE Trans. Autom. Control, AC-19*:153 (1974).

[40] P. Faurre, "Stochastic Realization Algorithms," in R. K. Mehra and D. G. Lainiotis, eds., *System Identification: Advances and Case Studies*, Academic Press, New York, 1976.

[41] H. Akaike, "Markovian Representation of Stochastic Processes by Canonical Variables," *SIAM J. Control, 13*(1):162–173 (Jan. 1975).

[42] H. Akaike, "Markovian Representation of Stochastic Processes and Its Application to the Analysis of Autoregressive Moving Average Processes," *Ann. Inst. Stat. Math.*, *26*:363–387 (1974).

[43] T. Kailath, "The Innovations Approach to Detection and Estimation Theory," *Proc. IEEE, 58*:680–695 (May 1970).

[44] D. Mcginn and D. H. Johnson, "Reduction of All-Pole Parameter Estimation Bias by Successive Autocorrelation," *Proc. ICASSP*, Boston, Apr. 1983, pp. 1088–1091.

[45] V. C. Klemma and A. J. Laub, "The Singular Value Decomposition: Its Computation and Some Applications," *IEEE Trans. Autom. Control, AC-25*(2):164–176 (Apr. 1980).

[46] S. Y. Kung, K. S. Arun, and D. V. Bhaskar Rao, "State Space and SVD Based Approximation Methods for the Harmonic Retrieval Problem," *J. Optical Soc. of America*, *73*:1799–1811 (Dec. 1983).

[47] T. J. Ulrych and R. W. Clayton, "Time Series Modeling and Maximum Entropy," *Phys. Earth Planet. Interiors, 12*:188–200 (1976).

[48] Y. H. Hu, "New Algorithms and Parallel Architectures for Toeplitz Systems, with Applications to Spectrum Estimation," Ph.D. dissertation, University of Southern California, Los Angeles.

[49] U. B. Desai and D. Pal, "A Realization Approach to Stochastic Model Reduction and Balanced Stochastic Realizations," *16th Annu. Conf. Inf. Sci. Syst.*, Princeton University, Princeton, N.J., Mar. 1982.

[50] S. Y. Kung and K. S. Arun, "Approximate Realization Methods for ARMA Spectral Estimation," *IEEE Int. Symp. Circuits Syst.*, Newport Beach, Calif., May 1983.

[51] J. White, "Stochastic State Space Models from Emperical Data," *Proc. IEEE Int. Conf. Acoust. Speech Signal Process.*, Apr. 1983, pp. 243–246.

[52] W. F. Gabriel, "Spectral Analysis and Adaptive Arrays Superresolution Techniques," Proc. IEEE, Vol. 68, pp. 654–666, June 1980.

[53] J. E. Shore, and R. W. Johnson, "Axiomatic Derivation of the Principle of Maximum Entropy and the Principle of Minimum Cross-Entropy," *IEEE Trans. Info. Theory, IT-26*(1):26–37 (Jan. 1980).

[54] J. E. Shore, "Minimum Cross-Entropy Spectral Analysis," *IEEE Trans. Acoustics, Speech and Signal Processing*, Vol. ASSP-29, no. 2, pp. 230–237, April 1981.

[55] S. Kullback and R. A. Leibler, "On information and sufficiency," *Annals of Mathematical Statistics*, Vol. 22, pp. 79–86, 1951.

[56] S. Y. Kung, D. V. Bhaskar Rao, and K. S. Arun, "New state space and singular value decomposition based approximate modeling methods for hormone retrieval," *IEEE ASSP Spectrum Estimation Workshop II*, Tampa, Florida, Nov. 1983, pp. 266–271.

4

High-Resolution Spectrum Analysis by Dominant-Mode Enhancement

NORMAN L. OWSLEY

Naval Underwater Systems Center
New London, Connecticut

4.1 INTRODUCTION

High-resolution power spectrum analysis methods have well-known applications to single-channel frequency analysis [1] and multiple-channel sensor array frequency–wavenumber analysis [2]. More recently, sensor array simultaneous frequency–wavenumber–range analysis has provided a further extension of fundamentally the same power spectrum analysis procedures [3,4]. Principal among these techniques are, first, the minimum variance, distortionless response (MV) filter, which is closely related to the maximum-likelihood procedure [1,2]; and second, the class of least-squares linear smoothing/prediction techniques referred to herein corporately as the maximum-entropy (ME) method [5–7]. It is typical of these techniques to assume no prior knowledge of the vector space dimensionality of either the signal or noise components of the data to be analyzed. Such prior knowledge of the signal takes the form of total bandwidth, angular, and radial extent of the signal for frequency, wavenumber, and range analyses, respectively. Prior knowledge that the noise can be prewhitened and the signal is known to have a dimensionality which is small relative to the analysis data dimensionality can be exploited to enhance the spectrum estimator resolution capabilities to an extent limited only by the observation time. Accordingly, an approach to data adaptive spectral analysis is described herein which can be characterized by a modal decomposition of the observation data covariance matrix. This approach, which can be expressed alternatively in terms of either orthonormal [8,9,10], singular-value [10], or Cholesky decompositions of the covariance matrix, is in contrast to either the Gram–Schmidt orthogonalization procedures [11] or related schemes as embodied in the lattice filter structure [12].

The use of eigenvector orthonormal decompositions for frequency–

wavenumber spectrum analysis has evolved from nonparametric and adaptive array processing schemes [13–15] to applications as high-resolution spectral estimators [9,10,15–22]. Frequently, these high-resolution techniques are viewed as ad hoc methods based on the notion of separability of signal-and-noise and noise-only processes into orthogonal vector subspaces. In fact, two distinct versions of these high-resolution spectrum estimators are derived herein from the formal expressions for the minimum-variance, distortionless response (MV), and maximum-entropy method (ME) spectrum estimators. The additional prior information that allows the extension of the MV and ME estimators to their enhanced high-resolution, dominant-mode forms is either that the noise is uncorrelated or can be prewhitened and the signal components of the analyzed data are spectrally narrow.

In this chapter the dominant-mode form of the analysis data covariance matrix is presented together with the expression for the readily derived inverse of this matrix. Next, the MV and ME spectrum estimators in modal decomposition form are developed along with 3-dB down-resolution expressions. Finally, the modal MV and ME estimators are applied to sensor receiving array processing for source range estimation.

4.2 SIGNAL MODEL

Let the covariance matrix for the stationary, zero-mean complex data N vector $\mathbf{x}(t)$ at discrete time t be given by the expectation

$$R = E\{\mathbf{x}(t)\mathbf{x}(t)'\} \tag{4.1}$$

where the prime indicates the complex-conjugate transpose. Of course, in practice, a time-averaged estimate of R can be used instead of R in terms of the ensemble average as defined in equation (4.1). This covariance matrix is assumed to consist of the signal and noise components,

$$R = P + \sigma^2 I_N \tag{4.2}$$

where P is a rank $K \leq L \ll N$ signal covariance matrix, σ^2 is the uncorrelated noise variance, and I_N is an $N \times N$ unit diagonal matrix. The signal covariance matrix has two representations. First, in terms of actual signal descriptors, we can write

$$P = \sum_{p=1}^{L} \sum_{q=1}^{L} \sigma_{pq}^2 \, \mathbf{D}(\boldsymbol{\theta}_p)\mathbf{D}(\boldsymbol{\theta}_q)' \tag{4.3}$$

and, second, in terms of a dominant mode(s) decomposition, we have

$$P = \sum_{k=1}^{K} \lambda_k \mathbf{M}_k M_k' \tag{4.4}$$

The signal is represented in (4.3) on an N-dimensional vector subspace by the spanning set of signal vectors $\{\mathbf{D}(\mathbf{e}_p): p = 1, 2, \ldots, L\}$. These L signal vectors are not necessarily linearly independent. The quantity σ_{pq}^2 is the cross correlation between the pth and qth signal complex envelopes, which, in turn, are assumed to be zero mean. In (4.4), the signal covariance matrix is written as an orthonormal expansion in terms of its dominant eigenvectors and K associated eigenvalues where $\lambda_1 \geq \lambda_2 \geq \cdots \geq \lambda_K$ are the rank-ordered eigenvalues of P and $M_1, M_2, \ldots, M_K$ are the corresponding orthonormal eigenvectors.

A narrowband (i.e., frequency-domain) representation of the signal is assumed which allows separability of the time and space description of the signal [23]. Two important examples of signal representations in each of these domains are given by defining the nth element of the signal vector $\mathbf{D}(\theta_p)$ to be

$$\mathbf{D}(\theta_p)\Big]_n = \exp\left(-j2\pi f_p n\Delta\right) \tag{4.5}$$

for spectral analysis of a time series in terms of the discrete Fourier frequency parameter f_p with a uniform temporal sampling interval Δ. The spatial spectrum analysis counterpart to (4.5) for a one-dimensional (linear) spatial sensor array on the X axis is

$$\mathbf{D}(\theta_p)\Big]_n = \exp\left(-j2\pi f\tau_{pn}\right) \tag{4.6a}$$

$$= \exp\left[-\frac{j2\pi f(r_p^2 + X_n^2 - 2r_p X_n \cos\beta_\rho)^{1/2}}{c}\right] \tag{4.6b}$$

$$\simeq \exp\left\{-\frac{j2\pi f[r_p - X_n \cos\beta_p + (\sin^2\beta_p/2r_p)X_n^2]}{c}\right\} \tag{4.6c}$$

The spatial model for the pth signal component at frequency f represented in (4.6) expresses the propagation time τ_{pn} for wavefront propagation at speed c in a homogeneous medium from the pth source at range r_p and bearing β_p to the nth of N sensors in a linear array. This geometry is illustrated in Figure 4.1. Equation (4.6c) is obtained from (4.6b) by taking a Taylor series expansion of the propagation time

$$\tau_{pn} = \frac{(r_p^2 + X_n^2 - 2r_p X_n \cos\alpha_p)^{1/2}}{c} \tag{4.7}$$

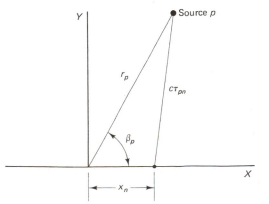

Figure 4.1   Linear array of N discrete sensors located at points $X_n, n = 1, 2, \ldots, N$, with a source at range r and bearing β.

with respect to X_n assuming that $r_p \gg X_n$. Estimation of the time delay

$$D_{m,n}(p) = \tau_{pm} - \tau_{pn} \qquad (4.8)$$

between the mth and nth sensors $\{m, n = 1, 2, \ldots, N\}$ and the corresponding set of L source range and bearing parameters is a problem of considerable current interest [24,25]. This problem is equivalent to estimating the geometric parameters β_p and r_p, $p = 1, 2, \ldots, L$, for the signal sources [30].

The estimation of either frequency or time delay is discussed herein in terms of operations on the estimated data covariance matrix given, for example, simply by

$$\hat{R} = \frac{1}{T} \sum_{t=1}^{T} \mathbf{x}(t)\mathbf{x}(t)' \qquad (4.9)$$

However, in the following sections, the ensemble-average form of the data covariance matrix as given in (4.2) is assumed and the effect of finite averaging time T is addressed empirically in the section, which considers specifically the range estimation example. Before proceeding, two forms of the data covariance matrix R and the inverse R^{-1} are presented. If the $N \times K$ modal matrix M of P with M_k as the kth column and the $K \times K$ diagonal matrix λ with λ_k as the kth diagonal element are defined, an enhanced data covariance matrix, $R(e)$, is defined as

$$R(e) = eM\lambda M' + \sigma^2 I_N \qquad (4.10)$$

with

$$R(e)^{-1} = \frac{1}{\sigma^2}\left[I_N - M\left(I_K + \frac{\sigma^2}{e}\lambda^{-1} \right)^{-1} M' \right] \qquad (4.11a)$$

$$= \frac{1}{\sigma^2}\left[I_N - \sum_{k=1}^{K} \left(1 + \frac{\sigma^2}{e\lambda_k} \right)^{-1} M_k M_k' \right] \qquad (4.11b)$$

The scalar parameter e is hereafter referred to as the modal enhancement factor. With respect to the original definition of R in (4.2), we have $R(1) = R$. However, once the signal eigendata are estimated they are readily inserted into (4.11). Next, consider the alternative form

$$R(e) = eCC' + \sigma^2 I_N \qquad (4.12)$$

wherein $P = CC'$ is a Cholesky factorization of the signal covariance matrix P obtained from R after the diagonal, uncorrelated noise matrix $\sigma^2 I_N$ has been removed. An algorithm for this procedure is discussed in a subsequent section. As above, the form

$$R(e)^{-1} = \frac{1}{\sigma^2}\left[I_N - C\left(C'C + \frac{1}{e}I_K \right)^{-1} C' \right] \qquad (4.13)$$

for the inverse of the data covariance matrix with signal enhancement is referenced in subsequent discussions.

4.3 SPECTRUM ANALYSIS

Two approaches to high-resolution spectrum analysis are presented in this section. Both methods result from a linearly constrained quadratic minimization problem formulation in conjunction with the additional information that, first, the signal vector subspace is of dimension K; second, the noise process is zero mean, independent, and identically distributed; and third, the spectral resolution is to be maximized. The maximization of resolution is equivalent to saying that the enhancement factor, e, is to be made large. This is because a large e emulates a high signal-to-noise ratio condition which is a principal factor in resolution improvement.

4.3.1 Minimum-Variance Distortionless Response (MV)

For this procedure it is desired to find a linear filter vector $\mathbf{W}$ for the analysis data vector which is a solution to the following constrained minimization problem.

Minimize:
$$\sigma^2_{MV} = E\{|\,\mathbf{W}'\mathbf{x}(t)|^2 : e\} \tag{4.14a}$$

$$= \mathbf{W}'R(e)\mathbf{W} \tag{4.14b}$$

Maximize: spectral resolution (i.e., $e \to \infty$)

Constraints: (1) Distortionless (unit) response requires that

$$1 = \mathbf{W}'\mathbf{D}(\theta) \tag{4.14c}$$

(2) Signal space of dimension K and uniform independent noise requires (4.10).

Solution:
$$\mathbf{W} = \mathbf{W}_{MV} \tag{4.15a}$$

$$= \frac{R(e)^{-1}\mathbf{D}(\theta)}{\mathbf{D}(\theta)'R(e)^{-1}\mathbf{D}(\theta)} \tag{4.15b}$$

$$= g(\boldsymbol{\theta}, e)\left[\mathbf{D}(\boldsymbol{\theta}) - \sum_{k=1}^{K} b(k, e)\mathbf{M}'_k\,\mathbf{D}(\theta)\mathbf{M}_k\right] \tag{4.15c}$$

where

$$g(\boldsymbol{\theta}, e) = \left[N - \sum_{k=1}^{K} b(k, e)\,|\,\mathbf{M}'_k\,\mathbf{D}(\boldsymbol{\theta})|^2\right]^{-1} \tag{4.16}$$

and

$$b(k, e) = \left(1 + \frac{\sigma^2}{e\lambda_k}\right)^{-1} \tag{4.17}$$

The MV power spectrum estimator is (4.14a) with $\mathbf{W} = \mathbf{W}_{\text{MV}}$, which yields

$$P_{\text{MV}}(\boldsymbol{\theta}) = \frac{1}{\mathbf{D}(\boldsymbol{\theta})' R(e)^{-1} \mathbf{D}(\boldsymbol{\theta})} \tag{4.18}$$

$$= \sigma^2 g(\boldsymbol{\theta}, e) \tag{4.19}$$

The enhanced minimum-variance (EMV) spectrum estimator is obtained from (4.18) by taking

$$\lim_{e \to \infty} b(k, e) = 1 \tag{4.20}$$

which gives

$$P_{\text{EMV}}(\boldsymbol{\theta}) = \frac{\sigma^2}{N - \sum_{k=1}^{K} |M_k' \mathbf{D}(\boldsymbol{\theta})|^2} \tag{4.21}$$

for maximum resolution. It is noted that (4.15b) is the solution to the minimization requirement subject only to the distortionless response constraint of (4.14c). When the signal and noise vector space dimensionality information is exploited, (4.15c) and (4.19) result. Finally, the resolution maximization requirement is satisfied when (4.20) is implemented and results in the estimator of (4.21).

4.3.2 Maximum Entropy (ME)

For smoothing and prediction it is desired to find a filter vector $\mathbf{W}$ which is to be applied linearly to the data vector $\mathbf{x}(t)$ in such a way that the nth element in $\mathbf{x}(t)$, say $x_n(t)$, is estimated in a least-squares sense by a linear combination of the other $N - 1$ elements of $\mathbf{x}(t)$. This objective in conjunction with both resolution maximization and the a priori knowledge that the data covariance matrix has the structure specified by (4.10) can be stated as the following constrained optimization problem.

Minimize: $\sigma_{\text{ME}}^2 = E\{|\mathbf{W}'\mathbf{x}(t)|^2 : e\}$ $\tag{4.22a}$

Maximize: spectral resolution (i.e., $e \to \infty$)

Constraints: (1) Smoothing/prediction at point n in data window requires

$$1 = \mathbf{W}'\mathbf{l}_n \tag{4.22b}$$

where $\mathbf{l}_n$ is a real N vector consisting of all zeros except a 1 for the nth element.

(2) Signal space of dimensionality K and uniform independent noise requires (4.10).

The solution to this problem invoking only the constraint of (4.22b) is

$$\mathbf{W}_{ME} = \frac{R(e)^{-1}\mathbf{l}_n}{\mathbf{l}'_n R(e)^{-1}\mathbf{l}_n} \tag{4.23}$$

and with the addition of constraint (2),

$$\mathbf{W}_{ME} = g(n, e)\left[\mathbf{l}_n - \sum_{k=1}^{K} b(k, e)M'_k\mathbf{l}_n M_k\right] \tag{4.24}$$

where

$$g(n, e) = \left(1 - \sum_{k=1}^{K} b(k, e)|M'_k\mathbf{l}_n|^2\right)^{-1} \tag{4.24}$$

Now let the data vector $\mathbf{x}(t)$ contain samples which are uniformly spaced at an interval Δ for either a time sequence for frequency analysis or a homogeneous spatial signal field for wavenumber analysis. Both processes are stationary. Spatial stationarity requires that there be no near-field sources for the spatial spectrum analysis application. This limits the present discussion to wavenumber analysis (i.e., bearing estimation). With these restrictions and $\hat{x}(t - n)$ defined as a least-squares estimate of $x_n(t) = x(t - n)$, the equation

$$\varepsilon_n(t) = x(t - n) - \hat{x}(t - n) \tag{4.26}$$

$$= x(t - n) + \sum_{k=0}^{N-1} W^*_k x(t - k) \tag{4.27}$$

gives the smoothing error $\varepsilon_n(t)$ for $0 < n \le N - 1$ and prediction error if $p = 0$. It is the entropy (uncertainty) in the residual error process $\varepsilon_n(t)$ which is maximized as a function of $\mathbf{W}$. The term W_k is the kth element of the filter weight N vector $\mathbf{W}_{ME}$ and the * indicates a complex conjugate. Let $x(t)$ be modeled as an autoregressive (AR) process generated by an all-pole filter excited by white noise, $\varepsilon_n(t)$, with variance $\sigma^2_{ME} = [\mathbf{l}'_n \mathbf{R}^{-1}(e)\mathbf{l}_n]^{-1}$. Now (4.27) has the z transform

$$E_n(z) = z^{-n}\left(1 + \sum_{\substack{k=0 \\ k \ne n}}^{N-1} W^*_k z^{n-k}\right)X(z) \tag{4.28}$$

$$= z^{-n}\sum_{k=0}^{N-1} W^*_k z^{n-k}X(z) \qquad (W_n = 1) \tag{4.29}$$

$$= A_n(z^{-1})X(z) \tag{4.30}$$

The corresponding ME power spectrum estimate for $x(t)$ is therefore

$$P_{ME}(\theta = \omega) = \frac{[\mathbf{l}'_n R(e)^{-1}\mathbf{l}_n]^{-1}}{|A_1(z^{-1})|^2}\bigg|_{z=e^{j\omega}} \tag{4.31}$$

$$= \frac{[\mathbf{l}'_n R(e)^{-1}\mathbf{l}_n]^{-1}}{|\mathbf{W}'_{ME}\mathbf{D}(\omega)|^2} \tag{4.32}$$

$$= \frac{\mathbf{l}'_n \mathbf{R}(e)^{-1} \mathbf{l}_n}{|\mathbf{l}'_n \mathbf{R}(e)^{-1} \mathbf{D}(\omega)|^2} \tag{4.33}$$

The kth element in $\mathbf{D}(\omega)$ is given by

$$d_k(\omega) = \exp \left[-j\omega(n-k)\Delta \right] \tag{4.34}$$

for frequency spectrum analysis with $\omega = 2\pi f$ and

$$d_k(\omega) = \exp \left[-j\omega(n-k)d \right] \tag{4.35}$$

for wavenumber spectrum analysis with d equal to a uniform sensor spacing interval and $\omega = 2\pi f \cos (\beta/C)$. Note that the ME power spectrum estimation process is not strictly realizable without accepting a delay of n samples. This is because the signal model generation process is as illustrated in Figure 4.2. The z^{-1} terms represent the unit-delay operation which is realizable, and the z terms represent a unit advance which is not realizable without a process delay. In actuality, a delay is implicit in the ME process realization which requires that $\mathbf{R}(e)^{-1}$ be estimated with a delay for time averaging.

The final step in the implementation of the enhanced ME process is to substitute the dominant-mode signal covariance matrix constraint expressed by (4.11b) into (4.33) to yield

$$P_{\mathrm{ME}}(\theta) = \frac{1}{g(n, e) \left| 1 - \displaystyle\sum_{k=1}^{K} b(k, e) \mathbf{l}'_n \mathbf{M}_k \mathbf{M}'_k \mathbf{D}(\theta) \right|^2} \tag{4.36}$$

with the enhanced version of (4.36), $P_{\mathrm{EME}}(\theta)$, given by taking $b(k, e) = 1$.

We can now generalize the use of $P_{\mathrm{MV}}(\boldsymbol{\theta})$ and $P_{\mathrm{ME}}(\boldsymbol{\theta})$ to more general forms for $\mathbf{D}(\boldsymbol{\theta})$. Clearly, for $P_{\mathrm{ME}}(\boldsymbol{\theta})$ to be a power (wavenumber) spectrum estimator in the strict sense, the uniform sampling interval $\Delta(d)$ specified above is required. However, this does not preclude the use in general of (4.24) for $\mathbf{W}_{\mathrm{ME}}$ as a filter for least-squares

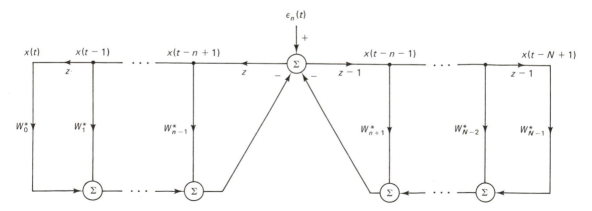

Figure 4.2 Generation of a nonrealizable all-pole process.

estimation of $x_n(t)$. Moreover, the use of (4.32) as a measure of the orthogonality between $\mathbf{W}_{\text{ME}}$ and $\mathbf{D}(\omega)$ is still perfectly valid. This orthogonality test point of view is useful whether the exponents in the elements of $\mathbf{D}(\omega)$ exhibit either a uniform sampling interval structure with a linear dependence on the transform variable as for frequency spectrum analysis and wavenumber analysis or a nonuniform sampling interval. Such a nonuniform sampling interval could result from a quadratic dependence on the transform variable such as in the range spectrum analysis application. To illustrate this, we take the simple case of $K = 1$ where it can be shown that $\mathbf{M}_1 = \mathbf{D}(\theta_0)/\sqrt{N}$. Using this in (4.37) gives

$$P_{\text{ME}}(\boldsymbol{\theta}) = \frac{1}{g \left| \mathbf{I}'_n \left[\mathbf{I} - \dfrac{D(1, e)}{N} \mathbf{D}(\theta_0)\mathbf{D}(\theta_0)' \right] \mathbf{D}(\boldsymbol{\theta}) \right|^2} \tag{4.37}$$

which exhibits a peak value at $\boldsymbol{\theta} = \boldsymbol{\theta}_0$. In fact, (4.37) becomes infinite when $\boldsymbol{\theta} = \boldsymbol{\theta}_0$ if the $\mathbf{M}_1$ mode is enhanced [i.e., $b(1, e) = 1$]. This is because, for this single-source case,

$$\lim_{e \to \infty} \boldsymbol{R}(e)^{-1} = \frac{1}{\sigma^2} \left[\mathbf{I}_N - \frac{1}{N} \mathbf{D}(\theta_0)\mathbf{D}(\theta_0)' \right]$$

$$= \boldsymbol{R}(\infty)^{-1} \tag{4.38}$$

and with $\boldsymbol{\theta} = \boldsymbol{\theta}_0$, it is easy to see that the signal transform vector $\mathbf{D}(\theta_0)$ is in the null space of $\boldsymbol{R}(\infty)^{-1}$, causing an infinite response in the power spectrum estimator.

4.4 RESOLUTION PERFORMANCE

The resolution performance of the MV and MEM procedures can be examined at high signal-to-noise ratio (SNR) and long averaging time, T, by using (4.10) for the estimated data covariance matrix. For a single Gaussian signal source with SNR $= a$, the enhanced covariance matrix is given by

$$R(e) = a e \mathbf{D}(\theta_0)\mathbf{D}(\theta_0)' + \boldsymbol{I}_N \tag{4.39}$$

with θ_0 equal to either the signal frequency (f), wavenumber (k), or range (r) depending on the particular application. Following the procedures of [1] and [26], the width of the spectrum response at the 3-dB down points due to this single component signal in a white-noise background can be obtained. Specifically, the values of δ for which

$$\frac{P_{\text{EMV}}[\theta_0 + (\delta/2)]}{P_{\text{EMV}}(\theta_0)} = 0.5 \tag{4.40a}$$

and

$$\frac{P_{\text{EME}}[\theta + (\delta/2)]}{P_{\text{EME}}(\theta_0)} = 0.5 \tag{4.40b}$$

TABLE 4.1   THE 3-dB RESPONSE WIDTHS FOR HIGH-RESOLUTION
ANALYSIS OF FREQUENCY, WAVENUMBER, AND RANGE SPECTRA
ASSUMING HIGH SIGNAL-TO-NOISE RATIO (SNR) AND LONG
AVERAGING TIME (T)

Type of analysis	Domain of application		
	Frequency	Wavenumber (line array)	Range ($N = 3$)
Conventional (EC)	$\sqrt{6}\left(\dfrac{1}{N\pi\Delta}\right)$	$\beta = \pi/2$: $\ \sqrt{6}\,\dfrac{\lambda}{N\pi d}$ $\beta = 0$: $\ 2\sqrt{\sqrt{6}\,\dfrac{\lambda}{N\pi d}}$	$3\dfrac{\lambda}{\pi}\left(\dfrac{r}{d\sin\beta}\right)^2$
Minimum variance (EMV)	$2\sqrt{\dfrac{3}{Nea}}\left(\dfrac{1}{N\pi\Delta}\right)$	$\beta = \pi/2$: $\ 2\sqrt{\dfrac{3}{Nea}}\,\dfrac{\lambda}{N\pi d}$ $\beta = 0$: $\ 2\sqrt{2\sqrt{\dfrac{3}{Nea}}\,\dfrac{\lambda}{N\pi d}}$	$\sqrt{\dfrac{6}{ea}}\dfrac{\lambda}{\pi}\left(\dfrac{r}{d\sin\beta}\right)^2$
Maximum entropy (EME)	$\dfrac{2}{Nea}\left(\dfrac{1}{N\pi\Delta}\right)$	$\beta = \pi/2$: $\ \dfrac{2}{Nea}\dfrac{\lambda}{N\pi d}$ $\beta = 0$: $\ 2\sqrt{\dfrac{2}{Nea}\dfrac{\lambda}{N\pi d}}$	$\dfrac{1}{ea}\dfrac{\lambda}{\pi}\left(\dfrac{r}{d\sin\beta}\right)^2$

are referred to as the 3-dB response widths for the enhanced MV and ME spectral
analysis processes, respectively. A third estimator,

$$P_{\text{EC}}[\theta_0 + (\delta/2)] = \mathbf{D}[\theta_0 + (\delta/2)]'R(e)\mathbf{D}[\theta_0 + (\delta/2)] \tag{4.41}$$

referred to as the enhanced conventional spectrum estimator, is also considered.
Equation (4.41) comes from the uniformly weighted data transform

$$P_{\text{EC}}[\theta_0 + (\delta/2)] = E\{|\,\mathbf{D}[\theta_0 + (\delta/2)]'\mathbf{x}(t)\,|^2 : e\} \tag{4.42}$$

The 3-dB response widths for the three spectrum estimators given above are summarized in Table 4.1. Table 4.1 includes δ for spectrum analysis, wavenumber analysis, and range analysis. In the range analysis case, the results are restricted to an
$N = 3$ sensor array with uniform spacing d between the sensors for both wavenumber and range analysis.

The response width δ is presented herein as a measure of a particular spectral
analysis method to resolve two closely spaced signal components of equal strength.

For a single component in the presence of uncorrelated noise, neither MV nor ME processing of any type can provide a better estimate of the signal frequency (wavenumber, range) location than can conventional Fourier processing in terms of unbiased estimator variance. In fact, it is possible for a high-resolution estimator to have a higher variance than the conventional Bartlett estimator in the single-component case when a finite limited averaging time is required [1]. It is only in the multiple-component case that the lack of resolution capability can lead to superior performance of the EMV and EME techniques because increased resolution reduces the component of total rms error due to bias. These concepts are illustrated by an example in the final section.

Finally, the dependence of δ on the SNR $= a$ for each of the three spectral analysis techniques is noted. The conventional process response width is independent of SNR, while the MV and ME widths vary inversely as the square root of the SNR and SNR, respectively.

4.5 COMPUTATION OF THE ENHANCED SPECTRUM ESTIMATORS

Two fundamental approaches are available for the realization of the enhanced minimum-variance (EMV) and maximum-entropy (EME) estimators [27]. The first approach suggests a stochastic gradient search algorithm for adaptive estimation of the K largest eigenvalues and corresponding eigenvectors. This method involves feedback of the modal filter outputs

$$y_k(t) = M'_k(t)\mathbf{x}(t) \qquad k = 1, 2, \ldots, K \qquad (4.43)$$

which are then correlated with the data vector $\mathbf{x}(t)$ in an attempt to maximize the average value of the instantaneous eigenvalue estimate [8]

$$|y_k(t)|^2 = \mathbf{M}'_k(t)\mathbf{x}(t)\mathbf{x}'(t)\mathbf{M}_k \qquad (4.44)$$

This objective is accomplished by adjustment of the eigenvector estimates at time t, $\mathbf{M}_k(t)$, $k = 1, 2, \ldots, K$, to maximize (4.44) subject to the orthogonality

$$\mathbf{M}'_k(t)\mathbf{M}_m(t) = \delta_{km} \qquad (1 \le k, m \le K) \qquad (4.45)$$

and normality

$$|\mathbf{M}_k(t)|^2 = 1 \qquad (1 \le k \le K) \qquad (4.46)$$

constraints. This approach is computationally superior to the second approach to be outlined subsequently because both storage and multiply/add requirements are proportional only to NK. However, the performance of this gradient search scheme degrades drastically with SNR, as might be anticipated with noisy feedback as a component in the realization.

The second scheme is an open-loop, feedforward realization utilizing a direct computation of the data covariance matrix estimate $\hat{R}$ using (4.9). Given $\hat{R}$ with acceptable dimensionality N, subroutine EIGCH for Hermitian matrices contained in the IMSL Library [19], for example, can be used to obtain all N eigenvalues and eigenvectors of $\hat{R}$. Alternatively, because N can be large and the dimensionality K of the signal space is typically small relative to N, numerical techniques based on the power method for eigensystem computation can be used to obtain only the first K eigenvalues and eigenvectors of $\hat{R}$. However, the storage and computation requirements in this direct method are proportional to N^2.

As a variation on the feedforward approach, $\hat{R}$ can be assumed to be of the form given by (4.12). An estimate of the diagonal component $\sigma^2 I$ can be obtained and subtracted from $R(e)$. Then the residual matrix

$$P(e) = R - \sigma^2 I_N \tag{4.47}$$

can be factorized using a Cholesky decomposition. This factorization is used in (4.13), which, in turn, is used in the MV and ME spectral estimators in the form

$$R^{-1}(e) = \frac{1}{\sigma^2} [I_N - C(C'C)^{-1}C'] \tag{4.48}$$

(i.e., with $e = \infty$). While computationally expedient, the difficulty with this method is that the estimation and removal of the diagonal matrix $\sigma^2 I_N$ can result in problems. One algorithm, which follows that given in [18], suggests iteratively subtracting an increasingly larger term from the diagonal of R until the resulting matrix becomes singular. This singular matrix is used as an estimate of P, its rank is determined, and Cholesky factorization is performed. In reality, an estimate of R obtained from a finite time average will not be characterized by $\lambda_k = \sigma^2$ for $K + 1 \leq k \leq N$ as would be true of the assumed model in (4.12). Such a modeling error would lead to a premature termination of the iterative diagonal removal algorithm above. The result is poor relative enhancement of the dominant-mode eigenvectors and excessive computation because the dimensionality of the factorization (i.e., the number of columns in the matrix C is too large and the signal modes are insufficiently isolated).

A final variation on the direct feedforward method is to estimate the orthonormal complement $\{\mathbf{M}_k: k = K + 1, K + 2, \ldots, N\}$ of the dominant modal vectors $\{\mathbf{M}_k: k = 1, 2, \ldots, K\}$. The complement form is based on the fact that

$$\sum_{k=1}^{N} \mathbf{M}_k \mathbf{M}'_k = I_N \tag{4.49a}$$

$$= \sum_{k=1}^{K} \mathbf{M}_k \mathbf{M}'_k + \sum_{k=K+1}^{N} \mathbf{M}_k \mathbf{M}'_k \tag{4.49b}$$

or

$$\sum_{k=1}^{K} \mathbf{M}_k \mathbf{M}'_k = I_N - \sum_{k=K+1}^{N} \mathbf{M}_k \mathbf{M}'_k \tag{4.50}$$

With mode enhancement, (4.11b) substituting (4.50) becomes

$$R(e = \infty)^{-1} = \frac{1}{\sigma^2}\left(I_N - \sum_{k=1}^{K} \mathbf{M}_k \mathbf{M}'_k\right)$$

$$= \frac{1}{\sigma^2}\sum_{k=K+1}^{N} \mathbf{M}_k \mathbf{M}'_k \qquad (4.51)$$

and the EMV and EME estimators can be written as [16–18,28]

$$P_{\text{EMV}}(\theta) = \frac{\sigma^2}{\displaystyle\sum_{k=K+1}^{N} |\mathbf{D}(\theta)'\mathbf{M}_k|^2} \qquad (4.52)$$

and

$$P_{\text{EME}} = \frac{\displaystyle\sum_{k=K+1}^{N} |\mathbf{l}'_n \mathbf{M}_k|^2}{\left|\displaystyle\sum_{k=K+1}^{N} \mathbf{l}'_n \mathbf{M}_k \mathbf{M}'_k \mathbf{D}(\theta)\right|^2} \qquad (4.53)$$

after (4.21) and (4.34), respectively. Equations (4.52) and (4.53) are intuitively satisfying when it is noted that the orthonormal complementary set of eigenvectors spans a vector subspace which is orthogonal to the subspace spanned by the dominant eigenvectors. Thus, when $\mathbf{D}(\theta)$ lies in the signal subspace we have $\mathbf{M}'_k \mathbf{D}(\theta) = 0$, $K + 1 \le k \le N$, and the response functions given in (4.52) and (4.53), theoretically at least, can become infinite. Obviously, the mechanism that precludes such an infinite response is the inability to estimate either the eigenvectors for the dominant modes or their orthonormal complements exactly due to finite averaging time. This issue is examined more completely in the following section.

4.6 HIGH-RESOLUTION RANGE ESTIMATION

To passively locate a source with a linear array of sensors, it is desired to estimate the two-dimensional, range-bearing power spectrum of the array obtained by processing the array sensor output data over the sequence of T samples. Three power spectrum estimators are considered: the uniformly weighted-averaged transform, referred to as the focused beamformer (FB) [29,30], the minimum-variance distortionless response (MV), and the maximum-entropy (ME) estimators [10]. The array focusing vector $\mathbf{D}(\theta) = \mathbf{D}(r)$ at known frequency f and bearing β is defined by (4.6b). The conventional focused beamformer response is formed as illustrated in Figure 4.3. The MV and ME response functions are generated in a functionally similar manner with the only difference being in the operations performed on the estimated correlation matrix $\hat{R}$ in the beamformer.

The range spectrum estimator is emphasized exclusively in this discussion with

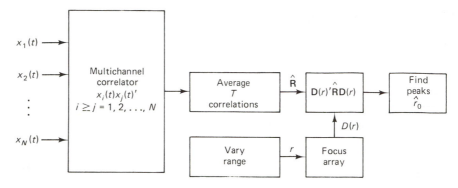

Figure 4.3 Realization of the focused beamformer for estimating the range r_0 of a source.

similar performance results obtainable for the bearing-angle dimension in terms of the wavenumber spectrum. The FB, MV, and ME range spectrum estimators are examined herein together with their enhanced dominant-mode decomposition spectra, referred to as the EFB, EMV, and EME, respectively. The issue of primary interest is a comparison of the various range estimator standard deviations to the minimum bound in the presence of both one and two sources.

For a single Gaussian source in a background of spatially uncorrelated and uniform Gaussian noise with zero mean and variance σ^2, the minimum bound on the variance of the range estimator is given by the Cramer–Rao (CR) variance [31]

$$\sigma_r^2 = \left[2N \left(\frac{2\pi}{\lambda} \right)^2 \frac{\sigma_s^2}{\sigma^2} \frac{1}{1 + N(\sigma_s^2/\sigma^2)} \left(\sum_{n=1}^{N} \alpha_n^2 - N\bar{\alpha}^2 \right) T \right]^{-1} \tag{4.54}$$

where N = number of sensors
     λ = signal wavelength
 σ_s^2/σ^2 = signal-to-noise ratio at the output of a sensor (SNR)

$$\alpha_n = \frac{r - x_n \cos \beta}{(r^2 + x_n^2 - 2x_n r \cos \beta)^{1/2}} \tag{4.55}$$

$$\bar{\alpha} = \frac{1}{N} \sum_{n=1}^{N} \alpha_n \tag{4.56}$$

and

T = number of samples in the observation interval

As a simple example for simulation of a linear array, the parameters $r = 4000\lambda$, $\beta = 90°$, and $N = 3$ are selected with a uniform spacing between sensors of $d = 75\lambda$. Figure 4.4 gives a comparison of the minimum range estimator standard deviation from (4.54) and the standard deviation of the location of the spectrum peak for the FB, MV, ME, EMV, and EME range power spectrum estimators as a function of signal-to-noise ratio. It is noted that the actual performance results are better than

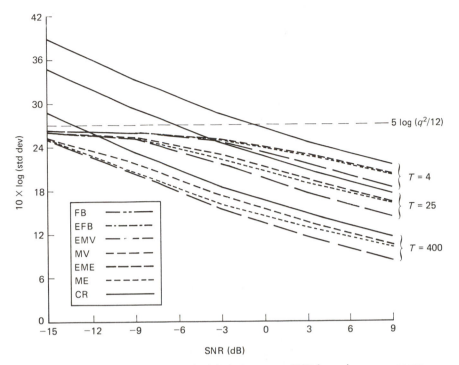

Figure 4.4   Range estimate standard deviation versus SNR for various range power spectrum estimators: $r = 4000$; $\beta = 90°$; $N = 3$.

minimum-bound results. This occurs for two reasons. First, a complex Gaussian (i.e., incoherent signal envelope) model is used both to obtain the result of (4.54) and to define the processors, whereas in the simulation performed, a constant-frequency coherent signal was used. Second, a range gated estimator was used. This results in the simulation results being consistent with the predicted minimum bound at either high SNR or long averaging time T. However, for low SNR and short averaging time, the range estimator variance appears to saturate. This effect occurs because the range gate region of allowable spectrum peaks is selected to contain only the correct response lobe for a representative high SNR, large T response function as illustrated in Figure 4.5. With a gate width q and either a low SNR or small T, the range estimate histogram generated in the simulation becomes nearly flat, with resultant range estimator variance approaching $q^2/12$. A similar problem which uses a time-delay gate to prevent selection of ambiguous lobes of a cross-correlation function is considered in [32]. The use of a range gate in this case is justified by noting that the location of the true response peak is invariant with respect to λ. This result says that, by frequency diverse range response function for a bandlimited signal esti-mation, the ambiguous response peaks can be resolved. The most significant feature about the simulation results shown in Figure 4.4 is that, in terms of range estimator rms error, none of the high-resolution spectrum estimators perform either better or

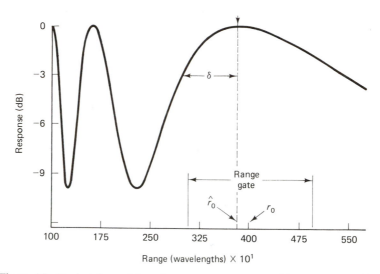

Figure 4.5 Typical focused beamformer range response with a range gate from 3000λ to 5000λ ($q = 2000\lambda$) and a source located at $r_0 = 4000\lambda$.

worse than the focused beamformer, with the exception of the maximum-entropy process. Both the maximum entropy (ME) and enhanced maximum entropy (EME) do marginally worse at high SNR with performance approaching that for the Gaussian signal when, in fact, a coherent signal is present. As SNR decreases, the ME performance approaches that for the FB and MV processes. This is consistent with predictions made by Baggeroer [33] for relative ME performance.

To evaluate the ability of any range power spectrum estimation technique to resolve the spectrum peak resulting from one source, from a second peak resulting from a nearby source, two metrics are appropriate. First, the separation of the 3-dB down points, δ, on the peak due to a single source as shown in Figure 4.4 is considered. Second, the rms range error for one source in the presence of a second source is established. The 3-dB down-range response peak widths for the techniques under consideration are as given in Table 4.1. The effective dominant-mode enhancement, e, for range estimation is a function of the SNR and observation time T. Figures 4.6(a) ($T = 400$) and (b) ($T = 25$) give the range response 3-dB down-peak width for the simulation experiment previously discussed. The theoretical ME and MV results from Table 4.1 are included for reference and are seen to be in excellent agreement with experimental results at high SNR (≥ -3 dB) and long averaging time ($T \geq 25$).

The ability of either an enhanced MV or ME process to improve resolution as indicated by 3-dB response peak width is clearly demonstrated in Figure 4.6 to be a function of both averaging time T and SNR. Table 4.2 gives the empirically determined effective enhancement factors for the MV and ME processes as a function of these two parameters. The results in Table 4.2 indicate that effective enhancement increases with averaging time, T. Furthermore, the results indicate that the obtain-

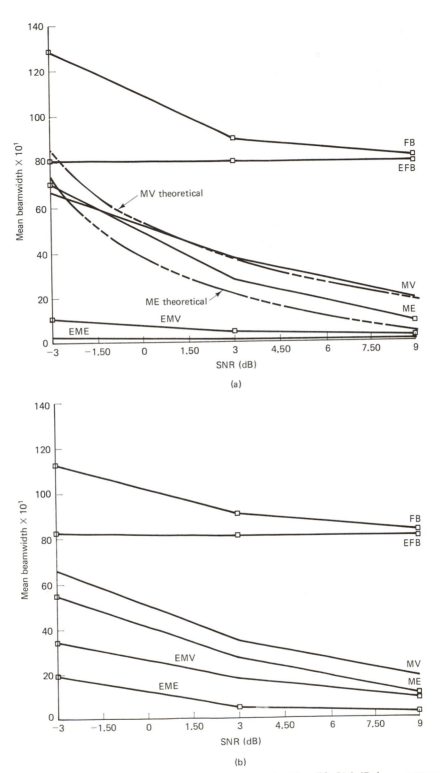

Figure 4.6 (a) 3-dB down-response width δ versus SNR for $T = 400$: (b) 3-dB down-response width δ versus SNR for $T = 25$.

TABLE 4.2a ENHANCEMENT FACTOR FOR THE MV PROCESS ($r = 2000\lambda$)

		SNR	
T	3	6	9
4	1.68	1.74	1.74
25	10.12	12.96	6.76
400	156.25	125.00	90.25

TABLE 4.2b ENHANCEMENT FACTOR FOR THE ME PROCESS ($r = 2000\lambda$)

		SNR	
N	3	6	9
4	1.16	1.39	1.29
25	5.20	4.86	4.00
400	18.66	12.33	9.00

able relative enhancement of the MV process is greater than that for the ME process. However, the absolute 3-dB peak widths for the ME and EME processes are still marginally less than for the MV and EMV processes, respectively. There is slightly better range estimator rms error performance exhibited by the MV and EMV processes compared to the corresponding ME approaches. Finally, it is important to note that both enhancement processes continue to exhibit impressive resolution capabilities with sufficient averaging time. This remains true even for SNR levels below -3 dB.

For the analysis of range estimator performance when two sources are present, let two sources be located at ranges r_1 and r_2. Also, let $x(t)$ be the N vector of array time data with $J = 2$. It is assumed that the signals and noise are uncorrelated. The covariance matrix of $\mathbf{x}(t)$ is

$$\mathbf{R} = E[\mathbf{x}(t)\mathbf{x}(t)'] \tag{4.57}$$

$$= \sigma_s^2[\alpha\mathbf{D}(r_1)\mathbf{D}(r_1)' + (1 - \alpha)\mathbf{D}(r_2)\mathbf{D}(r_2)'] + \sigma^2 I_N \tag{4.58}$$

where $0 < \alpha < 1$ indicates the relative power of the sources. Because $\mathbf{x}(t)$ is assumed to be zero mean, complex, and Gaussian, the probability density function of $\mathbf{x}(t)$ is

$$P(\mathbf{x}(t)/r_1, r_2) = \frac{1}{\pi^M |\mathbf{R}|} \exp\left[-\mathbf{x}(t)'\mathbf{R}^{-1}\mathbf{x}(t)\right]. \tag{4.59}$$

The minimum-variance bound for an unbiased estimator of the range parameter vector $\mathbf{r}$ defined by the transpose $\mathbf{r}^T = [r_1, r_2]$ is cov $(\mathbf{r}) = \mathbf{J}^{-1}(\mathbf{r})$, where $\mathbf{J}(\mathbf{r})$ is the 2×2 Fisher information matrix. The ijth entry of $\mathbf{J}(\mathbf{r})$ is given by

$$[\mathbf{J}(\mathbf{r})]_{ij} = -E\left[\frac{\partial^2 L(\mathbf{x}/\mathbf{r})}{\partial r_i\, \partial r_j}\right] \tag{4.60}$$

where

$$L(\mathbf{x}/\mathbf{r}) = -MN \log \pi - N \log |\mathbf{R}| - \sum_{n=1}^{T} \mathbf{x}(t)'\mathbf{R}^{-1}\mathbf{x}(t). \tag{4.61}$$

Using the identity

$$-E\left[\frac{\partial^2 L(\mathbf{x}/\mathbf{r})}{\partial r_i\, \partial r_j}\right] = \text{tr}\left(\mathbf{R}^{-1}\frac{\mathbf{R}}{\partial r_i}\mathbf{R}^{-1}\frac{\mathbf{R}}{\partial r_j}\right) \tag{4.62}$$

expressions can be derived for J_{11}, J_{22}, and J_{12} which are functions of the focusing vectors, signal to noise, number of sensors, and number of samples. For high signal-to-noise ratio and $X_n \ll r_i$ $(i = 1, 2)$ [3,34],

$$J_{11} \simeq 2N\left(\frac{2\pi}{\lambda}\right)^2 \left(\frac{\alpha\sigma_s^2}{\sigma^2}\right)^2 \left[N \sum_{n=1}^{N} t_{n1}^2 - \left(\sum_{n=1}^{N} t_{n1}\right)^2 \right] \frac{1 + \dfrac{(1-\alpha)M\sigma_s^2}{\sigma^2}(1-|g|^2)}{\mathrm{DV}} \tag{4.63}$$

$$J_{22} \simeq 2N\left(\frac{2\pi}{\lambda}\right)^2 \left(\frac{(1-\alpha)\sigma_s^2}{\sigma^2}\right)^2 \left[N \sum_{n=1}^{N} t_{n2}^2 - \left(\sum_{n=1}^{N} t_{n2}\right)^2 \right] \frac{1 + \dfrac{\alpha M\sigma_s^2}{\sigma^2}(1-|g|^2)}{\mathrm{DV}} \tag{4.64}$$

$$J_{12} \simeq 0$$

where

$$t_{ni} = \frac{r_i - X_n \cos\beta_i}{\sqrt{r_i^2 + X_n^2 - 2r_i X_n \cos\beta_i}} \qquad n = 1, 2, \ldots, N; \quad i = 1, 2 \tag{4.65}$$

$$g = \frac{1}{N} \sum_{n=1}^{N} \exp\left[-j2\pi \frac{c}{\lambda}(\tau_{1n} - \tau_{2n}) \right] \tag{4.66}$$

$$\mathrm{DV} = \left\{ \left(1 + \frac{\alpha N\sigma_s^2}{\sigma^2}\right)\left[1 + \frac{(1-\alpha)N\sigma_s^2}{\sigma^2}\right] - \frac{\alpha(1-\alpha)M^2|g|^2}{\sigma^4} \right\} \tag{4.67}$$

    The remaining terms are defined for the minimum-variance bound for a single source. Using J_{11} and J_{22}, it follows that the minimum-variance bounds for estimators of r_1 and r_2 are J_{11}^{-1} and J_{22}^{-1}, respectively. These expressions provide insight into the performance of multiple range estimation. For example, if $g = 0$, the minimum-variance bounds for r_1 and r_2 are the same as they would be in the single-source case in (4.54). If $|g| \neq 0$, it can be shown that the minimum-variance bounds for r_1 and r_2 will be larger. In fact, J_{11}^{-1} and J_{22}^{-1} are monotonically increasing functions of $|g|$.

    As an illustration of range estimation high-resolution performance with two sources present, a typical set of range response functions from a simulation run is presented in Figure 4.7. The simulation parameters are SNR = 9 dB, $T = 100$, $N = 5$, $d = 75\lambda$, $r_1 = 3500\lambda$, and $r_2 = 4000\lambda$. The single-source range estimate one standard deviation interval is shown above the individual source ranges centered at the true range value. All response functions resolve the two sources except the FB, MV, and EMV with $e = 1$. The enhanced processes use $K = 2$ for the number of dominant modes for which the eigenvalues and eigenvectors are estimated. The response level for the source at 4000λ is lower than for that at $r_1 = 3500\lambda$ because the reduced resolution at r_2 manifests itself more as a range distributed source but having the same total power. Clearly, the enhanced processes with $e = \infty$ exhibit substantial resolution capability. The inability to resolve the two sources by the FB, MV, and EMV $(e = 1)$ processes results in a critical rms bias error of 200 to 300λ in addition to suggesting a completely erroneous source range spectrum.

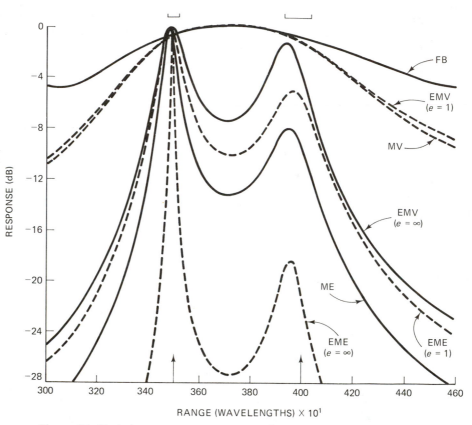

Figure 4.7 Typical range spectrum response for a two-source example with SNR = 9 dB, $T = 100$, $N = 5$, $d = 75\lambda$, $r_1 = 3500\lambda$, and $r_2 = 4000\lambda$.

4.7 CONCLUDING REMARKS

Two high-resolution spectral analysis procedures based on the well-known minimum-variance (MV) and maximum-entropy (ME) techniques have been presented. These procedures exhibit superior resolution performance relative to either the MV or ME spectrum estimators with ultimate performance limited only by observation time. Such performance should have substantial impact in spectral analysis applications which are estimator bias limited as opposed to estimator random fluctuation limited.

REFERENCES

[1] R. T. Lacoss, "Data Adaptive Spectrum Analysis Methods," *Geophysics, 36*(4):661–675 (Aug. 1971).

[2] J. Capon, "High-Resolution Frequency-Wavenumber Spectrum Analysis," *Proc. IEEE, 57*: 1408–1418 (1969).

[3] N. L. Owsley and G. Swope, "High Resolution Range Estimation with a Linear Array," *Proc. IEEE EASCON '82*, Washington, D.C., Sept. 1982.

[4] N. L. Owsley, "An Overview of Optimum Adaptive Control in Sonar Array Processing," in K. S. Narendra and R. V. Monopoli, eds., *Applications of Adaptive Control*, Academic Press, New York, 1980, pp. 131–164.

[5] J. P. Burg, "Maximum Entropy Spectrum Analysis," *37th Annu. Meet. Soc. Explor. Geophys.*, Oklahoma City, Okla., 1967.

[6] J. P. Burg, "The Relationship between Maximum Entropy and Maximum Likelihood Spectra," *Geophysics*, *37*(2):375–376 (Apr. 1972)

[7] A. Vanden Bos, "Alternative Interpretation of Maximum Entropy Spectral Analysis," *IEEE Trans. Inf. Theory* (Corresp.), *IT-17*:493–494 (July 1971).

[8] N. L. Owsley, "Adaptive Data Orthogonalization," *Proc. IEEE ICASSP*, Tulsa, Okla., Apr. 1978, pp. 109–112.

[9] N. L. Owsley, "Modal Decomposition of Data Adaptive Spectral Estimates," *Yale Univ. Workshop Appl. Adaptive Syst. Theory*, K. S. Narendra, ed., May 1981.

[10] N. L. Owsley and J. F. Law, "Dominant Mode Power Spectrum Estimation," *Proc. IEEE ICASSP*, Paris, Apr. 1982.

[11] D. Tufts and R. Kumaresen, "Data-Adaptive Principal Component Signal Processing," *Proc. 1980 IEEE CDC*, Albuquerque, N.M., Dec. 1980.

[12] L. J. Griffiths, "A Continuously Adaptive Filter Implemented as a Lattice Structure," *Proc. IEEE ICASSP*, Hartford, Conn., 1977.

[13] W. S. Liggett, "Passive Sonar: Fitting Models to Multiple Time Series," in J. W. R. Griffiths et al., eds., *Signal Processing*, Academic Press, 1973.

[14] N. L. Owsley, "A Recent Trend in Adaptive Spatial Processing for Sensor Arrays: Constrained Adaptation," in J. W. R. Griffiths, et al., eds., *Signal Processing*, Academic Press, New York, 1973.

[15] H. Mermoz, "Complementarity of Propagation Model Design with Array Processing," in G. Tacconi, ed., *Aspects of Signal Processing*, D. Reidel, Dordrecht, The Netherlands, 1977.

[16] A. Cantoni and L. Godara, "Resolving the Directions of Sources in a Correlated Signal Field Incident on an Array," *J. Acoust. Soc. Am.*, *67*(4):1247–1255 (Apr. 1980).

[17] G. Bienvenu and L. Koop, "Adaptive High Resolution Spatial Discrimination of Passive Sources," in L. Bjorno, ed., *Underwater Acoustics and Signal Processing*, D. Reidel, Dordrecht, The Netherlands, 1981.

[18] R. Klemm, "High-Resolution Analysis of Nonstationary Data Ensembles," *Signal Processing: Theories and Applications*, Elsevier-North Holland Publishing Co., Oct. 1980, pp. 711–714.

[19] International Mathematical and Statistical Libraries Inc., *Eigensystem Analysis*, 9th ed., Vol. 2, IMSL, Houston, 1982, Chap. E.

[20] D. Tufts and R. Kumarasen, "Singular Value Decomposition and Spectral Analysis," *Proc. IEEE ASSP Workshop Spectral Anal.*, McMaster University, Hamilton, Ont., Aug. 1981.

[21] D. Johnson, "Improving the Resolution of Bearing in Passive Sonar Arrays by Eigenvalue Analysis," *Proc. First ASSP Workshop Spectral Anal.*, McMaster University, Hamilton, Ont., Aug. 1981.

[22] W. Gabriel, "Adaptive Superresolution of Coherent RF Spatial Sources," *Proc. First ASSP Workshop Spectral Anal.*, McMaster University, Hamilton, Ont., Aug. 1981.

[23] S. Pasupathy and A. N. Venetsanopoulos, "Optimum Active Array Processing Structure and Space–Time Factorability," *IEEE Trans. Aerosp. Electron. Syst.*, AES-10(6):770–778 (Nov. 1974).

[24] Special Issue on Time Delay Estimation, Part II, *Trans. IEEE Acoust. Speech Signal Process.*, ASSP-29(3) (June 1981).

[25] L. Ng and Y. BarShalom, "Time Delay Estimation in a Multitarget Environment," *Proc. 21st IEEE CDC*, Orlando, Fla., Dec. 1982.

[26] O. B., Gammelsaeter, "Adaptive Beamforming with Emphasis on Narrowband Implementation," in L. Bjorno, ed., *Underwater Acoustic and Signal Processing*, NATO ASI Series C, D. Reidel, Dordrecht, The Netherlands, 1981, pp. 307–326.

[27] N. L. Owsley, "Adaptive Data Orthogonalization," *Proc. IEEE ICASSP*, Tulsa, Okla., Apr. 1978, pp. 109–112.

[28] R. Schmidt, "Multiple Emitter Location and Signal Parameter Estimation," *Proc. RADC Spectral Estimation Workshop*, Rome, N.Y., 1979.

[29] G. C. Carter, "Time Delay Estimation for Passive Sonar Signal Processing," *IEEE Trans. Acoust. Speech Signal Process.*, ASSP-29(3):463–470 (June 1981).

[30] N. L. Owsley and G. R. Swope, "Time Delay Estimation in a Sensor Array," *IEEE Trans. Acoust. Speech Signal Process.*, ASSP-29(3):519–523 (June 1981).

[31] W. J. Bangs and P. Schultheiss, "Space–Time Processing for Optimal Parameter Estimation," in J. W. R. Griffiths et al., eds., *Signal Processing*, Academic Press, New York, 1973, pp. 591–604.

[32] J. P. Ianniello, "Threshold Effects in Time Delay Estimation Using Narrowband Signals," *IEEE ICASSP 1982*, Paris, May 3, 1982, pp. 375–379.

[33] A. B. Baggeroer, "Confidence Intervals for Maximum Entropy Spectral Estimates," in *Aspects of Signal Processing*, Part I, NATO Advanced Study Institute, D. Reidel, The Netherlands, 1976.

[34] G. Swope, Ph.D. thesis, Rensselaer Polytechnic Institute, Aug. 1982.

5

Principles of High-Resolution Array Processing

G. Bienvenu

Thomson-CSF
Cagnes-sur-Mer, France

H. Mermoz

Ingénieur Général des Télécommunications
Six Fours Les Plages, France

5.1 INTRODUCTION

One of the main functions of an underwater passive listening system is the determination of the number of sources present as well as the characteristic parameters of these sources. The basic tool is spatial processing, which has to separate the sources, using their spatial properties. The traditional tool for signal analysis is the classical beamformer; with the objective of improving the performance, the adaptive beamformer for the sensor array was evolved. This method brings an improvement in array gain which is asymptotically bound by the signal-to-noise ratio of the source noises when measured on a particular sensor. More recently, more powerful methods called "high resolution" have appeared. The improvement in performance compared to previous processing is predicated on an additional assumption regarding the medium. Nevertheless, these methods carry the possibility of including free parameters for the medium model, making the assumptions somewhat more flexible.

Adaptive array processing, just as classical beamforming, relies only on assumptions about sources and on the propagation of the transmitted signals. The sources are point-like. They have, through the array, a perfect spatial coherence. Generally, the form of the wavefront, as it is translated by the sensors, is a known function of the position (direction, distance) of the source. The sensor transfer function is also assumed to be perfectly known. This latter condition is expressed by the notion of position vector $\mathbf{D}(f)$ of the source. When it is not known a priori, $\mathbf{D}(f)$ will only be referred to as the source vector. Let $\mathbf{r}(t)$ be the vector representing the set of signals received on the N sensors. This vector will also be referred to as a column matrix

$$\mathbf{r}^+(t) = [r_1(t), \ldots, r_n(t), \ldots, r_N(t)] \tag{5.1}$$

where $\mathbf{r}^+(t)$ is the transposed conjugate of $\mathbf{r}(t)$, and $r_n(t)$ is the signal received by sensor number n. The received signal correlation matrix is defined by

$$C(\tau) = E[\mathbf{r}(t)\mathbf{r}^+(t + \tau)] \tag{5.2}$$

where $E(\cdot)$ stands for the mathematical expectation. The cross-spectral density matrix (CSDM) of the received signal is denoted by $\boldsymbol{\Gamma}(f)$. $\boldsymbol{\Gamma}(f)$ is the Fourier transform of $C(\tau)$. Under the previous assumptions, the CSDM of one source alone has particular properties. It can be written as

$$\boldsymbol{\Gamma}(f) = \gamma(f)\mathbf{D}(f)\mathbf{D}^+(f) \tag{5.3}$$

where $\mathbf{D}(f)$ is the source position vector based on the transfer functions between the source and every sensor. This set of transfer functions is normalized so that, for example, the value of the square of the modulus of vector $\mathbf{D}(f)$ is N; $\gamma(f)$ is the spectral density of the signal received from the source. $\mathbf{D}(f)$ represents the wavefront shape and can be computed if the location $\boldsymbol{\theta}$ of the source is given a priori. The rank of matrix $\boldsymbol{\Gamma}(f)$ is unity, which is characteristic of full spatial coherence. The last assumption is that the sources are *not* cross-correlated either between one another or with the background noise.

Under these assumptions, the global CSDM at the sensor outputs is expressed by

$$\boldsymbol{\Gamma}(f) = \boldsymbol{\Gamma}_B(f) + \sum_{i=1}^{K} \gamma_i(f)\mathbf{D}_i(f)\mathbf{D}_i^+(f) \tag{5.4}$$

where $\boldsymbol{\Gamma}_B(f)$ is the CSDM of the background noise and K the number of sources.

High-resolution methods (HRM) need one more assumption as to the medium; namely, there exists some information about the spatial coherence of the background noise. As a result of this assumption, HRM do better than the adaptive array. First we present the principle of HRM using the basic assumptions, which are restrictive ones; then we tell how they can be extended to more flexible conditions.

5.2 BASIC PRINCIPLES OF THE HIGH-RESOLUTION METHODS

The basic assumptions of the HRM are:

1. For the sources, the same ones already used for adaptive arrays, which are presented in the introduction, with an additional assumption that the sources can be singled out because the number K of sources is smaller than the

number N of sensors of the array. This is a way of saying that some complexity of the array is needed to surmount the complexity of the medium.

2. For the background noise there is no spatial coherence; its CSDM is equal to

$$\boldsymbol{\Gamma}_B(f) = \sigma(f)\boldsymbol{I} \tag{5.5}$$

where $\sigma(f)$ is the unknown spectral density of background noise and the unit matrix $\boldsymbol{I}$ is its spatial coherence matrix. So that the global CSDM is expressed by

$$\boldsymbol{\Gamma}(f) = \sigma(f)\boldsymbol{I} + \boldsymbol{\Gamma}_s(f) = \sigma(f)\boldsymbol{I} + \sum_{i=1}^{K<N} \gamma_i(f)\mathbf{D}_i(f)\mathbf{D}_i^+(f) \tag{5.6}$$

5.2.1 CSDM Decomposition

HRM are based on an examination of the CSDM eigenvectors and eigenvalues. An eigenvector $\mathbf{V}(f)$ and its corresponding eigenvalue $\lambda(f)$ satisfy

$$\boldsymbol{\Gamma}(f)\mathbf{V}(f) = \sigma(f)\mathbf{V}(f) + \sum_{i=1}^{K<N} \gamma_i(f)\mathbf{D}_i(f)\mathbf{D}_i^+(f)\mathbf{V}(f) = \lambda(f)\mathbf{V}(f) \tag{5.7}$$

The matrix $\boldsymbol{\Gamma}(f)$ has

1. K eigenvectors $\mathbf{V}_i(f)$ ($i \in [1, K]$), which also are eigenvectors of the "sources alone" CSDM, $\boldsymbol{\Gamma}_s(f)$. They correspond to the nonzero K eigenvalues $\lambda_{si}(f)$ of $\boldsymbol{\Gamma}_s(f)$ whose rank is K. The corresponding eigenvalues are

$$\lambda_i(f) = \lambda_{si}(f) + \sigma(f)$$

These K eigenvectors form a basis in the K-dimensional subspace of the K position vectors $\mathbf{D}_i(f)$ and this subspace will be called the source subspace.
   One can write the two fundamental relations

$$\mathbf{D}_i(f) = \sum_{i=1}^{K} \alpha_{ij}\mathbf{V}_j(f) \tag{5.8}$$

and

$$\sum_{i=1}^{K} \gamma_i(f)\mathbf{D}_i(f)\mathbf{D}_i^+(f) = \sum_{i=1}^{K} [\lambda_i(f) - \sigma(f)]\mathbf{V}_i^+(f)\mathbf{V}_i(f) \tag{5.9}$$

2. $(N - K)$ eigenvectors $\mathbf{V}_{i0}(f)$ ($i \in [K + 1, N]$) orthogonal to the previous ones and fundamentally orthogonal to all the source position vectors; that is,

$$\mathbf{V}_{i0}^+(f)\mathbf{D}_j(f) = 0 \ \forall \ i \in [K + 1, N] \quad \text{and} \quad j \in [1, K] \tag{5.10}$$

The corresponding noise-only eigenvalues are all $\sigma(f)$ and consequently smaller than any of the signal-plus-noise eigenvalues. These $(N - K)$ eigenvectors form a basis that defines a noise-only orthogonal subspace.

5.2.2 The High-Resolution Methods

The HRM is based on the following observations:

1. From the eigenvalues of the CSDM one can get:
   a. The number of sources, which is the number of sensors minus the number of equal and smallest eigenvalues.
   b. The source signal subspace and the corresponding orthogonal subspace through a partitioning of the eigenvectors to those that do not correspond and to those that do correspond to the smallest eigenvalues.
   One must remark that to get these results, the only assumptions needed are those on the spatial incoherence of background noise and the coherence of the sources.
2. The source locations can be obtained using either the source subspace vectors, equations (5.8) and (5.9), or the orthogonal subspace, (5.10).

In both cases we use a model position vector $\mathbf{D}(f, \theta)$ where θ stands for the position of the source. This vector complies with the accepted assumptions. When one exploits the source subspace, one works on a reconstruction of the "source-alone" CSDM, given by

$$\sum_{i=1}^{K} [\lambda_i(f) - \sigma(f)] \mathbf{V}_i(f) \mathbf{V}_i^+(f) \tag{5.11}$$

where $\sigma(f)$ is given by one of the smallest eigenvalues of $\Gamma(f)$. The source parameters θ_i and $\gamma_i(f)$ can be obtained from the equation [1,2]

$$\sum_{i=1}^{K} [\lambda_i(f) - \sigma(f)] \mathbf{V}_i(f) \mathbf{V}_i^+(f) = \sum_{i=1}^{K} \gamma_i(f) \mathbf{D}(f, \theta_i) \mathbf{D}^+(f, \theta_i) \tag{5.12}$$

It should be mentioned that the classical methods such as conventional beamforming [3] (in the particular case of sources with orthogonal position vectors), maximum likelihood, and maximum entropy can be modified by using the source subspace eigenvectors only [4]. When the orthogonal subspace is used [5,6], the source positions result from a projection of the model position vector onto this orthogonal subspace, according to

$$G(\theta) = \sum_{i=K+1}^{N} |\mathbf{V}_{i0}^+(f) \mathbf{D}(f, \theta)|^2 \tag{5.13}$$

When θ varies, $G(\theta)$ produces a null every time θ equals the position θ_i of a source. The nulls of $G(\theta)$ yield the source locations. Knowing them, spectral densities are then at hand [7].

5.2.3 Practical Use of Estimated CSDM

The properties previously described are of course asymptotic ones, since in practice, one can only get an estimate $\hat{\Gamma}(f)$ of the received signal CSDM. An infinite time would be needed (plus perfect stationarity) to obtain $\Gamma(f)$. One can only hope to get estimates of the source parameters. The $(N - K)$ smallest eigenvalues are not, in practice, strictly equal. Therefore, the decision about the number of sources, or equivalently the source signal subspace and its orthogonal subspace, is relevant to detection theory. It has been shown [8] that under Gaussian assumptions about the actual signals and the use of the generalized likelihood ratio, it is possible to find a test for the number of sources. This test uses only the eigenvalues of $\hat{\Gamma}(f)$, the maximum-likelihood estimate of $\Gamma(f)$.

It has been also shown that the properties that characterize $\Gamma(f)$ remain valid for $\hat{\Gamma}(f)$. In particular, relations (5.8), (5.9), and (5.10) remain valid when $\gamma_i(f)$, $\mathbf{D}_i(f)$, and $\sigma(f)$ are, respectively, replaced by their maximum-likelihood estimates, and when $\lambda_i(f)$, $\mathbf{V}_i(f)$, and $\mathbf{V}_{io}(f)$ are replaced by the eigenvalues and eigenvectors of the CSDM estimate $\hat{\Gamma}(f)$.

5.2.4 Main Property of High-Resolution Method

The partitioning of the array data into two subspaces, that of the sources and its orthogonal complement, gives the HRM its main power; namely, its resolution is no longer limited by the signal-to-noise ratio $\gamma_i(f)/\sigma(f)$ as in the case of adaptive array. The resolution increases with the observation time up to infinity if the latter could be unlimited. Therefore, asymptotically as averaging time increases, two sources can be separated, how close and how weak, compared to the background noise, they may be. This property is illustrated in Figure 5.1, which shows results from a simulation using the orthogonal subspace. The receiving array is a linear one with 12 equispaced sensors. The sources, two in number, are infinitely remote with bearings θ: $0°$ and $-5.3°$. The sources' signal-to-noise ratios are both 0 dB. Since an

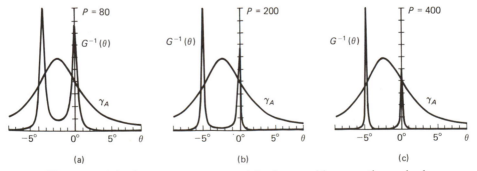

Figure 5.1   Adaptive array output spectral density γ_A and inverse orthogonal subspace method response $G^{-1}(\theta)$ versus bearing θ for two sources: (a) $P = 60$; (b) $P = 200$; (c) $P = 400$.

estimate of the CSDM is used, the function $G(\theta)$ of (4.13) yields minimum points instead of nulls in the source directions. The estimate of the CSDM is

$$\hat{\boldsymbol{\Gamma}}(f) = \frac{1}{P} \sum_{i=1}^{P} \mathbf{X}_i(f)\mathbf{X}_i^+(f) \tag{5.14}$$

where $\mathbf{X}_i(f)$ is the discrete Fourier transform of the incoming signal $\mathbf{r}(t)$, computed over adjacent portions of time whose duration is T. The observation time is, consequently, PT. As the bearing θ varies, Figure 5.1 shows the spectral density on the output of the adaptive array γ_A and the value of $1/G(\theta)$, which has peaks instead of minimums. One can see first that the adaptive array does not resolve the two sources, while the orthogonal subspace method does. Moreover, it is clear that the $G(\theta)$ response improves when the observation time increases, while the response of the adaptive array is stabilized.

Physical limitations of the "infinite" resolution power in the HRM come from a nonperfect knowledge concerning fluctuations of both the spatial coherence of the background noise and the shape of the source wavefronts. In the next paragraph we shall see how the assumptions about these two quantities can be made more flexible, so as to increase the validity domain of HRM.

5.3 GENERALIZATION OF HIGH-RESOLUTION METHODS

The first task to be achieved in order to use the HRM is to separate the source subspace from its orthogonal subspace because all of the HRM power lies in this partition. But it relies strictly on the assumed lack of spatial coherence of the background noise. It is then necessary to extend this assumption.

After the best possible partition the second task is the location of the sources. To do so, we need wavefront models as flexible as possible to adhere more closely to the real world.

5.3.1 Partition of Source Subspace and Orthogonal Subspace: Modeling of Background Noise Spatial Coherence (BNSC)

Influence of BNSC.   In an actual array environment, the background noise can be correlated between two sensors. As a consequence its CSDM is not the unity matrix. In the most general case the background noise CSDM can be written

$$\boldsymbol{\Gamma}_B(f) = \sigma(f)\boldsymbol{J}(f) \tag{5.15}$$

where $\boldsymbol{J}(f)$ is the normalized coherence matrix. Then the previous properties of the eigenvectors and eigenvalues are no longer valid.

As an example and to illustrate this remark, let us see what happens when the surface wind noise is dominant in the background noise for a passive underwater listening array. For two sensors at distance d in a horizontal plane, the cross-

TABLE 5.1

$d = 0.7$	2.67	2.22	1.38	1.11	0.76	0.61	0.58	0.55	0.54	0.53	0.52	0.52
$d = 0.3$	2.24	2.02	1.14	1.10	0.96	0.93	0.87	0.85	0.83	0.78	0.24	0.03

spectral density can be represented by

$$\gamma_d(f) = \sigma(f)\,\frac{2^m m!}{2\pi f d/c}\,y_m(2\pi f d/c) \tag{5.16}$$

where c is the sound velocity in the water and $y_m(\cdot)$ is the Bessel function of order m; the value of m depends on the sea state.

Signals of a horizontal 12 equispaced sensor array have been simulated. Table 5.1 shows the eigenvalues (for an infinite observation time) due to a noise field made only of background noise with $m = 0$, and for a distance d between two adjacent sensors equal to 0.3 and 0.7 wavelength. If the CSDM of the background noise were a unit matrix, all the eigenvalues would have been equal. Instead, Table 5.1 presents significant differences.

Figure 5.2 presents the projection of the vector $\mathbf{V}_M(f)$ (which corresponds to the maximum eigenvalue of the CSDM estimated by using (5.14) with $d = 400$) on

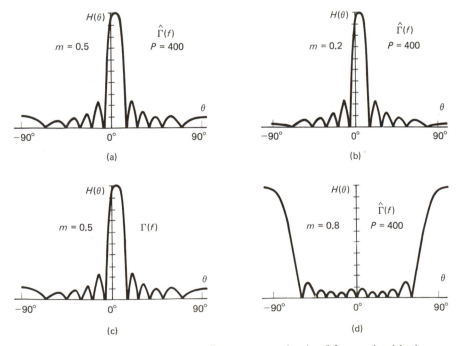

Figure 5.2 Maximum eigenvector diagrams versus bearing θ for correlated background noise and one source [$\Gamma(f)$ = theoretical CSDM, $\hat{\Gamma}(f)$ = CSDM estimate].

the model position vector $\mathbf{D}(f, \theta)$, as a function of the bearing

$$H(\theta) = |\mathbf{D}^+(f, \theta)\mathbf{V}_M(f)|^2$$

The noise field is made of background noise plus a source at bearing $4°$ with a signal-to-noise ratio of -15 dB. In this case, the eigenvector corresponding to the maximum eigenvalue of the theoretical CSDM is identical to the source position vector and develops the same diagram as the classical time-delay beamformer. The array is the same as previously mentioned with adjacent sensors one-half wavelength apart. Figure 5.2 involves four diagrams. The first one is theoretical. The three others correspond to different background noise spatial coherence given by $m = 0.5$, $m = 0.2$, and $m = 0.7$, respectively. For $m = 0.5$ the CSDM is a unit matrix accordingly with the assumptions. Actually, the $H(\theta)$ diagram is in this case identical to the theoretical diagram. For $m = 0.2$, $H(\theta)$ shows small differences with theory but is completely different for $m = 0.8$.

Figure 5.3 presents, for the same data, the results of the orthogonal subspace method. What is plotted is the inverse of $G(\theta)$ (5.13) so as to observe peaks instead of minimums. Again we observe some degradation when the background noise spatial coherence matrix is no longer a unit matrix ($m \neq 0.5$).

Adaptivity to background noise spatial coherence [9]. When the spatial coherence $\mathbf{J}(f)$ of background noise is not a unit matrix, it can be made obvious that if it is known, we are back to the previous problem. Actually, since $\mathbf{J}(f)$ is a Hermitian positive-definite matrix, it is possible to find a matrix $\mathbf{C}(f)$ such that

$$\mathbf{C}(f)\mathbf{J}(f)\mathbf{C}^+(f) = \mathbf{I} \tag{5.18}$$

When the original CSDM is transformed by $\mathbf{C}(f)$, one gets

$$\boldsymbol{\Gamma}_c(f) = \mathbf{C}(f)\boldsymbol{\Gamma}(f)\mathbf{C}^+(f)$$

$$= \sigma(f)\mathbf{C}(f)\mathbf{J}(f)\mathbf{C}^+(f) + \sum_{i=1}^{K} \gamma_i(f)\mathbf{C}(f)\mathbf{D}_i(f)\mathbf{D}_i^+(f)\mathbf{C}^+(f) \tag{5.19}$$

Noticing that $\mathbf{C}(f)\mathbf{D}_i(f)$ is a vector that can be written $\mathbf{D}_{ci}(f)$, and using (5.18), it

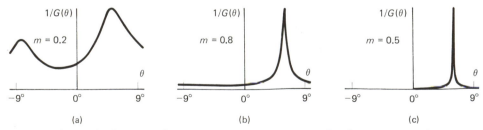

(a)   (b)   (c)

Figure 5.3 Inverse orthogonal subspace response versus bearing θ for the noise fields of Figure 5.2.

follows that

$$\boldsymbol{\Gamma}_c(f) = \sigma(f)\boldsymbol{I} + \sum_{i=1}^{K} \gamma_i(f)\mathbf{D}_{ci}(f)\mathbf{D}_{ci}^{+}(f) \qquad (5.20)$$

In this way, the background noise has been spatially "whitened"; the eigenvalues and vectors of the modified CSDM, $\boldsymbol{\Gamma}_c(f)$, are endowed with the theoretical properties previously mentioned. Of course, to get the source locations, it will be necessary to use a transformed position vector according to

$$\mathbf{D}_c(f, \boldsymbol{\theta}) = C(f)\mathbf{D}(f, \boldsymbol{\theta}) \qquad (5.21)$$

Moreover, it can be shown that it is not necessary to know exactly the spatial coherence of the background noise. It is sufficient to have it comply with a general structure or model, in which several parameters can remain unknown. The model of the CSDM is then written in the form $J(f, \mathbf{m})$, where $\mathbf{m}$ symbolizes the unknown parameters in the model.

Let $C(f, \mathbf{m})$ be the matrix such as

$$C(f, \mathbf{m})J(f, \mathbf{m})C^{+}(f, \mathbf{m}) = I \qquad (5.22)$$

and $\mathbf{m}_0$ the true value of the background noise parameter vector. The CSDM transformed by $C(f, \mathbf{m})$ is

$$\boldsymbol{\Gamma}_c(f, \mathbf{m}) = C(f, \mathbf{m})\boldsymbol{\Gamma}(f)C^{+}(f, \mathbf{m}) \qquad (5.23)$$

Consequently, we obtain

$$\boldsymbol{\Gamma}_c(f, \mathbf{m}) = \sigma(f)C(f, \mathbf{m})J(f, \mathbf{m})C^{+}(f, \mathbf{m})$$
$$+ \sum_{i=1}^{N} \gamma_i(f)C(f, \mathbf{m})\mathbf{D}_i(f)\mathbf{D}_i^{+}(f)C^{+}(f, \mathbf{m}) \qquad (5.24)$$

It is clear that if $\mathbf{m}$ is varied through different values, the background noise will become perfectly incoherent when it reaches the value $\mathbf{m}_0$. Consequently, if the eigenvalues of $\boldsymbol{\Gamma}_c(f, \mathbf{m})$ are plotted against $\mathbf{m}$, the $(N - K)$ smallest eigenvalues become equal at $\mathbf{m}_0$; it is therefore possible, in doing so, to derive the value $\mathbf{m}_0$ and to spatially whiten the background noise.

Of course, in practice, only an estimate $\hat{\boldsymbol{\Gamma}}(f)$ of $\boldsymbol{\Gamma}(f)$ is available and when the eigenvalues of

$$\hat{\boldsymbol{\Gamma}}_c(f, \mathbf{m}) = C(f, \mathbf{m})\hat{\boldsymbol{\Gamma}}(f)C^{+}(f, \mathbf{m}) \qquad (5.25)$$

are plotted against $\mathbf{m}$, what is observed instead of a perfect whitening is a sort of convergence of the smallest eigenvalues. This convergence is all the more acute and close to $\mathbf{m}_0$ as the observation time becomes larger. The value of $\mathbf{m}$ at convergence point yields simultaneously (1) an estimate $\hat{\mathbf{m}}_0$ of the parameters $\mathbf{m}_0$ of the background noise, and (2) the number of sources.

The source subspace and the orthogonal subspace are defined from the eigenvectors of the CSDM estimate transformed by $C(f, \hat{\mathbf{m}}_0)$. We get the estimate of the "source-alone" CSDM with

$$C^{-1}(f, \hat{\mathbf{m}}_0)\left\{\sum_{i=1}^{\hat{N}} [\hat{\lambda}_i(f) - \hat{\sigma}(f)]\hat{\mathbf{V}}_i(f)\hat{\mathbf{V}}_i^+(f)\right\}[C^+(f, \hat{\mathbf{m}}_0)]^{-1} \qquad (5.26)$$

where $\hat{N}$ = estimate of the number of sources
$\hat{\lambda}_i(f)$ = $\hat{N}$ largest eigenvalues of $\boldsymbol{\Gamma}_c(f, \hat{\mathbf{m}}_0)$
$\mathbf{V}_i(f)$ = $\hat{N}$ eigenvectors corresponding to the $\hat{N}$ largest eigenvalues of $\hat{\boldsymbol{\Gamma}}_c(f, \hat{\mathbf{m}}_0)$
$\hat{\sigma}(f)$ = mean value of the $(K - \hat{N})$ smallest eigenvalues

To check the feasibility of the method, simulations have been conducted on a very simple example using the same array previously described. The background noise spatial coherence model is that of the surface noise of (5.16). It is then a model with one parameter m. The signal field consists of two sources at bearing $0°$ and $4°$ with a signal-to-noise ratio of -10 dB; the spatial coherence parameter for the noise is $m_0 = 0.5$. Figure 5.4 shows (a) the plotting against m of the theoretical CSDM eigenvalues, (b) the same plotting for the estimate $\hat{\boldsymbol{\Gamma}}(f)$ according to (5.14) with $P = 80$, and (c) the same with $P = 400$.

With the theoretical CSDM, one can see that for $m = 0.5$ the minimum eigenvalues are equal and that the two larger eigenvalues correspond to the two sources. With the estimated matrices, one can see a convergence point when m passes across 0.5. Moreover, the quality of the convergence increases with the observation time P.

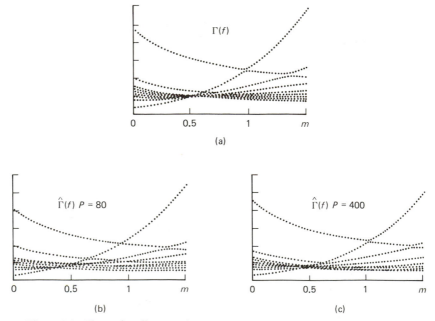

Figure 5.4 Eigenvalue diagrams for background noise with coherence parameter $m_0 = 0.5$ and two sources: $0°$ and $4°$, $\gamma(f)/\sigma(f) = -10$ dB: (a) $\Gamma(f)$; (b) $\hat{\Gamma}(f)$ $(P = 80)$; (c) $\hat{\Gamma}(f)$ $(p = 400)$.

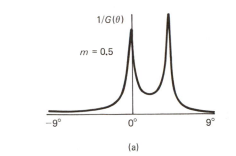

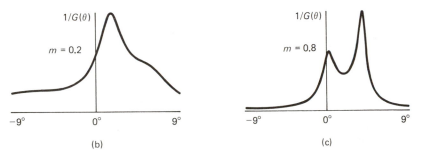

Figure 5.5 Inverse orthogonal subspace method inverse $G^{-1}(\theta)$ versus bearing θ for the noise field of Figure 5.4, for different values of the background noise coherence parameter m.

Above the noise eigenvalues, two eigenvalues corresponding to the sources show clearly. Figure 5.5 gives the results of the orthogonal subspace method (with the use of the inverse of $G(\theta)$ as previously done) with $P = 400$ and with $m = 0.5$, $m = 0.2$, and $m = 0.8$. The best results are given of course for $m = 0.5$. Therefore, thanks to the parametric modeling of the background noise spatial coherence, the fields of application of the HRM are considerably widened. They are more likely to fit an actual array environment.

The following section concerns the assumptions on the wavefronts of the sources themselves to utilize HRM when $\mathbf{D}(f, \boldsymbol{\theta})$ is widely unknown.

5.3.2 Exploitation of the Source Subspace [3]

The CSDM of the "sources alone" as it results from (5.26) is nonregular (with zero eigenvalues) but of course definite, nonnegative. Its rank is the number of sources where the source vectors are assumed to be linearly independent. Actually, it is the sum of the single-source CSDM matrices.

Every source CSDM is itself the dyadic product of the corresponding source vector. The rank of a one-source CSDM is unity. Therefore, the problem is to know whether or not we can get the source vectors starting from the source CSDM.

Necessity of modeling. Let us remark that an array as such can learn, at the best, only the individual source vectors bound to the sources. To go from the knowledge of a wavefront to that of a location of the sources in the medium, additional information is needed about the transfer functions between a given point source and the sensors. In other words, a model of the medium should be able to yield a wavefront as a function of a source position. If we already know the wavefront shape from the sources CSDM, it is required to find the coordinates and spectrum of a source through an identification of the measured wavefront, with the wavefront given by the model. It is out of the question to get rid of a model of the medium to ensure the localization.

But even before this step of localization, one has to know whether the first problem, namely, the description of the wavefronts from the CSDM, does not already require some additional information about the wavefronts and consequently about the medium. That is, can the unknown source vectors be computed from the CSDM only? The strict answer is "no." Nevertheless, we shall see later that it can be modified to a "no but ...." For the moment the CSDM alone can yield the source vectors only as functions of a number of unknown scalar parameters. This number is $K(K - 1)/2$, and it is possible to say that what is lacking for the CSDM to fully yield the wavefronts is an arbitrary unitary matrix of order and rank equal to K. Nothing can suggest the arbitrary unitary choice of this matrix except some a priori knowledge or assumption about the wavefronts or about the medium complexity: in other words, some elements of a model.

One-step computation of the source location. One possible approach is to solve in one step the source vectors and location problem. On one hand, we have an expression of the source vectors yielded by the experimental CSDM but only as functions of $K(K - 1)/2$ unknown scalar parameters. On the other hand, we build a model that will yield another expression of the source vectors, where the unknown parameters are

—the K spectral densities
—the $3K$ source coordinates
—"free" scalars, which are used to describe the medium in the frame of an a priori general structure, but without presuming numerical values

These parameters are the same for all the sources that are assumed to be Z in number.

Therefore, for identifying the expressions of the K source vectors (every one with N components), we have

—on the "model" side, $4K + Z$ unknown scalars

—on the CSDM side, $K(K - 1)/2$ unknown scalars

Since the number of available scalar equations is KN, the identification can give the value of all the unknown scalars if

$$KN = 4K + Z + \frac{K(K - 1)}{2} \tag{5.27}$$

Thus the number Z of the "medium descriptive" parameters, in the general frame of the chosen model, cannot be larger than

$$Z = K\left(N - \frac{K + 7}{2}\right) \tag{5.28}$$

This number increases with N; in other words, the number of sensors in the array limits the capacity of medium description. It increases also with K, the number of sources, which is supposed to be significantly smaller than N. Therefore, the medium is better described when the number of sensors and the number of sources is large. For example, for a 50-sensor array with 11 sources we obtain that $Z = 410$, and with 21 sources we obtain that $Z = 720$. Yet this can only be an approximation because other conditions have to be fulfilled for some parameters. For example, the coordinates have to be real numbers (note that the KN equations are complex).

Two-step computation of the source locations.   Another possible approach is based on the observation that the *amount of a priori* knowledge to be accepted in order to express the source vectors themselves is much less and more credible than the amount needed for a model of the medium. For example, one can accept without significant hesitation that every source vector can be represented as a *coherent* sum of plane and/or spherical waves. Thus, before presuming anything about the medium model itself, one can formulate a general structure of the source vectors. The *total* number of free parameters should match the system of KN equations with $K(K - 1)/2$ unknown parameters already on the CSDM side. This total number is consequently

$$KN - \frac{K(K - 1)}{2} = K\left(N - \frac{K - 1}{2}\right) \tag{5.29}$$

Thus for every source vector, there would be, in principle,

$$N - \frac{K - 1}{2}$$

free parameters available for its representation by a series of plane and/or spherical waves. Using a priori information we may have about the relative complexity of the K wavefronts, we may attribute more elementary waves to some of them than to the others.

Thus we end up, through the same kind of identification as that mentioned above in our discussion of one-step computation, with a numerical expression for all the source vectors, requiring only a minimum of a priori assumptions, with each source vector modeled as a summation of elementary wavefronts. Then, in the second step, we have to use a model of the medium, which should, of course, be *imperatively compatible with all the source vectors previously computed.*

Since every source "absorbs"—as in one-step computation—four scalar parameters, the number Z' of "medium descriptive" scalars is given by

$$KN = 4K + Z' \tag{5.30}$$

It must be noticed that Z' is a bit larger than the Z of one-step computation, since by putting a constraint (even a small one) on the general form of the wavefronts, we have, in some way, "set free" more parameters for the medium.

Obviously, the same concept holds in the problems of pure medium description with controlled sources (in position and spectral density). Such sources "absorb" no parameters at all, and set free more of them for the medium description.

5.3.3 Orthogonal Subspace Exploitation

With the orthogonal subspace method it is possible, as previously done, to use propagation parameters to get a model of the wavefronts. The model of the source position vector then depends on these parameters, and can be written

$$\mathbf{D}(f, \boldsymbol{\theta}, \mathbf{Z})$$

where $\mathbf{Z}$ is the vector representing the propagation parameters. To find the source position, we have now to look for the minimums of the following function not only for $\boldsymbol{\theta}$ but also for $\mathbf{Z}$:

$$G(\boldsymbol{\theta}, \mathbf{Z}) = \sum_{i=N+1}^{K} |\hat{\mathbf{V}}_{i0}^{+}(f)\mathbf{D}(f, \boldsymbol{\theta}, \mathbf{Z})|^2 \tag{5.31}$$

5.3.4 Experimental Results

To illustrate the method, experimental results are presented here from an underwater acoustic environment, with a very simple array. The array is linear, horizontal, and made of five equispaced sensors 1 m apart. The CSDM has been estimated around 850 Hz using a 5-Hz spectral resolution and an integration over 10 spectrums. There are three sources involved. Figure 5.6 shows the plotting of the sound field against bearing angle for the classical beamformer (C), the adaptive array (A),

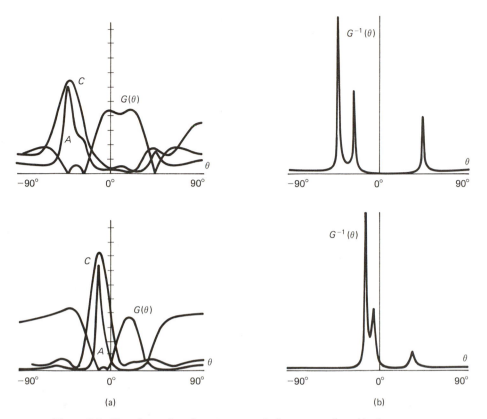

(a)   (b)

Figure 5.6 Experimental underwater acoustical array results with three sources responses versus bearing θ of (a) classical beamforming C, adaptive array A, and orthogonal subspace method $G(\theta)$, and (b) $1/G(\theta)$.

and the orthogonal subspace method $G(\theta)$, assuming that the background noise is incoherent.

Figure 5.7 shows another situation, still with three sources but with an attempt to model the background noise spatial coherence. The model is that of ocean surface noise (5.16). Again Figure 5.7(a) shows the plotting of the classical beamformer (C), the adaptive array (A), and the orthogonal subspace method $G(\theta)$ when the background noise is assumed to be not incoherent. Figure 5.7(b) shows the evolution of the five eigenvalues as functions of the parameter m of the model. Figure 5.7(c) shows the inverse $1/G(\theta)$ for $m = 0$, $m = 0.4$, and $m = 1.5$. One can observe the two smallest eigenvalues' convergence for $m = 0.4$. Clearly, the three larger eigenvalues correspond to the three sources. Thus correspondingly in Figure 5.7(c), the best result is for $m = 0.4$. These "at sea" experiments illustrate the superiority of high-resolution methods over classical and adaptive array.

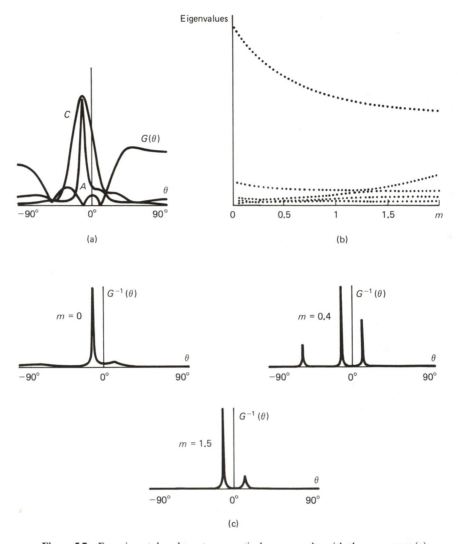

Figure 5.7 Experimental underwater acoustical array results with three sources: (a) responses versus bearing θ of classical beamforming C, adaptive array A, and orthogonal subspace method $G(\theta)$ with background noise coherence parameter m supposed equal to 0.5; (b) eigenvalue diagram versus background noise coherence parameter m: the two smallest eigenvalues coincides for $m = 0.4$; (c) inverse orthogonal subspace method response $G(\theta)$ versus bearing θ for $m = 0$, the best value 0.4, and 1.5.

5.4 IMPLEMENTATION OF THE HIGH-RESOLUTION METHOD

In this section the general structure of the HRM implementation is described, and the complexity of the calculations to be performed is evaluated.

5.4.1 General Structure

The general structure of the HRM processing scheme is shown on Figure 5.8. It can be divided in three main steps. The first and the third steps are straightforward, whereas the second step, which is the focus of HRM, requires new types of calculation.

Step 1: Input data preprocessing. The first step is devoted to preprocessing the input data in order to extract the quantities needed by the HRM, namely, the CSDM estimates. The operations involved are conventional.

First, the Fourier transform of the signals from the N sensors is performed. Next, if L is the number of frequency cells of the spectral analysis, we find L identical processing paths in parallel. That parallelism is maintained throughout the processing scheme.

The second step of the input data preprocessing consists in estimating the CSDM by using relation (5.14):

$$\hat{\boldsymbol{\Gamma}}_n(f_e) = \frac{1}{P} \sum_{i=nP}^{(n+1)P} \mathbf{X}_i(f_e)\mathbf{X}_i^+(f_e)$$

The N components of the vector $\mathbf{X}_i(f_e)$ are the output samples of the frequency cells centered on f_e ($e \in [1, L]$). It is noted that $\mathbf{X}_i(f_e)$ is a column matrix. Therefore, these calculations are carried out by matrix multiplication and accumulation.

CSDM estimation is not usually performed in conventional array processing, except for adaptive array processing in the frequency domain using direct inversion of the CSDM estimate.

Step 2: HRM specific calculations. The main HRM calculations are of a new type in array processing and typically belong to the linear algebraic matrix computation domain. They consist of the eigensystem decomposition of the CSDM estimates. Different methods are available. But perhaps a particularly well-suited method for Hermitian matrices is the following: first, the input matrix is reduced to tridiagonal form $\hat{\boldsymbol{\Gamma}}_T$ by the Householder method, and second, the QR method is used. That procedure is widely described in the literature [10]. Its main steps are now reviewed.

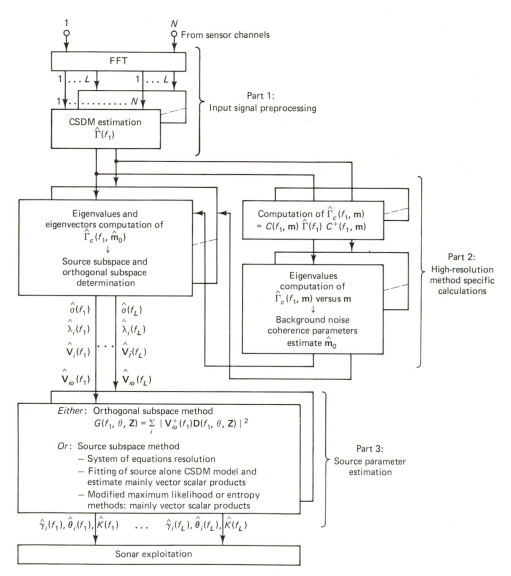

Figure 5.8 General structure of high-resolution-method processing scheme.

The reduction to tridiagonal form is carried out by $(N - 2)$ steps with identical structures:

$$\hat{\boldsymbol{\Gamma}}_T = \boldsymbol{U}_{N-2} \cdots \boldsymbol{U}_1 \hat{\boldsymbol{\Gamma}} \boldsymbol{U}_1 \cdots \boldsymbol{U}_{N-2} = \boldsymbol{U}\hat{\boldsymbol{\Gamma}}\boldsymbol{U} \tag{5.32}$$

Matrices $\boldsymbol{U}_i$, and therefore $\boldsymbol{U}$, are unitary and Hermitian. They have to be computed at each step.

The QR method is a recursive algorithm that reduces the matrix $\hat{\Gamma}_T$ to a right-triangular matrix as the number of iterations grows to infinity. The tridiagonal form is conserved. The right-triangular reduction is obtained by unitary similarity transformations, as the previous tridiagonal form reduction; thus the eigenvalues and eigenvectors of $\hat{\Gamma}$ are easily deduced from those of the final right-triangular matrix. The computation of the eigensystem of a right-triangular matrix is straightfoward. At the first iteration $\hat{\Gamma}_T$ is reduced to a right-triangular matrix $\boldsymbol{R}_1$ by $(N - 1)$ steps with identical structures

$$\boldsymbol{R}_1 = \boldsymbol{U}_{N-1}^1 \cdots \boldsymbol{U}_1^1 \hat{\Gamma}_T = \boldsymbol{Q}_1 \hat{\Gamma}_T \tag{5.33}$$

Matrices $\boldsymbol{U}_i^1$, and thus $\boldsymbol{Q}_1$, are unitary and Hermitian. It is deduced that

$$\hat{\Gamma}_1 = \boldsymbol{R}_1 \boldsymbol{Q}_1$$

At the second iteration, $\hat{\Gamma}_1$ is reduced to a right-triangular matrix $\boldsymbol{R}_2 = \boldsymbol{Q}_2 \hat{\Gamma}_1$, from which $\hat{\Gamma}_2 = \boldsymbol{R}_2 \boldsymbol{Q}_2$ is deduced, and so on. At iteration p it can be shown that

$$\hat{\Gamma}_p = \boldsymbol{R}_p \boldsymbol{Q}_p = \boldsymbol{Q}_p^+ \cdots \boldsymbol{Q}_1^+ \hat{\Gamma}_T \boldsymbol{Q}_1 \cdots \boldsymbol{Q}_p = \boldsymbol{W}_p^+ \hat{\Gamma}_T \boldsymbol{W}_p \tag{5.34}$$

Iterations are stopped when a given precision is reached. Matrices U and $\boldsymbol{W}_p$ being unitary, eigenvalues of $\hat{\Gamma}$ are equal to those of $\hat{\Gamma}_p$, which are on its main diagonal. Eigenvectors $\hat{\Gamma}$ are given by $U\boldsymbol{W}_p V$ if V is an eigenvector of $\hat{\Gamma}_p$.

Eigensystem computation is used in the two main steps of HRM processing.

1. The first step is the estimation of the parameters $\mathbf{m}_0(f_e)$ of the background noise spatial coherence. The following operations have to be performed for some given number of values of the modeling parameter $\mathbf{m}$:
   a. Complex matrix multiplication to compute the matrix [relation (23)]

   $$\hat{\Gamma}_{cn}(f_e, \mathbf{m}) = \boldsymbol{C}(f_e, \mathbf{m})\hat{\Gamma}_n \boldsymbol{C}^+(f_e, \mathbf{m}) \tag{5.35}$$

   b. Eigenvalue computation for the matrix $\hat{\Gamma}_{cn}(f_e, \mathbf{m})$
   c. Calculation of a simple function of the eigenvalues to determine the estimation $\hat{\mathbf{m}}_0(f_e)$
2. The second step is the construction of the source signal subspace and of the orthogonal subspace. This operation needs the eigenvalues and eigenvectors of the matrix $\hat{\Gamma}_{cn}[f_e, \hat{\mathbf{m}}_0(f_e)]$: the source subspace is spanned by the $\hat{K}(f_e)$ eigenvectors associated with the $\hat{K}(f_e)$ largest eigenvalues, and the orthogonal subspace by the remaining $[N - \hat{K}(f_e)]$. $\hat{K}(f_e)$ is determined by a simple function of the eigenvalues [8].

Step 3: Source parameter estimation.   Source parameter estimation for each frequency cell can be obtained using different methods. The source subspace or the orthogonal subspace can be utilized. In each case, the complexity of the calculations are more or less dependent on the use of unknown propagation parameters $\mathbf{Z}$ in the model of the source position vector.

1. Orthogonal subspace method. Source position estimates $\hat{\theta}_i$ are given by the values of θ associated with the $\hat{K}(f_e)$ maximums (versus θ and $\mathbf{Z}$) of [relation (5.31)]

$$G^{-1}(f_e, \theta, \mathbf{Z}) = \left[\sum_{i=\hat{K}(f_e)+1}^{N} |\mathbf{V}_{i0}^{+}(f_e)\mathbf{D}(f_e, \theta, \mathbf{Z})|^2 \right]^{-1} \qquad (5.36)$$

In cases where the number of parameters in θ and $\mathbf{Z}$ is small, particularly when the model does not include propagation parameters, the maximums can be determined by scanning the values of $G^{-1}(f_e, \theta, \mathbf{Z})$ versus θ and $\mathbf{Z}$ with increments for the θ and $\mathbf{Z}$ components that depend on resolving power. This operation is based on scalar products of complex vectors. It is very similar to classical beamforming, but needs more calculations due to the $[N - \hat{K}(f_e)]$ scalar products and to the small increments on θ required because of the high-resolution capability.

2. Source subspace methods. With the source subspace, several methods can be used. The method of [4], which does not include propagation parameters $\mathbf{Z}$, is also based on a search for the maximums of a function involving scalar products of complex vectors.

The method of [1] does not use propagation parameters either. It utilizes a fitting method between the source-only CSDM estimate and the source-only CSDM model.

Reference [3] shows that it is possible to include propagation parameters in the model, which leads to a system of $N\hat{K}(f_e)$ equations with $N\hat{K}(f_e)$ unknowns if the number Z of unknown propagation parameters is equal to $Z = \hat{K}(f_e)[N - \hat{K}(f_e) + 7/2]$. This method requires the solution of a system of equations which are generally nonlinear. The complexity of the computation depends on the structure of the source position vector model.

5.4.2 Evaluation of the Complexity of the Calculation

After proper transformation, the input signals are supposed to be complex signals in the frequency bandwidth $(-B, B)$ which are sampled at a rate of 2.5B.

Input data preprocessing. Spectral analysis is achieved by using the averaged periodogram procedure. Let $\Delta f = 1/T$ be the spectral resolution needed. If L is the number of useful frequency cells (positive frequencies), we have $L = B/\Delta f$. If the spectral analysis is carried out with a period equal to $T/2$, N FFTs of $2.5L$ points must be performed each $T/2$ seconds.

The CSDM estimates $\hat{\Gamma}(f_e)$ are computed for each frequency cell. As $\hat{\Gamma}(f_e)$ is a Hermitian matrix, the estimation procedure needs L identical operations of $PN[(N + 1)/2]$ complex multiplications and additions. This computation is performed every $PT/4$ seconds.

HRM specific calculations.   Computations are composed primarily of eigenvalue and eigenvector decomposition of matrices. The complexity of the calculations depends on the dimension of matrices that is on the number N of sensors, and on the required precision, and on the matrices themselves. Typically, calculation of eigenvalues for 32 sensors requires about 10^5 multiplications and additions and the calculation of both eigenvalues and eigenvectors about 2×10^5 multiplications and additions. These quantities vary as about $N^{2.5}$. For stability, 32-bit floating-point computation has to be used.

1. Background noise spatial coherence parameter estimation. Computations must be performed for L frequency cells and M values of the parameters:

Calculation of matrix $\hat{\Gamma}_{cn}(f_e, \mathbf{m})$ needs about $N^2(3N + 1)/2$ complex multiplications and additions.

Calculation of the eigenvalues of $\hat{\Gamma}_{cn}(f_e, \mathbf{m})$ needs about $20N^{2.5}$ multiplications and additions.

The background noise parameter estimation needs negligible calculations compared to preceding ones. These ML identical operations are performed with a period equal to the stationarity time of the background noise, which is generally greater than the estimation period $PT/4$ of $\hat{\Gamma}(f_e)$.

2. Determination of the source subspace and of the orthogonal subspace. The following operations have to be performed for the L frequency cells, within a period equal to $PT/4$:

Computation of matrix $\hat{\Gamma}_{cn}(f_e, \hat{\mathbf{m}}_0)$: $N^2(3N + 1)/2$ complex multiplications and additions.

Computation of eigenvalues and eigenvectors of that matrix: about $20N^{2.5}$ multiplications and additions.

Source number estimation: the corresponding calculations are negligible.

Source parameter estimation.   The complexity of the calculations needed by the source parameter estimation is more difficult to evaluate. It depends on the number of unknown propagation parameters. But they are generally composed primarily of inner products of complex vectors. As an example, one of the simplest methods is the orthogonal subspace method with no propagation parameters. Using the scanning procedure for each frequency cell, about $[N - \hat{K}(f_e)](N + 1)\Theta$ complex multiplications and additions are required, where Θ is the number of increments of θ. If propagation parameters are included to improve system performance, many more calculations must be performed. In this case, more appropriate search methods may be used.

5.5 CONCLUDING REMARKS

High-resolution methods appear to have an application to the underwater passive listening problem because of their improved resolving capability compared to those of either conventional or adaptive beamforming. However, more accurate modeling of the medium and more parameters can be required and consequently estimated.

These new methods lead to new types of architecture which are characterized by two main features. The first feature is that, compared to conventional beamforming, the complexity of computation is greatly increased. Whereas in a conventional processor, beamforming is carried out immediately after the input FFT is performed, in high-resolution methods the additional operations of cross-spectral density matrix estimation and eigenvalue-eigenvector matrix decomposition are required. The second feature is that most of the calculations belong to the field of matrix operations, including scalar products of complex vectors, complex matrix multiplications, and Hermitian matrix eigensystem decomposition. Thus new parallel computation methods such as systolic array processing [11] can yield an appropriate solution for this new kind of sensor array space-time processing.

ACKNOWLEDGMENT

The work presented here has been supported in part by the Direction des Recherches, Etudes et Techniques, Paris, France.

REFERENCES

[1] W. S. Ligget, "Passive Sonar: Fitting Models to Multiple Time Series," *Proc. NATO ASI Signal Process.*, Academic Press, Loughborough, U.K., 1972, pp. 327–345.

[2] H. Mermoz, "Imagerie, corrélation et modèles," *Ann. Télécommun.*, *31*(1–2):17–36, (jan.–fév. 1976).

[3] N. L. Owsley, "A Recent Trend in Adaptive Spatial Processing for Sensor Arrays," *Proc. NATO ASI Signal Process.*, Academic Press, Loughborough, U.K., 1972, pp. 131–164.

[4] N. L. Owsley, and J. F. Law, "Dominant Mode Power Spectrum Estimation," *Proc. ICASSP 82*, Paris, May 3–5, 1982, pp. 775–778.

[5] G. Bienvenu, "Influence of the Spatial Coherence of the Background Noise on High Resolution Passive Methods," *Proc. ICASPP 79*, Washington, D.C., Apr. 2–4, 1979, pp. 306–309.

[6] R. Schmidt, "Multiple Emitter Location and Signal Parameter Estimation," *Proc. RADC Spectrum Estimation Workshop*, Oct. 1979, pp. 243–258.

[7] G. Bienvenu and L. Kopp, "Source Power Estimation Method Associated with High Resolution Bearing Estimation," *Proc. ICASSP 81*, Atlanta, Ga., Mar. 1981, pp. 153–156.

[8] G. Bienvenu and L. Kopp, "Optimality of High Resolution Array Processing Using Eigensystem," *IEEE Trans. Acoust. Speech Signal Process.*, Oct. 1983, pp. 1235–1248.

[9] G. Bienvenu and L. Kopp, "Adaptivity of Background Noise Spatial Coherence for High Resolution Passive Methods," *Proc. ICASSP 80*, Denver, Colo., Apr. 9–11, 1980, pp. 307–310.

[10] J. N. Franklin, *Matrix Theory*, Prentice-Hall, Englewood Cliffs, N.J., 1968.

[11] K. Bromley, I. J. Symanski, J. M. Speiser, and J. H. Whitehouse, "Systolic Array Processor Developments," *Proc. CMU Conf. VLSI Syst. Comp.*, Carnegie-Mellon University, Pittsburg, Pa., Oct. 19–21, 1981.

6

Remarks on Matrix Eigenvalue Computations

B. N. PARLETT

University of California at Berkeley
Berkeley, California

6.1 INTRODUCTION

Eigenvalue calculations are beginning to play an important role in modern signal processing, as the earlier chapters in this volume indicate. On the other hand, there are scientists and engineers who spend considerable sums of money (megadollars) and much of their professional lives on the computation of eigenvectors and eigenvalues of the matrices that concern them. Structural engineers need to know natural frequencies and normal modes; quantum chemists hope to derive chemical properties of molecules from their possible states and associated energies; oceanographers can validate their models by comparing predicted tides against observed tides. VLSI circuits for computing eigenvalues may help some of these groups to extend their analyses far beyond today's range. At the end of the chapter we examine briefly the implications of the new technology for future algorithms. Now let us turn to the present.

What is the situation with today's serial and vector computers? In the following sections we attempt to give the essential features without getting trapped in details and pedantry. The subject must be divided up according to two obvious but inadequately appreciated criteria. Matrices are either *large* or *small* (see Section 6.3) and their eigenvalues are either *real* or *complex*. Moreover, the property that produces real eigenvalues in the overwhelming number of cases is the *symmetry* of a real matrix.

In this chapter capital letters denote matrices, lowercase letters denote vectors. We try to reserve symmetric letters for symmetric matrices. Lowercase Greek letters denote numbers.

6.2 REDUCTION TO STANDARD FORM

A good number of the symmetric matrices that do get fed to eigenvalue programs should never have been created in the first place. We now explain this remark.

Many applications require solutions $(\lambda, \mathbf{z})$ to

$$(H - \lambda M)\mathbf{z} = 0$$

where H and M are real, symmetric matrices and one of them (or some linear combination of them) is positive definite. The well-intentioned user dutifully reduces H and M to a single symmetric matrix according to the precepts laid down in the past by the eigenvalue experts in their books and research papers (Wilkinson and Reinsch [20]; Parlett [12], hereafter called SEP). A favorite technique when M is positive definite is to compute the Cholesky factorization of M, namely,

$$M = LL^t \qquad (t \text{ denotes transpose})$$

where L is lower triangular, then build up

$$A = L^{-1}H(L^t)^{-1}$$

so that

$$(A - \lambda I)\mathbf{y} = 0 \qquad \text{and} \qquad \mathbf{y} = L^t\mathbf{z}$$

Today we realize that this procedure, although not always fatal, is almost always bad. Why is it bad? Briefly, because usually A is very much larger than H and M (in norm) and usually only a few eigenvalues are wanted. These few are the smallest eigenvalues of A. When A is large (say 1000×1000), that is bad news. First, there are fundamental limitations on the relative accuracy with which the smallest eigenvalue of a matrix can be computed. Second, they are more difficult to compute than the big ones. Third, any structure in the pattern of zero elements of H and M is lost during the construction of A.

Suppose that only eigenvalues of (H, M) close to σ are wanted. The right procedure is to define, but not form, the matrix $A = M^{1/2}(H - \sigma M)^{-1}M^{1/2}$. No inversion, please. To form $\mathbf{x} := A\mathbf{v}$ for any vector $\mathbf{v}$, just solve linear systems of the form $(H - \sigma M)\mathbf{x} = M^{1/2}\mathbf{v}$ for $\mathbf{x}$, however painful. With this capability the Lanczos algorithm can be used to find the largest eigenvalues of A. These are simply related to the wanted eigenvalues of (H, M). See [6] or [SEP, p. 325] for more details. Now we return to the consideration of a single matrix.

6.3 THE POWER OF TODAY'S PROGRAMS

Definition.   An $n \times n$ matrix is *small* (in a given computer system) if it and its matrix of eigenvectors can be held as conventional square arrays in the fast (random access) store or memory. Otherwise, it is *large*.

This is a useful distinction. There is a sharp difference between the techniques used for the two cases (large and small), and also between the very problems that give rise to the matrices. On conventional number-crunching systems (CDC 7600, 6600; IBM 370/195, 165) a 100×100 matrix is small. On a small minicomputer system a 50×50 might well be large. Nevertheless, we may safely say that 1000×1000 is large and we must note that quantum physicists would like to compute the spectra of matrices of size 2^p with p rising to 20, 30, 40, and beyond! *Most numerical analysts consider that the eigenvalue problem is solved for small matrices.* There is one open problem of a theoretical nature (convergence of shifted QR for nonnormal matrices), and I am not aware of any computational bottlenecks, complaints, or unfulfilled requests in this category.

A few figures may give substance to this opinion. On an IBM 370/165 *all* eigenvalues and *all* eigenvectors of an 80×80 random symmetric matrix can be computed in 5.25 s. It is interesting to note that it takes nearly 1 second (0.88 s) to compute the largest eigenvalue alone. On the rather old-fashioned CDC 6400, the 5.25 must be increased to 32.64, over half a minute. One should add that those 80 eigenvectors will be orthogonal to almost working accuracy. The eigenvalues will be accurate to within a few units of the last digit of the largest stored eigenvalue. It would be interesting to learn of important computing tasks for which this speed is not adequate. Real-time applications are the only ones that come to mind.

Another aspect to ponder is that it is just not worth trying to exploit special patterns of zero elements in *small* matrices (their so-called sparsity structure). The only exception is when the matrix has very narrow bandwidth (e.g., tridiagonal, pentadiagonal). In other words, there is enough overkill in the techniques used for *small* matrices that there is no point in exploiting any special properties of the matrix (except for tridiagonality and the Toeplitz property of constant diagonals.)

Another fact that surprises many people is that it is as easy (i.e., as quick) to compute *all* the eigenvalues of a small, full, real symmetric matrix (with $n \geq 30$) as to multiply that symmetric matrix by another one! Briefly, "eigenvalues are easier than multiplication." Exactly n^3 scalar multiplications are required to form the product.

Suppose that we drop the adjective "symmetric" in the preceding paragraph. The assertion is then false, but not by much. Of course, the eigenvalues may well be complex. Nevertheless, they can *all* be computed in less time than it takes to form five matrix-matrix products (i.e., $5n^3$ scalar multiplications). Again, n must be large enough so that its leading powers dominate the others, say $n = 20$, but still be small.

This state of affairs would have been unbelievable in 1955. What made the difference? Special and cooperative efforts between 1958 and 1970 by a small group of experts whose work is now widely available through program libraries: EISPACK [5,16] for eigenvalue problems, LINPACK [4] for linear equations and least-squares problems, IMSL [9] for everything, and NAG ([11], European effort) for everything. The packages are well documented and have been severely tested.

I have suggested that the accuracy of the output of, say, an EISPACK program is as good as could be desired. This may puzzle some of you who are familiar with the extreme difficulty of computing the Jordan canonical form when it contains blocks larger than 1×1.

The explanation of this puzzle is important. The EISPACK codes produce eigenvalues and eigenvectors of matrices that equal the given matrix almost to working precision. If B is $n \times n$, the output really belongs to $B - E$, where E is unknown but $\| E \| \leq n \cdot \varepsilon \cdot \| B \|$. This is an intrinsic limitation arising from the use of a computer instead of exact arithmetic. The numerical analysts have ensured that the uncertainty in the output that can be attributed to their algorithms is certainly not more than n times the uncertainty arising from storing the matrix to working precision in the first place. So our attention shifts to the effect on eigenvalues of uncertainty in the matrix elements.

Definition.   The *condition* of a simple eigenvalue λ is the secant of the angle between the row and column eigenvectors of λ. For more details, see [19, Chap. 2]. This number is also the norm of the spectral projector onto λ's eigenspace and is the coefficient of the leading term in a perturbation series for the change in λ. The condition of an eigenvalue λ associated with a Jordan block is ∞.

All eigenvalues of all symmetric matrices are perfectly conditioned (i.e., the condition number is 1). The change in an eigenvalue is bounded by the (spectral) norm of the change in the matrix.

The EISPACK and related programs achieve almost all that is possible in the given computing environment for small matrices. It is true that the uncertainty for ill-conditioned problems can be very great, perhaps leaving no significant figures. Nevertheless, the accuracy of well-conditioned computed eigenvalues is not impaired by the presence of ill-conditioned eigenvalues of the same matrix.

6.4 THE SITUATION FOR LARGE PROBLEMS

The picture changes sharply when the matrix (the input) and the results (the output) can no longer be held in the random access memory. Such problems are vital to a variety of users, as mentioned in the introduction. Some good methods have been developed, research is active, but there is nothing like EISPACK available. Indeed, some of the good methods are buried inside special-purpose packages: for example, NASTRAN, ADINA for structural analysis, and the Quantum Chemistry Program Exchange at Indiana University (see [3]). The general user should consult Mike Heath's catalog [18] begun in 1982.

Two aspects of the problem need emphasis.

1. It is important to exploit any *sparsity structure* that the original problem may enjoy.
2. The efficiency of a technique depends quite strongly on the *computer system* (virtual memory or explicit transfer between primary and secondary storage).

The information demanded from large problems is less varied than for small. The symmetric case dominates. I have never heard of a need for the complete spectral factorization (all eigenvalues and all eigenvectors), but I do know of one application using the whole spectrum! The dominant demand (by far) is for some eigenvalues (maybe 3, maybe 50) at the left end of the spectrum (near 0) together with the associated eigenvectors. In 1978 an engineering company spent $12,000 of computer time for 30 eigenpairs of a problem with $n = 12,000$. This figure excludes the cost of program development.

There is a beautiful way to exploit sparsity without developing a different method for every important pattern of zero elements. The *user* supplies a program that takes any n vector $\mathbf{v}$ as input and produces the n vector $\mathbf{u} = A\mathbf{v}$ as output. The method must then compute eigenvalues and eigenvectors by supplying suitable vectors $\mathbf{v}$ to the program. The linear operator A is never known to the solution method. No transformations on A are possible. In this way the burden is on the user to exploit what is known about A to make the computation of $\mathbf{u}$ ($= A\mathbf{v}$) as efficient as possible. In practice, A will not be a simple matrix. A typical situation involves a diagonal matrix D, a structured matrix H, and a shift parameter σ. In fact, $A = D(H - \sigma D^2)^{-1} D$. The output $\mathbf{u}$ is found in three steps:

1. Form $\mathbf{w} = D\mathbf{v}$
2. Solve $(H - \sigma D^2)\mathbf{x} = \mathbf{w}$ for $\mathbf{x}$
3. Form $\mathbf{u} = D\mathbf{x}$

Please note that step 2 itself might employ some iterative method for solving the system whenever it is inconvenient to factor $H - \sigma D^2$. For big problems it is sometimes necessary to use secondary storage for step 2. This simple but essential transformation of the $(H - \lambda D^2)\mathbf{z} = 0$ problem to standard form $(A - \lambda I)\mathbf{y} = 0$ is always used in structural analysis. It is not surprising that vector computers (particularly the Cray-1 and Cyber 205) are attractive devices for problems where the vectors are long and full but the matrices are sparse.

A very stable way to solve $(H - \lambda M)\mathbf{z} = 0$ is to employ the QZ algorithm, which is widely available in package form. However, this procedure will find all the eigenvalues, whether wanted or not, and it will destroy symmetry. Moreover, it requires two $n \times n$ arrays in fast memory. Consequently, it is used only for small problems, as described in the next section.

Here ends our overview. In the next section we describe, in broad terms, the principal methods. Now we give some general advice to eigenvalue hunters. If your matrices are small then, in the United States, use EISPACK and its guide [5,16]. Most tasks require between N^3 and $10N^3$ arithmetic operations. If your matrices are large, consult M. Heath's catalog [18] of software for sparse matrices.

6.5 METHODS FOR SMALL MATRICES

All the methods found in EISPACK make explicit two-sided transformations on the given matrix A. Such techniques exploit well the large random access store that is available to a serial CPU. Moreover, for both symmetric (no surprise) and nonsymmetric (surprise) the transformations are exclusively orthogonal congruences. In symbols,

$$B \to PBP^t$$

where P is orthogonal (i.e., $PP^t = P^tP = I$).

Let us expand on this point. In pure linear algebra we learn that any similarity transformation, $B \to FBF^{-1}$, may be used to simplify our problem because the eigenvalues are preserved. However, when the elements of B are uncertain and round-off error is present, the nature of F must be taken into account. It turns out that any uncertainty in B is multiplied by approximately $\text{cond}(F) := \| F \| \cdot \| F^{-1} \|$, the condition number of F for inversion. Now $1 \le \text{cond}(F) < \infty$, so computer usage dictates that we confine our similarity transformations to those with modest values of $\text{cond}(F)$. The safest strategy is to have $\text{cond}(F) = 1$. The beauty of an orthogonal matrix P is that

1. $P^{-1} = P^t$ (inversion is trivial).
2. $\text{cond}(P) = 1$. Real symmetric matrices can actually be diagonalized by suitable orthogonal similarity transformations, but the best that can be achieved in general for nonnormal matrices is a triangular form.

A final comment on terminology. A congruence is a transform $B \to FBF^t$, so when F is orthogonal a congruence is also a similarity.

Jacobi methods (see [12, p. 202]) employ a sequence of simple P's (called plane rotations) each of which annihilates (or greatly diminishes) one pair of off-diagonal elements, say the pair in position (i, j) and (j, i). The symmetric matrix is gradually reduced to diagonal form. Of course, zeros that are created at one step are lost at a later step. Nevertheless, the process converges quite quickly if any reasonable strategy is used to choose the sequence of pairs (i, j).

For diagonally dominant matrices Jacobi is ideal, but for standard cases Jacobi methods are two or three times slower (see [SEP, Chap. 9] for data) than the favorite method, called *Householder-QR*. This judgment applies to conventional serial computers. This method has two phases (see [12, p. 202]).

1. A sequence of $(n - 1)$ orthogonal congruences reduces a symmetric matrix A to tridiagonal form T, and also reduces a nonsymmetric matrix B to Hessenberg form H. Typically,

$$H = \begin{bmatrix} * & * & * & * & * \\ * & * & * & * & * \\ 0 & * & * & * & * \\ 0 & 0 & * & * & * \\ 0 & 0 & 0 & * & * \end{bmatrix}$$

2. A different sequence of P matrices is used that diminishes the subdiagonal elements of T or H without destroying any zero elements. The last subdiagonal element diminishes very quickly. When it is negligible, the program simply

ignores the last row and column of the matrix (this is called automatic defla-
tion). At the end, T has become diagonal. Even in the nonsymmetric case the
eigenvalues lie on the diagonal of the resulting upper triangular matrix.

For the nonspecialist, the important fact is that it requires $(2/3)n^3$ floating-
point multiplications to get T and $(5/3)n^3$ to get H (see [20]). Phase II requires
about $10n^2$ in the symmetric case and about $(5/2)n^3$ in the nonsymmetric case.
There is really no question of anything going wrong or the program failing.
The techniques are stable and convergent and the codes have been tested extensively.
One technical point should be mentioned. When a real matrix has complex
eigenvalues, then H will be real, all steps in phase II will be real, but not all the
subdiagonal elements will tend to zero. The final form is merely *block* triangular
with a 2×2 diagonal block to each conjugate pair of complex eigenvalues. This is a
very efficient way to keep the arithmetic real and yet determine complex quantities.
If eigenvectors are wanted, then (in the symmetric case) the product of all the
P matrices can be accumulated, as each is created, to yield an accurate, orthogonal
system of eigenvectors, however close some eigenvalues may be. The cost of accumulating these products dominates the rest of the calculation.
Sometimes only a few eigenvalues are wanted and other techniques may be
called for. In the symmetric case, if more than one-fourth of the spectrum is wanted
to full accuracy, it is quicker to use Householder-QR to find them all than to find
the wanted eigenvalues one by one (see [16, p. 80]). In the nonsymmetric case the
situation is even more extreme. If only one eigenvalue is wanted, then, within the
scope of EISPACK programs, it is necessary to find all of them. This situation
occurs in the famous stability index problem: The best way we know to determine
one of the eigenvalues with largest real part of a small matrix is to compute *all* the
eigenvalues and then inspect them. Rival methods spend so much effort making sure
that they have the correct eigenvalue that they are not in general competitive. The
stability index is the real part of this eigenvalue. All this is quite surprising!

6.6. INVERSE ITERATION

There is one technical problem that is frequently misinterpreted. Imagine that a user
seeks all the eigenvalues in a given interval of a 100×100 symmetric tridiagonal
matrix. Suppose that there are just three of them. They can be found by a simple
method called bisection. The challenge is to compute the corresponding eigenvectors. The favorite technique is to use one step of inverse iteration: namely, solve
$(T - \lambda)\mathbf{s} = \mathbf{b}$ for $\mathbf{s}$, where λ is the computed eigenvalue and $\mathbf{b}$ is a vector carefully
chosen to depend on λ and to avoid being orthogonal to the wanted eigenvector.
We assume that λ is accurate to working accuracy so that $T - \lambda$ is singular to
working precision. Nevertheless, the only *computer* solution to $(T - \lambda)\mathbf{x} = 0$ will be
0, so it is necessary to use some tiny but *nonnull* right-hand side $\mathbf{b}$. The EISPACK

program TINVIT that embodies this technique produces an excellent approximation to λ's eigenvector. This can be verified a posteriori since

$$\frac{\| (T - \lambda)s \|}{\| s \|} = \frac{\| b \|}{\| s \|}$$

is available and should be as small as $\varepsilon \cdot \| T \|$, where ε is the round-off unit.

What is the trouble? Suppose that two eigenvalues λ_1 and λ_2 agree to nearly all their significant figures. In such cases the b vector changes very little when λ_2 replaces λ_1 and, as a consequence, the computed eigenvector s_2, although not too close to s_1 is, unfortunately, far from *orthogonal* to s_1. Of course, both s_1 and s_2 are excellent approximations to eigenvectors of T, as their residuals show. The trouble is that s_1 and s_2 lie in a plane that is an eigenspace to almost working precision. Indeed, any linear combination of s_1 and s_2 is just as good an approximation as is s_2.

In this way began the myth that it is *intrinsically* difficult to compute orthogonal eigenvectors for very close eigenvalues, whereas in truth, it is only the technique of inverse iteration that experiences this difficulty. *The defect is that b is too smooth a function of λ.* Moreover, the situation is not difficult to remedy once it has been detected. Indeed, the latest version of TINVIT does take the trouble to produce nearly orthogonal sets of eigenvectors. However, the elegance and simplicity of the original method have been sacrificed to include judgments and orthogonalizations. As mentioned in Section 6.5, the accumulation of the product of all the QR transforms will produce a full set of eigenvectors orthogonal to working precision.

6.7 METHODS FOR LARGE SYMMETRIC MATRICES

Similarity transformations tend to destroy any sparsity structure in A and, to my knowledge, explicit transformations are not used for large problems. It seems that all interested numerical analysts have been converted to the virtues of the *Lanczos algorithm*. Yet the majority of engineers use their favorite versions of the method of simultaneous iteration, which they call *subspace iteration*. Both these methods will be considered in somewhat more detail presently. The computational chemists use either subspace iteration or special techniques that work well on their special problems but are not general (see [3]).

There is nothing comparable to EISPACK yet. Nor is it clear that there ever will be, considering the strong influence of the computer system on efficiency. It must be emphasized that both the Lanczos method (LAN) and subspace iteration (SI) are general terms allowing for significant variations in implementation. This is in contrast to the QR algorithm, which now has little room for variation. The choice of shift strategy in QR is settled, whereas it is still something of an art in SI.

Many, but not all, large matrices are sparse (i.e., fewer than 20 nonzeros per row). Such matrices can often be held in fast memory provided that some compact data structure is used to hold the nonzero elements. Nevertheless, near the frontier of our current abilities are important computations in which every product $A\mathbf{u}$ demands transfers between primary and secondary storage. Quantum chemists produce simple symmetric matrices with $n \approx 10^4$ and 10^7 nonzero elements that are held on files or tapes. What helps the chemists is that crude approximations to the few wanted eigenvectors are known a priori; this is a tremendous asset. Many chemists work on medium-sized computer systems and their difficulties lie with features of the operating system rather than with the numerical method.

The software developed by most serious eigenvalue hunters is embedded in their special-purpose packages and it is difficult to see how they could take advantage of a superior, but general-purpose eigenvalue program for large problems. They would almost certainly have to understand it in detail and then integrate it into their systems. This is a task to be undertaken only with the prospect of order-of-magnitude improvements in reliability, speed, or storage requirements.

6.8 SUBSPACE ITERATION TO SOLVE $(K - \lambda M)\mathbf{z} = 0$

This is a fancy version of block inverse iteration with occasional shifts (see [SEP, Chaps. 14 and 15] for more details). The user must select a set of p orthonormal starting vectors. Let them be the columns of an $n \times p$ matrix X. The choice of p is not easy, but keep $p \leq 50$. The subspace of the title is the range or column space of X. The inner loop of SI is, in essence:

1. Solve $(K - \sigma M)Y = MX$ for $n \times p$ matrix Y.
2. Form $p \times p$ matrices $\bar{K} = Y^t K Y$ and $\bar{M} = Y^t M Y$.
3. Solve the complete $p \times p$ small eigenvalue problem $(\bar{K} - \sigma \bar{M})G = \bar{M}GD$, where D is diagonal and G is the $p \times p$ matrix of eigenvectors.
4. Set $X = YG$.
5. Test for convergence of the elements of D.

There are a number of significant modifications of this scheme, but they are not important for this survey (see [1] and [12, Chap. 14] for more details).

When the interesting elements of D have converged, the corresponding vectors of X are satisfactory approximations to eigenvectors.

Typical values are $n = 500$, $p = 20$. Sometimes p is big enough to necessitate the use of secondary storage in step 1. If p is chosen too small, the number of steps to convergence is increased signficantly.

SI is working implicitly with the matrix $(K - \sigma M)^{-1}M$, which is not symmetric but is self-adjoint with respect to the M inner product, and that is sufficient to make the eigenvalues real. The M inner product is defined by $(\mathbf{x}, \mathbf{y}) = \mathbf{y}^t M \mathbf{x}$.

*6.9 THE LANCZOS ALGORITHM**

Let $A = A^t$, $\mathbf{r}$ be a random starting vector, $\mathbf{q}_0 = 0$, and set $\beta_1 = \|\mathbf{r}\|$. The inner loop of the basic Lanczos algorithm is:

for $k = 1, 2, \ldots$ **until convergence, do**

1. $\mathbf{q}_k \leftarrow \mathbf{r}/\beta_k$.
2. $\mathbf{r} \leftarrow A\mathbf{q}_k$.
3. $\mathbf{r} \leftarrow \mathbf{r} - \mathbf{q}_{k-1}\beta_k$.
4. Put $\mathbf{q}_{k-1}$ into secondary storage.
5. $\alpha_k \leftarrow \mathbf{q}_k^t\,\mathbf{r}$.
6. $\mathbf{r} \leftarrow \mathbf{r} - \mathbf{q}_k \alpha_k$.
7. $\beta_{k+1} \leftarrow \|\mathbf{r}\|$.
8. Test for convergence.

Notice the sequential nature of this loop. It is full of vector operations, linear combinations, and dot products. The numbers α_i and β_i are considered as elements of a tridiagonal matrix

$$
T_k := \begin{bmatrix}
\alpha_1 & \beta_2 & & & \\
\beta_2 & \alpha_2 & \beta_3 & & \\
& \beta_3 & \alpha_3 & & \\
& & & & \beta_k \\
& & & \beta_k & \alpha_k
\end{bmatrix}
$$

which grows by one row and column at each step.

Let $Q_k = (\mathbf{q}_1, \mathbf{q}_2, \ldots, \mathbf{q}_k)$. The most important relation is

$$
AQ_k - Q_k T_k = 0 = \beta_{k+1}\mathbf{q}_{k+1}\mathbf{e}_k^t + F_k
$$

where F_k accounts for round-off errors and is always small. Here $\mathbf{e}_k^t = (0, 0, \ldots, 0, 1)$ has k elements; all other vectors are of length n. In exact arithmetic

$$
Q_k^t Q_k = I_k
$$

but this fails completely in practice. Nevertheless, this failure merely slows down the algorithm, it does not prevent convergence.

What happens is that some eigenvalues of T_k, as k increases, settle down at eigenvalues of A. If ϑ and s are an eigenpair of T_k, that is, if

$$
T_k \mathbf{s} = \mathbf{s}\vartheta, \qquad \|\mathbf{s}\| = 1
$$

then the approximate eigenvector belonging to ϑ is $\mathbf{y} := Q_k \mathbf{s}$. There is a computable

*See [SEP, Chap. 13] for more details.

error bound on ϑ. Let $\mathbf{s}(k)$ be the last element of $\mathbf{s}$. There is an eigenvalue λ of A satisfying

$$|\lambda - \vartheta| \leq \frac{(\beta_{k+1}|\mathbf{s}(k)| + \|F_k\|)}{\|\mathbf{y}\|} \approx \beta_{k+1}|\mathbf{s}(k)|$$

In most applications β_k never gets small, so that convergence follows the reduction in $|\mathbf{s}(k)|$ as $\mathbf{k}$ increases. To compute $\mathbf{s}(k)$ it is only necessary to find an eigenvector of a $k \times k$ tridiagonal matrix. In fact, there are even quicker ways of finding $\mathbf{s}(k)$. Again many important details have been omitted in order to enhance the main features.

The matrix A would often be the one described in Section 6.4, so step 2 ($\mathbf{r} \leftarrow A\mathbf{q}_k$) dominates the cost of a Lanczos step. The power of the algorithm comes from the fact that no information is discarded, in contrast to subspace iteration, in which one matrix X is overwritten by another one at each step. The convergence theory of LAN is well developed and the outer eigenvalues of A often converge after a mere $2\sqrt{n}$ steps.

LAN only requires three or four n-vectors in fast memory together with the data for effecting $\mathbf{r} \leftarrow A\mathbf{q}$. However, when eigenvectors are wanted, the Lanczos vectors $\mathbf{q}_1, \mathbf{q}_2, \ldots, \mathbf{q}_k$ must be retrieved from secondary storage to form

$$\mathbf{y} = Q_k \mathbf{s} = \mathbf{q}_1 \mathbf{s}(1) + \mathbf{q}_2 \mathbf{s}(2) + \cdots + \mathbf{q}_k \mathbf{s}(k)$$

Work continues on the task of turning Lanczos into a foolproof, black-box program. Meanwhile a good, documented program is available under the name LASO2 from the National Energy Software Center, Argonne National Laboratory, Argonne, IL 60439. It was written by D. Scott.

6.10 SINGULAR VALUES (SMALL MATRICES)

For small matrices there are very good programs available in most computing centers. The best known is in the LINPACK library (not EISPACK). The singular-value factorization (SVF) of a real $m \times n$ matrix B is $B = U\Sigma V^t$ where the $m \times r$ matrix U and the $n \times r$ matrix V are each orthonormal; that is, $U^t U = V^t V = I_r$, and $\Sigma = \mathrm{diag}(\sigma_1, \ldots, \sigma_r)$ with $\sigma_i > 0$. Here r is the rank of B. Clearly,

$$B^t B = V\Sigma^2 V^t$$

from which it follows that the $\{\sigma_i^2\}$ are the nonzero eigenvalues of $B^t B$ and the columns of V are the corresponding eigenvectors.

There is no need to form $B^t B$ explicitly. The LINPACK program is *equivalent to* the Householder-QR method applied to $B^t B$, but in fact only the array B is transformed explicitly (see [4, Chap. 11] for details). There are two phases.

1. A sequence of $(n - 1)$ simple two-sided orthogonal transformations

$$B \rightarrow P^t B Q$$

is used to reduce B to bidiagonal form.

$$\begin{bmatrix} x & x & & \\ & x & x & \\ & & x & x \\ & & & x \end{bmatrix}$$

2. A different sequence of P and Q matrices is used to diminish the nonzero off-diagonal elements without destroying any zero elements. The singular vectors (as the columns of U and V are called) are found by accumulating the product of the P and Q matrices as they are used.

This is a stable, compact algorithm guaranteed to work. The SVF offers the most informative and reliable technique for solving linear least-squares problems, *but* it is also the most expensive and constitutes overkill for a good number of applications. If $m \geq n$, the number of multiplications required to compute U and V as well as Σ is approximately

$$2m^2(n - m/3) + 5m^2 n$$

when $n = m$ this yields $\approx 6n^3$ (i.e., six matrix-matrix multiplications).

Notice that U and V are built up by operations of the form $U \leftarrow UP$, $V \leftarrow VQ$, where P and Q are orthogonal, to working precision. Thus U and V will be orthonormal to working precision and consequently, the left and right singular vectors will be orthogonal, however close singular values may be.

6.11 SINGULAR VALUES (LARGE MATRICES)

This is a topic still under investigation. There is no standard, readily available software. Least-squares problems do arise with $n \approx 300,000$ and $m \approx 2,000,000$, but with few nonzero elements in B [10]. The full SVD is not necessary for the least-squares problems, so other, cheaper methods are used. One such alternative is to compute the QR factorization (not to be confused with the QR algorithm) with column interchanges. Simplest of all, in the full-rank case, is the use of the Cholesky factor C of $B^t B$ (i.e., $B^t B = C^t C$, where C is $n \times n$ and upper triangular) to solve the normal equations. In exact arithmetic BC^{-1} will be orthogonal and its columns give an orthonormal basis of B's column space.

There are some clever parallel algorithms for computing the Cholesky factor of a positive-definite matrix [21]. How well they would work when $n \approx 300,000$ is not clear! For small problems and parallel techniques see [24].

6.12 LARGE UNSYMMETRIC PROBLEMS

There is at present no efficient, stable method for extracting some or all eigenvalues of large, sparse unsymmetric matrices, and very little good, readily available software. This is a topic of active research [13–15], but the economic incentives for progress seem weak beside those for the symmetric problem. See [18] for a review of this research. One important special case that warrants attention concerns perturbations of symmetric problems.

6.13 QUESTIONS FOR THE FUTURE

It is not surprising that the methods that have emerged as champions for serial computers exploit heavily the characteristics of these machines: ready access by the processor to a large, random access memory and the facility for complicated nesting of loops in programs. It would be surprising if such methods were well suited for parallel processors.

It seems that research on parallel algorithms is focused on surpassing serial machines on standard *small* problems ($n \leq 100$) such as matrix-matrix multiplication or solution of linear equations. Yet these same tasks are ignored by the conventional matrix experts, who feel that nearly all needs are being satisfactorily met for small problems. Their research efforts are directed to large problems; either to perfecting general methods or to devising special techniques for important special cases, such as solving quadratic λ matrices, $(\lambda^2 A + \lambda B + C)\mathbf{x} = 0$, or use of the Cray-1 in place of serial computers.

All the work I have read on eigenvalue techniques for parallel algorithms is devoted to treating full matrices (see [8], for example). Yet most large matrices are sparse and one would expect any efficient algorithm to exploit such structure, if only to conserve storage.

Image-processing and pattern-recognition tasks do generate full matrices. Consider briefly the problem of computing the largest 50 singular values (and vectors) of a 1000×500 matrix B. The data comprise $10^6/2$ numbers. (If these numbers are only known to one decimal digit, it seems unfortunate to begin by squandering $10^6/2$ words to store them.) General techniques, such as use of the normal equations or the clever parallel algorithm of Finn, Luk, and Pottle [7] and Kung and Gal-Ezer [23], seem to be overkill here. They do all the work required to find all 500 singular values! Is it permissible to incorporate prior, or outside knowledge? A good guess at the indices of the 10 or 20 columns of B with greatest norm would be a significant help.

It would permit a serial machine to use the block Lanczos algorithm, block size 10 or 20, with the operator $B^t B$. There is no need to form the product $B^t B$ in order to compute $B^t(BX)$. Here 20 or 10 iterations should suffice for convergence of the 50 largest eigenvalues of $B^t B$ and the number of arithmetic operations is approximately the same as needed just to form $B^t B$. How can parallel algorithms respond efficiently to a change in the number of singular values requested? Very small changes in the assumptions concerning a large problem can strongly affect the efficiency of rival techniques.

Here is one difficulty. How can systolic algorithms with a very uniform flow of information respond efficiently to changes in the specified task or the information available? If they cannot respond flexibly, can they be effective for big calculations?

ACKNOWLEDGMENT

The work of the author was supported in part by the Office of Naval Research under Contract N00014-76-C-0013.

REFERENCES

[1] K.-J. Bathé, and S. Ramaswarmy, "An Accelerated Subspace Iteration Method," *Comput. Methods Appl. Mech. Eng.*, *23*:313–331 (1980).

[2] K.-J. Bathé, and E. Wilson, *Numerical Methods in Finite Element Analysis*, Prentice-Hall, Englewood Cliffs, N.J., 1976.

[3] E. R. Davidson, "The Iterative Calculation of a Few of the Lowest Eigenvalues and Corresponding Eigenvectors of Large Real Symmetric Matrices," *J. Comp. Phys.*, *17*:87–94 (1975).

[4] J. J. Dongarra et al., *LINPACK User's Guide*, SIAM, Philadelphia, 1979.

[5] EISPACK.* The associated guide is published as *Lecture Notes in Computer Science*, Vol. 51, by B. S. Garbow et al, Springer-Verlag, New York, 1977.

[6] T. Ericsson and A. Ruhe, "The Spectral Transformation Lanczos Method in the Numerical Solution of Large, Sparse, Generalized, Symmetric Eigenvalue Problems," *Math. Comp.*, *34*:1251–1268 (1980).

[7] A. M. Finn, F. T. Luk, and C. Pottle, Systolic Array Computation of the Singular Value Decomposition (preprint), 1981.

[8] D. Heller, "A Survey of Parallel Algorithms in Numerical Linear Algebra," *SIAM Rev.*, *20*:740–777 (1978).

[9] International Mathematical Statistical Library (IMSL), Sixth Floor—GNB Building, 7500 Bellaire, Houston, TX 77036.

[10] P. Maissl, *A Priori Prediction of Roundoff Error Accumulation in the Solution of a Super-Large Geodetic Normal Equation System*, NOAA Prof. Paper 12, NOAA, Rockville, Md., 1980.

[11] Numerical Algorithms Group (NAG), 7 Banbury Road, Oxford OX2 6NN, England.

[12] B. N. Parlett, *The Symmetric Eigenvalue Problem*, Prentice-Hall, Englewood Cliffs, N.J., 1980. (Called **SEP** in this paper.)

*Requests for information and for the EISPACK tape should be addressed to Argonne Code Center, Building 221, Room C-235, Argonne National Laboratory, Argonne, IL 60439.

[13] B. N. Parlett, and D. Taylor, *A Lookahead Lanczos Algorithm for Unsymmetric Matrices*, Math. Comp., 44 (to appear).

[14] A. Ruhe, "The Two-sided Arnoldi Algorithm for Nonsymmetric Eigenvalue Problems," in B. Kagstrom and A. Ruhe, eds., *Matrix Pencils*, Springer-Verlag, New York, 1982.

[15] Y. Saad, Chebyshev Acceleration Techniques for Solving Nonsymmetric Eigenvalue Problems, *Math. Comp.*, 41 (to appear).

[16] B. T. Smith, et al., *Matrix Eigensystem Routines—EISPACK Guide*, Lecture Notes in Computer Science No. 6, Springer-Verlag, New York, 1974.

[17] G. W. Stewart, A Bibliographic Tour of the Large Sparse Generalized Eigenvalue Problem, in J. R. Bunch and D. J. Rose, eds., *Sparse Matrix Computations*, Academic Press, 1976, pp. 113–130.

[18] M. Heath, ed., *Sparse Matrix Software Catalog* (1982), available from Mathematics and Statistics Research Dept., Oak Ridge National Laboratory, Oak Ridge, TN.

[19] J. H. Wilkinson, *The Algebraic Eigenvalue Problem*, Oxford University Press, New York, 1965.

[20] J. H. Wilkinson, and C. Reinsch, 1971. *Handbook for Automatic Computation*, Vol. II: *Linear Algebra*, Springer-Verlag, New York, 1971.

[21] R. P. Brent, and S. P. Luk, "Computing the Cholesky Factorization Using a Systolic Architecture," *Proceedings of the 6th Australian Computer Science Conference*, Sydney, Feb. 1983.

[22] L. Snyder, "The Role of the CHiP Computer in Signal Processing," *Proceedings of USC workshop on VLSI and Modern Signal Processing*, Nov. 1982, p. 133.

[23] S. Y. Kung and Gal-Ezer, in this volume.

[24] R. Schreiber, "A Systolic Architecture for Singular Value Decomposition," *Proc. of 1st International Colloq. on Vector and Parallel Computing in Scientific Applications*, INRIA, Paris (1983).

Part II

CONCURRENT ARRAY PROCESSORS: ARCHITECTURES AND LANGUAGES

INTRODUCTION TO PART II

The theme for Part II is very large scale integrated (VLSI) architectures for parallel processing. The practicality of most theoretical research on signal processing, including the work discussed in Part I, will ultimately be determined by its computational feasibility. Particularly, for real-time data processing, it depends critically on the parallel processing capabilities, in both speed and volume, offered by the state-of-art computing machines. From the device front, the high-density, fast-speed VLSI devices and the emerging computer-aided-design (CAD) facilities have precipitated a timely architectural design revolution. While VLSI holds the promise of high parallelism by offering almost unlimited hardware at very low cost, there are several inherent technological constraints with respect to communication, design complexity, testability, and so on. Consequently, there is an urgent need for a new VLSI system design methodology. For the purpose of maximizing VLSI's potential for signal processing applications, the system design should embrace a broad spectrum of disciplinaries and holistically coordinate language design, architectural structure, and practical applications.

Until recently, computation-intensive tasks have been handled by high-performance supercomputers, including pipelined computers, array processors, and multiprocessor systems. The development of these computer systems has involved a

121

thorough exploration of parallel computing, efficient programming and resource optimization. However, the general-purpose nature of these machines leads to a complicated system organization and severe system operation overheads. These machines are not suitable for real-time signal processing, where a very high through-put rate is a must.

A solution to the high-speed requirement of signal processing is found in the use of special-purpose computers. There are two types of special-purpose computers. One type is characterized by inflexible and highly dedicated structures. The other allows some flexibility, such as programmability and reconfigurability. Hard-wired dedicated processors have been applied to real-time signal processing because they offer high processing speed. However, the special nature of hard-wired equipment often results in long development schedules with attendant high costs. With the advent of modern algorithm/architecture analysis, the programmable/reconfigurable parallel signal processors have become not only feasible but actually, in many instances, more economical than hard-wired machines. Therefore, Part II places a substantial emphasis on these parallel processors.

More specifically, Part II explores various aspects of the analysis of parallel algorithms, the design of parallel architectures, and the formulation of parallel languages. These fundamental studies, in our opinion, should provide the theoretical footing of the future VLSI computing systems for signal processing.

Facing the challenge of the revolutionary VLSI technology, modern signal processing systems should incorporate innovative but matured concepts and methods into their architecture/language design. The basic requirement in a top-down-design methodology is fundamental understanding of algorithm analysis. Although this is a well-developed research area from a sequential algorithmic point of view, much innovation is needed for the case of parallel processing. New ideas are needed in parallelism extraction, communication, data dependency, numerical performance, and commonality between algorithms.

For an example, as long as interconnection in VLSI remains restrictive, the locality of a recursive algorithm will be of great concern. An increase in efficiency can be expected if the algorithm arranges for a balanced distribution of work load while observing the requirement of locality (i.e., short communication paths). These properties of load distribution and information flow serve as a guideline to the designer of VLSI algorithms and eventually lead to new designs of architecture and language. The first such consideration for VLSI algorithms is the design of systolic array. Its original definition is given in its first publication, "Systolic Array (for VLSI)" by H. T. Kung and C. E. Leiserson (in *Sparse Matrix Proceedings*, SIAM, Philadelphia, 1978): "A systolic system is a network of processors which rhythmi-cally compute and pass data through the system. Physiologists use the word 'sys-tole' to refer to the rhythmically recurrent contraction of the heart and arteries which pulses blood through the body. In a systolic computing system, the function of a processor is analogous to that of the heart. Every processor regularly pumps data in and out, each time performing some short computation, so that a regular flow of data is kept up in the network."

For readers in the signal processing community, it should be particularly enlightening to see how to apply such VLSI algorithm analyses to signal processing architecture design. Due to the cost of communication in VLSI, a communication-oriented analysis and taxonomy of parallel algorithms will be most effective. An algorithm is said to be recursive if it repeats the same type of operations on sequentially available input data. In a parallel recursive algorithm, the input and output data are labeled with space as well as time indices. The time index of the output is always that of the input incremented by 1. The space indices of the output and input data are related by a regular pattern. A parallel algorithm is said to be locally communicative if the space indices of the input data elements involved in a same recursive operation are separated by no more than a given limit. In this taxonomy it is found that a great majority of signal processing algorithms belong to a common (locally recursive) group. The commonality of this class of algorithms can be exploited to facilitate the design of VLSI array processors for signal processing applications. In particular, the locality feature shared by these algorithms permits not only localizable data transactions but also localizable controls in the processor array. More precisely, this implies that the activation of all the computing processes in the array may be managed locally by a scheme similar to dataflow computing. As a matter of fact, the concept of dataflow computing is not necessarily novel or surprising to the signal processing research community. For example, it naturally arises in many signal flow graphic representations such as FFT and digital filter flow diagrams. Very interestingly, this additional flexibility lends itself to the design of the wavefront processor, a locally interconnected and self-timed array which can be programmed to process in parallel all the locally recursive algorithms.

Chapter 7, by Kung, offers an evolutionary perspective on the transition of VLSI concurrent array processors, from transversal filtering to systolic and wavefront array processing. The algorithm and architecture analysis is exemplified first by a one-dimensional ARMA filter design, and then extended to two-dimensional systolic and wavefront arrays. The paper stresses a holistic VLSI system design methodology that closely coordinates language design, architecture structure, and their practical applications. The advantages of (self-timed) data-driven computing and its supporting language in wavefront/systolic-type software are also highlighted. A major advance in the software and CAD techniques will be indispensable to the development of massively parallel array processors. In mapping parallel algorithms onto hardware, a top algorithmic analysis level should provide a powerful abstraction of the space-time activities in its full parallelism, while a lower level should specifically define the hardware design. This issue is addressed in Chapter 7, in which an HIFI (Hierarchical and Interactive Flow-graphic Integration) design is introduced. The HIFI design involves a close interplay between the recursive algorithm decompositions, abstract notations, flow-graph (structural) and functional (behavior) descriptions, transformation procedures, bi-directional mapping between graphic and textural codes, and silicon compilations.

To ease the bottleneck caused by computation-intensive problems, many special-purpose architectures have been proposed. Underlying these designs, there are some factors critical to their usefulness and cost-effectiveness. In particular, a very fundamental issue is the trade-off between a dedicated and a programmable processor array. Chapter 8, by Fisher and Kung, addresses this issue, first from a general perspective and then in a special-case example. The general discussion covers the broad field of implementation technology, applications, algorithms, architecture, hardware design and CAD, fabrication turnaround, and system integration. In the case study, the focus is shifted to the design of a building-block processor, a programmable systolic chip, which helps illustrate the role of the aforementioned general factors in the design consideration.

To meet the need of different signal processing environments, extra flexibility in special-purpose architecture is very desirable, and in many circumstances, necessary. It includes programmability, configurability, decomposability, and fault tolerance. Chapter 9, by Snyder, presents a configurable highly parallel (CHiP) computer for signal processing applications. The CHiP is composed of a collection of parallel processing elements placed at regular intervals in a two-dimensional lattice of programmable switches. The switches, which have local memory, allow the processor array to be dynamically connected into arbitrary topologies. The paper demonstrates the feasibility and convenience of a chip computer in hosting the FFT-type or general systolic-type algorithms. Most important, the CHiP machine can be efficiently implemented in VLSI and is suitable for wafer-scale integration, due to its reconfigurability and fault tolerance characteristics.

Highly parallel VLSI systems must provide some form of fault tolerance to be useful since defects in VLSI silicon are not uncommon. Chapter 10, by Kuhn, gives some innovative ideas for providing fault tolerance in VLSI concurrent array processors. It presents an interstitial fault tolerance (IFT) approach, a natural technique for incorporating fault tolerance into locally interconnected arrays. In general, fault tolerance implies redundancy in two forms: increased hardware or increased computation time. In seeking to use VLSI technology to implement highly parallel systems first and fault-tolerant systems second, time redundancy as used in the IFT technique becomes more appropriate. Contrasting to the previous wafer-scale integration techniques, the IFT is effective for both linear and two-dimensional arrays.

In many VLSI paper designs, it is common to make the assumption that the number of input ports exceeds some feature size of the input data. This may not necessarily be the case in real applications. Chapter 11, by Heller, addresses the important issue of mapping a large computing job into a smaller processor array. The paper surveys several state-of-the-art approaches. The most efficient partitioning methods require use of several array processor modules. In this case, programmability and flexibility of control of the array become highly desirable. This research area is just budding and many open problems are yet to be addressed.

Algorithms suitable for VLSI take advantage of the concurrency present in both the application problem and the target architecture while observing the requirements of locality (i.e., short communication paths) and balanced work-load distribution. The definition of a VLSI algorithm requires a precise notation capable of reflecting the concurrent nature of the algorithm. The abstractions of "function" and "state" should allow transitions to be computed on the whole state in a side-effect-free manner. In addition, the specification of data flow in locally synchronized algorithms calls for an adequate communication mechanism. Furthermore, the notation ought to be problem oriented (i.e., adaptable to a given level of abstraction in the design process). An important advantage of a precise notation is that it may be executable and thus provide problem-oriented diagnostics of prototypical implementations of a VLSI algorithm.

An executable VLSI algorithmic notation encouraging a functional style of specification is proposed in Chapter 12, by Cremers and Hibbard. The notation is based on the computational concept of a "data space," introduced by the authors several years ago as a formal model of abstract machines. A data space is specified in terms of an ensemble of data-defined functions of a rather general kind of associative "store" and an ensemble of expression-defined functions of a "processor." The notation allows the definition of recursive types and is applicative in the sense that during the computation of state transition as a function on the whole state, there are no side effects on the state. For local synchronization the syntactic concepts are "subspaces" intended as an encapsulation of computational substructures, and "synchronized types" that provide a data-flow-style communication mechanism. The notation has been implemented, and results are reported of using it as a vehicle for executable specification and high-level testing in various VLSI applications.

Chapter 13, by Chen and Mead, introduces a hierarchical methodology for specifying and verifying VLSI systems. It proposes to represent concurrent algorithms by a system of space-time recursive equations, with a denotational semantics (fixed-point semantics). The language has a wide range of applications and the same notation is used for systems ranging from the level of transistors up to the level of communicating processes. In other words, the framework can be viewed as a concurrent programming notation when describing communicating processes, and as a hardware description notation when specifying integrated circuits.

Digital signal processing is based on a well-established body of mathematical theory which is explicitly used during program development. Due to the central role of the concepts signal and system, a well-structured signal processing program is one organized as a collection of "signal" and "system" abstractions. Chapter 14, by Kopec, reviews three representations of discrete-time signals as data objects in a program. In particular, a representation with abstract signal objects is introduced in the new signal representation language, SRL, which satisfies a set of desirable signal representation criteria.

For signal processing applications, there is an extensive literature on digital filter designs. Recently, there has been renewed interest in filter structures for the purpose of extracting parallelism. To this end, Chapter 15, by Dewilde, Deprettere, and Nouta, proposes an approach using precedence graphs to identify systematically the maximum parallelism attainable for a given structure. In addition to the high throughput obtained by multilevel pipelining and parallelism, high quality of numerical conditions is achieved by using exclusively orthogonal arithmetics. It also demonstrates a doubly piped cordic module: globally between processing elements

(PEs) and locally at the bit level. Application of this cordic module to orthogonal filters or wavefront arrays results in a maximal throughput rate.

Most signal processing techniques involve intensive arithmetic computations. Therefore, it is important to select appropriate arithmetic units for specific applications. Chapter 16, by Ahmed, gives a broad survey of the various arithmetic unit architectures available for VLSI digital signal processing and compares them in different bases. The chapter should serve to offer a basic background for making a selection among several promising options currently being developed.

7

VLSI Signal Processing: From Transversal Filtering to Concurrent Array Processing

SUN-YUAN KUNG

University of Southern California
Los Angeles, California

7.1 INTRODUCTION

VLSI microelectronics technology and the emerging computer-aided-design (CAD) methodologies are spurring a new revolution in signal processing. The area of VLSI signal processing is bound to become a major focus of future technological research activity. To provide a bridge from signal processing theory and algorithm to VLSI processor architecture and implementation, it is critical to have a fundamental understanding of the basic computational requirements of modern signal processing and of the technology constraints of VLSI. This will require a cross-fertilization of the fields of computer science/engineering and signal engineering.

First, the ever-increasing demands for performance, sophistication, and real-time signal processing strongly indicate the need for tremendous computation capability, in terms of both volume and speed. The availability of low-cost, high-density, fast VLSI devices makes high-speed, parallel processing of large volumes of data practical and cost-effective [1]. This makes feasible ultrahigh throughput rates and presages major technological breakthroughs in real-time signal processing applications.

On the other hand, it is quite obvious that the full potential of VLSI can be realized only when its application domains are discriminatingly identified. For this purpose, it may be noted that traditional computer architecture designs are no longer suitable for the design of highly concurrent VLSI computing processors. For example, large design and layout costs suggest the utilization of a repetitive modular structure. The major technological constraint, however, is that communication in VLSI systems is very expensive in terms of area, power, and time consumption. Hence the communication has to be restricted to localized interconnections [1].

127

The traditional design of parallel computers and languages is deemed unsuitable for real-time (signal processing) systems [18], since it suffers from the heavy supervisory overhead incurred in scheduling communication, and storage tasks, which severely hampers the crucial throughput rate. For large-scale multiprocessor systems, VLSI imposes a local communication constraint, in order to reduce interdependence and ensuing waiting delays that result from excessive communication. This locality constraint prevents the utilization of global data transaction and global synchronization. The resultant use of distributed control, localized communication, and data-driven computing has become increasingly attractive from both the hardware and software standpoints. Such considerations have given rise to several VLSI-oriented special-purpose architectures for signal processing applications.

This chapter discusses the algorithmic analysis underlying the design evolution of these architectures and the trade-offs involved in their VLSI implementations. In Section 7.2 we first review several potential VLSI-oriented architectures for signal processing, including the transversal filter, systolic array, data-flow processor, and wavefront array. In Section 7.3 the main issue in the top-down-design methodologies, the mapping of algorithms into computer architectures, is explored. We further illustrate the evolution of the design philosophy from the conventional ARMA canonical form [2], the systolic array, to the (data-driven) wavefront array. In Section 7.4 we explore the extension to two-dimensional wavefront array processing. Based on a computational wavefront notion, we illustrate a holistic design approach which allows close coordination between the language and architecture design issues.

7.2 VLSI ARRAY PROCESSORS FOR SIGNAL PROCESSING

7.2.1 Transversal Filtering

Apart from FFT computing, perhaps the most useful digital signal processing techniques are FIR and IIR digital filters. An FIR filter for convolution, with transfer function

$$H(z) = \sum_{i=1}^{M} b_i z^{-i}$$

is often implemented in a tapped-delay line, sometimes called a transversal filter [3], as shown in Figure 7.1. Note that the conventional design requires global communications (for the summing operation). In VLSI design, however, it is preferred to have a more modular system with only localized interconnections, such as systolic [5,6] or wavefront arrays [7].

7.2.2 Systolic Array

"A systolic system is a network of processors which rhythmically compute and pass data through the system" [6]. For example, it is shown in [5] that some basic "inner product" processing elements (PEs) ($Y \leftarrow Y + A * B$) can be locally connec-

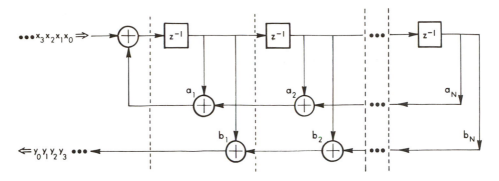

Figure 7.1   Transversal filter or tapped-delay-line design.

ted to perform FIR filtering in a manner similar to the transversal filter. Furthermore, two-dimensional systolic arrays (of the inner product PEs) can be constructed to efficiently execute matrix multiplication, L-U decomposition, and other matrix operations.

The basic principle of systolic design is that all the data, while being "pumped" regularly and rhythmically across the array, can be effectively used in all the PEs. The systolic array features the important properties of modularity, regularity, local interconnection, and highly pipelined and highly synchronized multiprocessing. They require no control and overlap I/O and computation, and hence speed up computer-bound computation without increasing I/O requirements [6].

However, there are several unresolved controversial issues regarding systolic arrays. First, systolic arrays require global synchronization (i.e., global clock distribution). This may cause clock-skew problems in high-order VLSI system implementations. In addition, (pure) systolic arrays tend to equalize the time units for different operations. As an example, for the convolution systolic arrays in [5], a local data transfer consumes the same time delay as a multiply-and-add (i.e., one full time unit). This often results in unnecessary wastage of processing time (see Section 7.3), since the data transfer time as needed practically is almost negligible. Another issue of concern is parallel language description and ease of programmability for complex data flows. These aforementioned problems motivate a revisit to a well-known asynchronous data-driven scheme, commonly adopted in data-flow machines.

7.2.3 Data-Flow Multiprocessor

A data-flow multiprocessor [8,9] is an asynchronous, data-driven multiprocessor which runs programs expressed in data-flow notations. Since the execution of its instructions is "data driven" (i.e., the triggering of instructions depends only on the availability of operands and resources required), unrelated instructions can be executed concurrently without interference. The principal advantages of data-flow multiprocessors over conventional multiprocessors are a simple representation of concurrent activity, relative independence of individual PEs, greater use of pipelining, and reduced use of centralized control and global memory.

For example, it is shown in [9] that a data-flow multiprocessor can perform matrix multiplication in $O(N)$ time, which is in the same order as the systolic array. However, for a general-purpose data-flow multiprocessor, the interconnection and memory conflict problems remain very critical [9]. Such problems can be eliminated if the notion of regularity and locality is elegantly inserted into data-flow processing [7]. In fact, this motivation leads to the notion of wavefront array processing.

7.2.4 Wavefront Array

A wavefront array is a programmable array processor [7] which combines the features of the asynchronous, data-driven properties in data-flow machines and the regularity, modularity, and local communication properties in systolic arrays. For VLSI array processors, it is very desirable to have an effective space-time programming language. The need for a powerful description tool is further aggravated when more complex algorithms, such as eigenvalue or singular-value decompositions, are encountered. Although the original systolic concept focuses on the data movements between processors in the array, unfortunately, it does not naturally lend itself to a simple programming language structure. In contrast, the concept of wavefront processing offers a simple solution to the space-time description of data movements in one- or two-dimensional arrays. More precisely, the wavefront model utilizes the asynchronous, data-driven property of data-flow machines and the locality and regularity of systolic arrays. This leads to the development of the matrix data flow language (MDFL) [7], which is capable of tracing complex sequences of interactions and data movements. Some examples will be given in the following sections.

7.3 FROM ALGORITHM ANALYSIS TO ARCHITECTURE DESIGN

A VLSI system design cycle should include the following research phases: (1) applicational specification, (2) theory and algorithm, (3) computing structure and language, (4) processor architecture, (5) CAD circuit layout and fabrication, and (6) insertion of chips into system. Therefore, a top-down design should start with a full understanding of the problem specification, signal mathematical analysis (parallel and optimal), algorithmic analysis, and then mapping of algorithms into suitable architectures.

7.3.1 Analysis of Recursive Algorithms

A recursive algorithm is said to be of local type if the space index separations incurred in two successive recursions are within a given limit. Otherwise, if the recursion involves globally separated indices, the algorithm will be said to be of global type; and it will always call for globally interconnected computing structures.

A typical example of global-type recursion is the FFT computations. The principle of the (decimation-in-time) FFT is based on successively decomposing the data, say $\{x(i)\}$, into even and odd parts. This partitioning scheme will result in a necessarily global communication between data. More precisely, the FFT recursions can be written as (using the "in-place" computing scheme [2])

$$x^{(m+1)}(p) = x^{(m)}(p) + w_N^r x^{(m)}(q)$$

$$x^{(m+1)}(q) = x^{(m)}(p) - w_N^r x^{(m)}(q)$$

with p, q, and r varying from stage to stage. The "distance" of the global communication involved will be proportional to $|p - q|$. At the last stage, the maximum distance is $|p - q| = N/2$. (For example, the maximum distance will be 512 units for a 1024-point FFT.) Obviously, an FFT cannot be mapped into local-type computing structures, such as systolic or wavefront arrays.

A more interesting example is the design of ARMA (IIR) filters. Generally, an IIR filter is defined by a transfer function

$$H(z) = \frac{\displaystyle\sum_{i=1}^{N} b_i z^{-i}}{1 + \displaystyle\sum_{i=1}^{N} a_i z^{-i}}$$

Convolution (or transversal filtering) is simply a special case when $a(z) = 1$. The corresponding recursive algorithm is often given in terms of the difference equation:

$$y(k) = \sum_{m=1}^{N} x(k - m)b(m) + \sum_{m=1}^{N} y(k - m)a(m) \tag{7.1}$$

This direct form involves global indexing, and thus leads to global-type realization in the canonical form design [3,4], shown in Figure 7.2(a). However, this design can easily be converted into a local-type one; therefore, so can the algorithm represented by equation (7.1).

A modified design is shown in Figure 7.2(b). To verify that Figure 7.2(b) yields the same transfer function as Figure 7.2(a), we simply check the transfer functions $V^{(N)}(z)/U^{(N)}(z)$ and $W^{(N)}(z)/U^{(N)}(z)$. Obviously, they remain unchanged by the modification. Thus, by induction, so do the transfer functions $V^{(i)}(z)/U^{(i)}(z)$, $W^{(i)}(z)/U^{(i)}(z)$, for $i = N - 1, N - 2, \ldots, 1$, and the proof is completed. Note that, in the modified direct form, there are always delays inserted between all the summing nodes; therefore, only localized communications are needed.

7.3.2 Systolic Array for ARMA (IIR) Filtering

We show below that the modified form is, in fact, equivalent to the systolic array design [see Figure 7.3(a)], except for a rescaling of time unit. Since the half-time delay unit $(z^{-1/2})$ is not common, our next step in Figure 7.3(a) is to rescale the time unit by setting $z^{-1/2} = z'^{-1}$ (i.e., to renormalize the time unit). Therefore, the delay z^{-1}, the multiplier, and the adder in each single section (defined by means of dashed

Figure 7.2 (a) Direct-form design of **ARMA (IIR)** filter (global interconnection required); (b) modified direct form with localized interconnection.

132

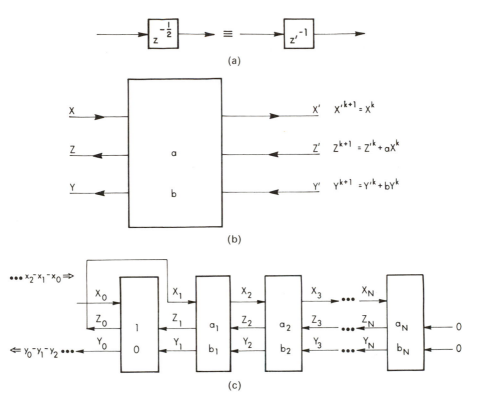

Figure 7.3   (a) $z^{-1/2} = z'^{-1}$, rescaling of time unit; (b) systolic processing element (note the one-unit time delay from X to X); (c) systolic array processor for ARMA (IIR) filter.

lines) in Figure 7.2(b) are all merged into an "inner product" SPE [5] as shown in Figure 7.3(b). An overall systolic array configuration for an IIR filter is shown in Figure 7.3(c). Note that due to the rescaling of time units, the input data $\{x_i\}$ have to be interleaved with "blank" data [see Figure 7.3(c)], and the throughput rate becomes $0.5T^{-1}$.

Now, having a localized array, we can, in turn, derive the localized version of the algorithm in (7.1). Assigning the superscript $\{k\}$ as the time variable and the subscript $\{n\}$ as the space variable for the data movements corresponding to the systolic array above, we have the following local-type recursions [setting $a(0) = b(0) = 0$]:

$$Z_n^{k+1} = X_n^{k+1}a(n) + Z_{n+1}^k \tag{7.2a}$$

$$Y_n^{k+1} = X_n^{k+1}b(n) + Y_{n+1}^k \tag{7.2b}$$

$$X_{n+1}^{k+1} = X_n^k \tag{7.2c}$$

except for $n = 0$, the leading PE, where

$$X_1^{k+1} = X_0^k + Z_0^k \qquad (7.2d)$$

Initial conditions:

$$Z_n^0 = 0 \qquad Y_n^0 = 0$$

Boundary conditions:

$$Z_{N+1}^k = 0 \qquad Y_{N+1}^k = 0$$

Input:

$$X_0^{2k-1} = x(k) \qquad (7.3a)$$

Output:

$$Y_0^{2k} = y(k) \qquad (7.3b)$$

Interestingly, the correctness of the systolic array can easily be verified by taking the z transform of (7.2) and (7.3), followed by some trivial algebra.

Note that the throughput rate for the systolic processor above is $0.5T^{-1}$ (i.e., one output data for every two time units). This rate is slower than that of the direct form design ($1.0T^{-1}$), because the data transfer ($X \rightarrow X'$) alone is required to consume the same full time unit as multiply-and-add, which is wasteful and unnecessary. There are two solutions to this problem: one is using multirate systolic array and the other is wavefront array based on asynchronous data-driven computing.

7.3.3 Multirate Systolic Array

A multirate systolic array is a generalized systolic array, where different data streams may be pumped through the array at different rates, thus allowing different basic operations to consume different time units. For the ARMA filter design example, we can assign Δ as the time unit for data transfer and T for multiply-and-add. Consequently, in the z-transform representation, there will be two different variables introduced: $z_1^{-1} = \Delta$ and $z_2^{-1} = T$, and in the circuit representation in Figure 7.2(b), we replace $z^{-1/2}$ on the feedforward path (for X) by z_1^{-1}, and $z^{-1/2}$ on the feedback paths (for Y and Z) by z_2^{-1}. These modifications lead to a multirate systolic array as shown in Figure 7.3(a) and (b). Here each datum X is pumped rhythmically at the rate of $1/\Delta$ and data Y and Z are pumped at the rate of $1/T$. Since the transfer function of the array will be $H(z'')$, where $z''^{-1} = z_1^{-1} + z_2^{-1}$, the input/output sequences $\{x_0, x_1, x_2, \ldots\}$ and $\{y_0, y_1, y_2, \ldots\}$ are pumped at a corresponding rate of $1/(T + \Delta)$.

By a multirate system realization theory (see, e.g., Kung and Gal-Ezer [18]), corresponding to the same transfer function $H(z'')$, there are many nonunique multirate array structures. This is worth noting because certain structures will be numerically better in their digital implementation [18].

In fact, a multirate systolic array is nothing but a synchronized version of the wavefront array (see the next section). A potential application of such an array is in mixed bit-wise and word-wise systolic/wavefront processing. A good example is the (synchronized) cordic wavefront array for QR decomposition [12]. Here, in a square array, the vertical wavefront propagation will be word by word, while the horizontal propagation will be bit by bit. (In this case, the bits to be pumped horizontally will be the cordic control bits for angle rotation [19].)

7.3.4 Wavefront Array for ARMA (IIR) Filter

When the issue of system synchronization becomes critical, one has to resort to the wavefront approach, using asynchronous data-driven computing. Our strategy now is to replace the globally synchronized computing in the systolic array by an asynchronous data-driven model, as shown in Figure 7.4. Therefore, at each node in Figure 7.4, the operation is executed when and only when the required operands are available. Since it uses asynchronous processing, the timing references become unnecessary. An immediate advantage of this model is that the data transfer $(X \to X')$ will use only negligible time compared to the time needed for the arithmetic processing. More precisely, the throughput rate achieved by the WAP is approximately $1.0T^{-1}$ (i.e., twice that of the pure systolic array in Figure 7.3).

Now we have another way to represent the recursion. Here we reassign the superscripts k (in parentheses) as the wavefront number (or the recursion number) since the timing index is no longer applicable. The new representation of (7.1) is [setting $a(0) = b(0) = 0$]

$$Y_n^{(k)} = X^{(k)}b(n) + Y_{n+1}^{(k-1)} \tag{7.4a}$$

$$Z_n^{(k)} = X^{(k)}a(n) + Z_{n+1}^{(k-1)} \tag{7.4b}$$

$$X^{(k)} = X(k) + Z_0^{(k)} \tag{7.4c}$$

Under the wavefront notion, $X^{(k)}$ is initiated at the leading PE($n = 0$) according to (7.4c), and then propagated rightward across the processor array, activating operations (7.4a) and (7.4b) in all the data-driven PEs. As is shown in (7.4), the updated data $\{Y_{n+1}^{(k)}, Z_{n+1}^{(k)}\}$ are fed back leftward ready for the next recursion (or wavefront).

Another important feature is that the data-driven model allows a simple data-flow language for the description of data movement. When programmed in MDFL, the description (for nonboundary PEs) goes as follows:

```
BEGIN
    FETCH   X,  RIGHT;
    FLOW    X,  LEFT;
    MUL     X,  b, D;
    MUL     X,  a, E;
    FETCH   Y,  RIGHT;
    ADD     D,  Y, Y;
    FETCH   Z,  RIGHT;
```

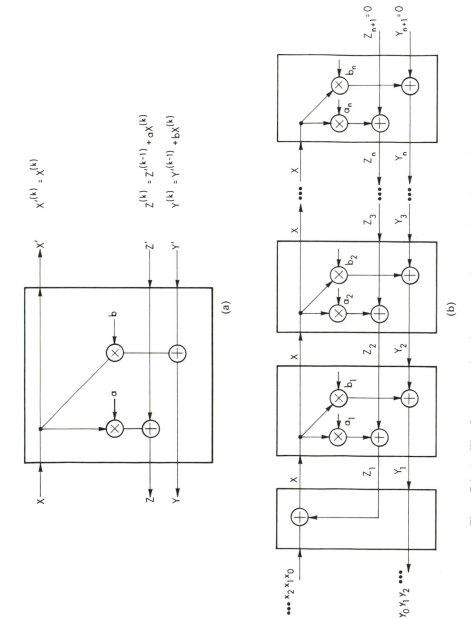

$$X'^{(k)} = X^{(k)}$$

$$Z^{(k)} = Z'^{(k-1)} + aX^{(k)}$$
$$Y^{(k)} = Y'^{(k-1)} + bX^{(k)}$$

(a)

(b)

Figure 7.4 (a) Wavefront processing element (asynchronous, data-driven model, i.e., operations take place only on availability of appropriate data); (b) wavefront array for ARMA (IIR) filter.

136

```
            ADD,    E, Z, Z;
            FLOW    Y, LEFT;
            FLOW    Z, LEFT;
       END
```

Note that the above is written in a local level, that is, taking the perspective of one processor encountering a sequence of activities. Once the notion of computational wavefront is formally introduced, it is possible to use a global level of MDFL, taking the perspective of one wavefront passing across all the processors. It is much easier to view the algorithm as a series of wavefronts; global MDFL reduces the burden on the programmer.

7.4 WAVEFRONT ARRAY PROCESSING

From a top-down point of view, one should exploit the fact that a great majority of signal processing algorithms possess the aforementioned recursiveness and locality properties. Indeed, a major portion of the computational needs for signal processing and applied mathematical problems can be reduced to a basic set of matrix operations and other related algorithms [10,11]. More challengingly, the inherent parallelism of these algorithms calls for parallel computing capacity much higher than what one-dimensional array can possibly furnish. The first major architecture breakthrough is the two-dimensional systolic arrays for matrix multiplication, inversion, and so on, and the reader is referred to [6] for an overall review.

In this section we concentrate on the algorithmic analysis, which will lead to a coordinated language and architecture design, needed for a two-dimensional array processor. In fact, the algorithmic analysis of the matrix operations will lead first to a notion of two-dimensional computational wavefronts. All of the algorithms above are decomposable into a sequence of recursions, and can be mapped into corresponding wavefronts on a homogeneous computing network. (A wavefront corresponds to a state of recursion.) Successively pipelining a sequence of computational wavefronts leads to a continuously advancing wave of data and computational activity.

7.4.1 Sequencing of Mathematical Recursions

We shall illustrate the sequencing of mathematical recursions by means of a matrix multiplication example. Let

$$A = a_{ij} \qquad B = b_{ij} \qquad C = A \times B$$

all be $N \times N$ matrices. The matrix A can be decomposed into columns A_i and the matrix B into rows B_j and therefore

$$C = A_1 B_1 + A_2 B_2 + \cdots + A_N B_N$$

The matrix multiplication can then be carried out in N recursions:

$$C_{i,j}^{(k)} = c_{i,j}^{(k-1)} + a_i^{(k)}b_j^{(k)} \tag{7.5a}$$

$$a_i^{(k)} = a_{ik} \tag{7.5b}$$

$$b_j^{(k)} = b_{kj} \tag{7.5c}$$

for $k = 1, 2, \ldots, N$ and there will be N sets of wavefronts involved.

Pipelining of computational wavefronts [7]. The computational wavefront for the first recursion in matrix multiplication will now be examined. Suppose that the registers of all the PEs are initially set to zero:

$$C_{ij}^{(0)} = 0 \qquad \text{for all } (i, j)$$

The entries of A are stored in the memory modules to the left (in columns), and those of B in the memory modules on the top (in rows). The process starts with PE $(1, 1)$:

$$C_{11}^{(1)} = C_{11}^{(0)} + a_{11} * b_{11}$$

is computed. The computational activity then propagates to the neighboring PEs $(1, 2)$ and $(2, 1)$, which will execute in parallel:

$$C_{12}^{(1)} = C_{12}^{(0)} + a_{11} * b_{12}, \qquad \text{and} \qquad C_{21}^{(1)} = C_{21}^{(0)} + a_{21} * b_{11}$$

The next front of activity will be at PEs $(3, 1)$, $(2, 2)$, and $(1, 3)$, thus creating a computational wavefront traveling down the processor array. This computational wavefront is similar to electromagnetic wavefronts (they both obey Huygens' principle), since each processor acts as a secondary source and is responsible for the propagation of the wavefront. It may be noted that wave propagation implies localized data flow. Once the wavefront sweeps through all the cells, the first recursion is over (see Figure 7.5).

As the first wave propagates, we can execute an identical second recursion in parallel by pipelining a second wavefront immediately after the first one. For example, the (i, j) processor will execute

$$C_{ij}^{(2)} = c_{ij}^{(1)} + a_{iz} * b_{zj}$$

and so on. The pipelining is feasible because the wavefronts of two successive recursions will never intersect (Huygens' wavefront principle), as the processors executing the recursions at any given instant will be different, thus avoiding any contention problems. Here let us be reminded that it is possible to have wavefronts propagating in several different fashions. The only critical factor is that the order of task sequencing must be correctly followed. This rule is ensured by the data-driven nature of wavefront processing. Therefore, the actual propagation pattern is practically of no consequence. As a matter of fact, when local clocks are used, the entire wavefront pattern may be crooked and yet yield accurate sequencing and computation.

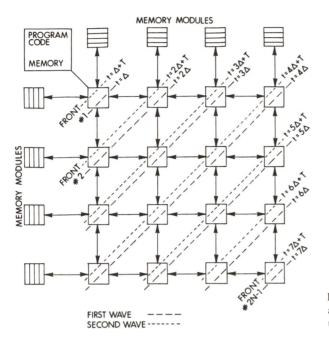

MEMORY MODULES

PROGRAM CODE

MEMORY

MEMORY MODULES

FRONT #1

FRONT #2

FRONT #2N-1

FIRST WAVE ----
SECOND WAVE ------

Figure 7.5 Two-dimensional wavefront array. Δ, unit time of data transfer; T, unit time of arithmetic operation.

7.4.2 L-U Decomposition

In the L-U decomposition, a given matrix C is decomposed into

$$C = A \times B \tag{7.6}$$

where A is a lower- and B an upper-triangular matrix. The recursions involved are

$$C_{ij}^{(k)} = C_{ij}^{(k-1)} - a_i^{(k)} b_j^{(k)} \tag{7.7a}$$

$$a_i^{(k)} = \frac{1}{c_{kk}^{(k)}} C_{i,k}^{(k-1)} \tag{7.7b}$$

$$b_j^{(k)} = C_{(k,j)}^{k-1} \tag{7.7c}$$

for $k = 1, 2, \ldots, N$; $k \le i \le N$; $k \le j \le N$.

Verifying the procedure by tracing back (7.7a), we note that

$$C = c_{ij}^{(0)} = \sum_{k=1}^{N} a_i^{(k)} b_j^{(k)} = AB$$

where $A = \{a_{mn}\} = \{a_m^{(n)}\}$ and $B = \{b_{mn}\} = \{b_n^{(m)}\}$ are the outputs of the array processing. [Compare with (7.6).]

Comparing with (7.5) and (7.7) is basically a reversal of the matrix multiplication recursions. In fact, in wavefront processing for the L-U decomposition, (7.7) will exhibit a similar wavefront propagation pattern as Figure 7.6, except for the following.

1. Just like the matrix multiplication, the data $a_i^{(k)}$ and $b_j^{(k)}$ are to be propagated rightward and downward, respectively. However, $a_i^{(k)}$ has to be derived from an arithmetic operation, causing extra delay; see (7.7b), [while $b_j^{(k)}$ is directly available from the previous recursion; see (7.7c)].

2. In a very straightforward scheme, the second recursion may start at the PE (2, 2) [the third at PE (3, 3), and so on]. However, there is a hardware advantage to have all recursions initiated in the PE (1, 1). In any event, the active area of array processing shrinks from one recursion to another. (This inevitably causes some wastage in the processor utilization [7].)

7.4.3 Eigenvalue Decomposition by Wavefront Processing

For many signal processing applications, such as high-resolution beamforming spectral estimation, image data compression, and so on [20,21], eigenvalue and singular-value decompositions have emerged as extremely powerful and efficient computational tools. For symmetric eigenvalue problems, according to Parlett [22], "the QL and QR algorithms ... have emerged as the most effective way of finding all the eigenvalues of a small symmetric matrix." The question is whether or not the QR algorithm may retain that same effectiveness when mapped into a parallel algorithm on a square or linear multiprocessor array. In the following section, we offer an answer to this question using the computational wavefront notion.

7.4.4 Linear Array Tridiagonalization of a Symmetric Matrix

In the QR algorithm a "full matrix is first reduced to tridiagonal form by a sequence of transformations and then the QL [QR] algorithm swiftly reduces the off diagonal elements until they are negligible. The algorithm repeatedly applies a complicated similarity transformation to the result of the previous transformation, thereby producing a sequence of matrices that converges to a diagonal form" [22]. The basic tridiagonalization of a symmetric matrix $A = \{a(i, j)\}$ is implemented by means of the similarity transform:

$$W = Q * A * Q^T$$

where W is tridiagonal and Q is orthogonal. Usually, Q consists of the product of $N - 2$ orthogonal matrices $Q^{(p)}$ such that $Q = Q^{(N-2)} * Q^{(N-3)} * \cdots Q^{(2)} * Q^{(1)}$ and $Q^{(p)}$ causes the $(N - p - 1)$ lower elements in the pth column of A to be set to zero. Similarly, $[Q^{(p)}]^T$ causes the $N - p - 1$ rightmost elements in the pth row of A to be set to zero. The same constraint of localized communications discussed above motivates use of Givens rotations on the matrix for tridiagonalization. In essence, the operator $Q^{(p)}$, described above, is again broken down into a sequence of finer operators, $Q^{(q,\,p)}$ where each operator annihilates the element $a(q, p)$. Thus $Q^{(p)} = Q^{(p+2,\,p)} * Q^{(p+3,\,p)} * \cdots * Q^{(N,\,p)}$. Each operator $Q^{(q,\,p)}$ is of the form

$$
\begin{array}{c}
\text{columns: } \quad q-1 \quad\;\; q \qquad\qquad\qquad \text{rows:} \\[4pt]
Q^{(q,\,p)} =
\begin{bmatrix}
1 & & & & & & \\
& 1 & & & & & \\
& & C(q, p) & S(q, p) & & q-1 \\
& & -S(q, p) & C(q, p) & & q \\
& & & 1 & & \\
& & & & & \\
& & & & & 1 \\
\end{bmatrix}
\end{array}
$$

Of major importance are the following facts:

1. The premultiplication of A by $Q^{(q,\,p)}$ modifies only rows $q-1$ and a of A. The elements of those two rows assume the following values after applying the rotation.

$$
\begin{bmatrix} a'_r(q-1) \\ a'_r(q) \end{bmatrix} = \begin{bmatrix} C(q, p) & S(q, p) \\ -S(q, p) & C(q, p) \end{bmatrix} \begin{bmatrix} a_r(q-1) \\ a_r(q) \end{bmatrix}
$$

where a'_r represents the row vector containing the elements of row k of matrix A. Note that $a'(q, p) := 0$.

2. The effect of postmultiplying $A_2 = (Q * A)$ by Q^T is to modify the elements of columns $q-1$ and q of A_2 to assume the following values:

$$
a''_c(q-1)a''_c(q) = a'_c(q-1)a'_c(q) \begin{bmatrix} C(q, p) & -S(q, p) \\ S(q, p) & C(q, p) \end{bmatrix}
$$

where $a''_c(k)$ represents column vectors of matrix A_2. As A was symmetric, this operation is largely a repetition of many of the row operations effected in the $Q * A$ process. The exceptions are the four elements located at the junction of rows and columns q and $q-1$.

3. The sequencing of operations is not quite as rigid and therefore allows for pipelining of wavefronts.

When taking the wavefront viewpoint of the operations, two types of waves are discernible. The first is an advancing wave, related to the row operations up to the diagonal elements and referred to as the "row wavefronts." The second involves computation in the junction regions and the column operations and can be seen as a reflected wavefront along the diagonal. These are dubbed the "column wavefronts." (Taking advantage of the symmetry of the problem, we can delete those PEs above the main diagonal, retaining a triangular array without losing any information.)

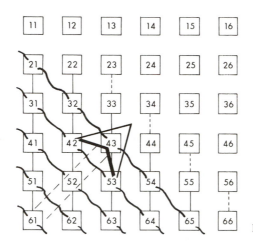

Figure 7.6 Row-wavefront propagation.

1. *Row-wavefront propagation.* The wavefront nature can be seen in Figure 7.6, which traces the fronts of activity relating to the row operations involved in annihilation of the elements of the first column. These fronts will be called row wavefronts.

2. *Column-wavefront propagation.* Figure 7.7 shows the somewhat similar sequencing of the column wavefronts and their propagation. The first column wavefront can be initiated *when and only when* the first row wave reaches the end of its travel (i.e., the last two elements of the last two rows). Its first task corresponds to iterating columns N and $N - 1$ through operator $[Q^{(N, 1)}]^T$. The column wave can advance by one stage when the row wave has operated on the last elements of rows $N - 1$ and $N - 2$. In the evolution of the computations, *row operations applied to rows q and q − 1 must terminate before the corresponding column operations are initiated.* This is due to the fact that column operations require data that are the outcome of the row operations. By the same token, the column operations corresponding to annihilation of the $(N - p - 1)$ elements of row p (column wave p) must

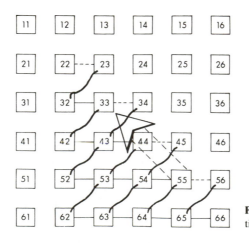

Figure 7.7 Column-wavefront propagation.

terminate before the row operations relating to the annihilation of column $p + 1$ [row wavefront $(p + 1)$] may commence.

From Figures 7.6 and 7.7, one can also see that essentially at most two PEs in each column are actively executing rotation oriented operations at any time instance. We therefore propose to apply the same procedure described above, utilizing a linear or bilinear array of processing elements. The computational tasks are assigned to the PEs so that first-row PEs perform parameter generation and row rotations, while second-row PEs execute column rotations.

Although the physical configuration of the processor array has changed from square to linear, the nature of the computational wavefront has not, and the theoretical propagation of computational activity is retained. Thus we have a square array virtual configuration mapped into a bilinear array actual machine. We also note that the bilinear array will yield the same $O(N^2)$ execution time as the square (or triangular) array, thus proving that the latter is unnecessary.

Among the most popular methods for determining the eigenvalue of a symmetric tridiagonal matrix is the iterative diagonalization scheme. It employs a series of similarity transformations which retain the symmetricity and bandwidth of the matrix, while reducing the off-diagonal norm and converging to a diagonal matrix, the elements of which are the sought eigenvalues. This algorithm can be easily mapped into a wavefront processing on a linear array. For details, we refer to [12,13].

Due to space limitation, we have to defer the detailed discussion on wavefront processing for eigenvalue and singular-value decomposition [13] and Toeplitz system solver [14,15] to the full presentation. Also of interest is its application to high-resolution beamforming problems (see, e.g., [20]).

7.4.5 Applications of Wavefront Processing

The notion of wavefront processing is applicable to all algorithms that possess recursivity and locality. Such algorithms can be roughly classified into three groups:

1. *Basic matrix operations* [7,13]: such as (a) matrix multiplication, (b) L-U decomposition, (c) L-U decomposition with localized pivoting, (d) Givens algorithm, (e) back substitution, (f) null-space solution, (g) matrix inversion, (h) eigenvalue decomposition, and (i) singular-value decomposition

2. *Special signal processing algorithms* [7,14,15]: (a) Toeplitz system solver, (b) linear convolution, (c) recursive filtering, (d) circular convolution filtering, and (e) DFT

3. *Other algorithms* [12]: (a) PDE (partial difference equation) solution, (b) sorting, and so on

7.4.6 Wavefront Language and Architecture

The wavefront concept provides a firm theoretical foundation for the design of highly parallel array processors and concurrent languages, and it appears to have two distinct features.

1. As to the architecture aspects, the wavefront notion leads to a wavefront-based architecture which preserves Huygens' principle and ensures that wavefronts never intersect. More precisely, the information transfer is by mutual convenience between each PE and its immediate neighbors. Whenever the data are available, the transmitting PE informs the receiver of the fact, and the receiver accepts the data when it needs it. It then conveys to the sender the information that the data have been used. This scheme can be implemented by means of a simple handshaking protocol [7,12,18]. The wavefront architecture can provide *asynchronous waiting* capability, and consequently, can cope with timing uncertainties, such as local clocking, random delay in communications, and fluctuations of computing times [12,17–19]. In short, the notion lends itself to an (asynchronous) data-driven flow computing structure that conforms well with the constraints of VLSI [7,16].

2. As to the language aspect, the wavefront notion helps greatly reduce the complexity in the description of parallel algorithms. The mechanism provided for this description is a special-purpose, wavefront-oriented language, termed matrix data flow language (MDFL) [7]. The wavefront language is tailored toward the description of computational wavefronts and the corresponding data flow in a large class of algorithms (which exhibit the recursivity and locality mentioned earlier). Rather than requiring a program for each processor in the array, MDFL allows the programmer to address an entire front of processors. In contrast to the heavy burden of scheduling, resource sharing, and the control of processor interactions often encountered in programming a general-purpose multiprocessor, the wavefront notion can facilitate the description of parallel and pipelined algorithms and drastically reduce the complexity of parallel programming.

A complete list of the MDFL instruction repertoire and the detailed syntax can be found in earlier publications [7,12]. For an example, a complete (global) MDFL program for matrix multiplication follows:

GLOBAL MDFL PROGRAM FOR MATRIX MULTIPLICATION

Array size: $N \times N$

Computation: $C = A \times B$

  kth wavefront: $c_{ij}^{(k)} = c_{ij}^{(k-1)} + a_{ik} b_{kj}$ $k = 1, 2, \ldots, N$

Initial: Matrix A is stored in the memory module (MM) on the left (stored row by row). Matrix B is in MM on the top and is stored column by column.

Final: The result will be in the C registers of the PEs.

```
1:      BEGIN
        SET COUNT  N;
        REPEAT;
          WHILE WAVEFRONT IN ARRAY DO
5:        BEGIN
              FETCH  A, LEFT;
              FETCH  B, UP;
```

```
                         FLOW   A, RIGHT;
                         FLOW   B, DOWN;
                            (* Now form C: = C + A × B)
     10:                 MULT   A, B, D;
                         ADD    C, D, C;
                    END;
                 DECREMENT COUNT;
              UNTIL TERMINATED;
     15:      ENDPROGRAM.
```

The power and flexibility of the MDFL programming on the WAP is best demonstrated by the broad range of algorithms that can be programmed in MDFL [7]. For more programming examples, see [7,12].

7.5 SOFTWARE TECHNIQUES AND CAD TOOLS FOR DESIGNING VLSI ARRAYS

A major advance in the software and CAD techniques will be indispensable to the development of massively parallel array processors. From an architectural perspective, modularity, communication, and system clocking are critical to the design of VLSI arrays. From an algorithmic point of view, a large class of algorithms in signal processing applications possess the properties of recursiveness and locality. Such a joint algorithmic and architectural consideration forms a basis for an advanced software and CAD techniques for VLSI array processors.

The algorithmic properties naturally lead to the systolic and wavefront arrays which effectively utilize a large number of modular and locally interconnected VLSI processors. Next, for a flexible and cost-effective array design, it is necessary to thoroughly examine the software requirements pertaining to the array processors. One major issue is how to express parallel algorithms in a powerful and easily understandable notation and yet possible to compile into efficient VLSI circuits and/or machine codes. For this, we propose a methodology which allows a hierarchical and recursive mapping from algorithms to Signal Flow Graph (SFG) network arrays, using an abstract and powerful notation. We also propose to look into the transformation procedures, which systematically convert SFG's into systolic or wavefront arrays and eventually their VLSI hardware implementations.

The main issue considered in the HIFI design system [24] is how to express parallel algorithms in a notation which is easy to understand by humans and possible to compile into efficient VLSI circuits and/or array processor machine codes. Essentially, the structural properties of parallel recursive algorithms point to the feasibility of a hierarchical and iterative flow-graphic design method of VLSI array processors. The HIFI design method will be based on this hierarchical, iterative and recursive mapping. First the algorithmic specification notation is required to be able to express the parallel space-time activities occurring in signal-processing algorithms in a clear and simple way. From this algorithmic specification, a detailed structural specification, based on instances of predefined primitive modules, will be derived.

As mentioned earlier, a taxonomical analysis points to two major classes of parallel algorithms for signal processing: i.e. the "locally recursive" and the "perfect shuffle" classes, both using the divide and conquer recursive computing schemes. This structural property motivates the idea of HIFI, which involves a close interplay between the recursive algorithm decompositions, abstract notations, flow-graph (structural) and functional (behavior) descriptions, transformation procedures, bi-directional mapping between graphic and textural codes, simulations and verifications, and silicon compilations.

The HIFI design system will offer the designer the definitional mechanisms for (1) temporal decomposition and (2) structural decomposition. Using these two basic decomposition techniques, the designer can do stepwise refinements in the hierarchical design approach [24]. The final HIFI design/description system will offer the following desirable features:

1. *Expressiveness*: The HIFI system must support a notation for describing the parallel space-time activities occurring in signal processing algorithms in a natural way.
2. *Freedom from detail*: Information hiding, i.e. hierarchical specifications and a separation between virtual and real machine design, will allow the designer to focus his attention at the appropriate level of detail.
3. *Simplicity*: The definitional kernel of the design system, i.e. the design objects and the operations allowed on them, must be simple. However, care must be taken to ensure that the operations are sufficiently powerful.

In addition to the algorithmic principles, the HIFI description language highlights an applicative functional specification. Here the term "applicative" simply means that there is no history sensitivity. (In another word, there is no storage and, therefore, no side-effects.) In fact, the HIFI design uses an extended functional programming (FP) system. (An FP system is a simple applicative system, in which "programs" are simply functions without variables and is founded on a fixed set of combining forms called functional forms.) In the HIFI description, The nodes the SFG network are of Formal Functional Programming (FFP) type [25], enhanced with a notion of state (i.e. history sensitivity). In a history-sensitive language, a program can affect the behavior of a subsequent one by changing some store which is saved by the system [25]. This motivates the model of applicative state transition system (AST). In the HIFI approach, a complete SFG network should then be viewed as a collection of AST systems, allowing the designer to think in terms of multiple units of tasks in parallel.

Speaking in a loose (but rather illustrative) term, we can say that an SFG representation, consisting of time-freezed (instant) nodes and delay arcs, suffices to represent the collection of AST systems. Furthermore, the SFG displays explicitly the structural information of a VLSI array. The focal issues of the HIFI design methodology are summarized below [27].

7.5.1 Mapping Algorithms onto Arrays

In most signal processing applications, several basic algorithms will be repetitively used with a large volume of data, including FFT, correlation, and matrix multiplication, inversion, and least-square solver. What is needed therefore is a thorough taxonomy analysis on the parallel algorithms most commonly encountered in the applications. For example, two prominent classes may be identified: one is the "locally recursive" class; the other is the "perfect shuffle" class. Both these classes belong to the important divide and conquer recursive computing schemes. This will eventually lead to a hierarchical and recursive description method.

Algorithmically, we have stressed the recursive nature of the signal processing algorithms and the SFG offers a tool directly expressing the sequencing of the recursions. More precisely, the input/output of nodes in an SFG indicates the order of computations within the same recursion, while a delayed arc denotes the separation (and ordering) between two consecutive recursions. This viewpoint matches well with that of Backus [25], in which he advocates a liberation of programming from the von Neumann style, (using the so-called Applicative State Transition systems).

Therefore, the HIFI approach offers an effective starting point for the design automation and software/hardware techniques. This is because (1) SFG provides a powerful (although mathematical) abstraction to express parallelism, and yet (2) transforming from SFG to (the more realistic) systolic/wavefront arrays is straightforward [27].

7.5.2 Flow-graphic Network based on SFG and AST Systems

A graphic notation based on the SFG is adopted. The nodes are conceptually delay-free and are specified by applicative functional specifications. The time-delay arcs and functional nodes combine to form a collection of applicative state transition (AST) systems [25, 26, 28]. Together they fully express the sequencing within and between recursions. Note that the delay operator D is now viewed as a sequencing operator, (as opposed to a timing device), and can be implemented either as globally clocked delay or self-timed data-driven separator (cf. the Equivalence Transformation Theorem in [27], Section V).

7.5.3 Recursive Structures and States

The recursive property of the algorithms leads naturally to a useful notion of "state" (history sensitivity), compatible to what is used in the AST systems. Briefly speaking, an applicative type mapping is often conveniently decomposed into a recursively defined "sub-mappings". The notion "state" represents the necessary and sufficient information summarizing the history of the activities. With reference to Fig. 2, the state is naturally associated with the values or objects carried by the delay-edges.

The same recursive decomposition scheme introduces certain primitive interconnection structures. Moreover, the semantics of the notations for such structures can be made very concise and precise. (The semantics of the structural primitives will also permit further structural refinements.) For an example, the concept of local recursiveness suggests locally interconnected structural primitives, applicable to most signal processing algorithms such as convolution, correlation, LU and QR decompositions.

7.5.4 Hierarchical Description

To ease the description of large-scale complicated networks, a hierarchical flow-graph design methodology proves to be very effective. A node is hierarchically specified using either a behavioral (applicative type) description or a flow-graph (structural type) description [29,30]. In order to have an executable specification, the primitive level nodes must be specified in terms of executable behavior modules. By varying the level of primitive modules the designer can select the level of detail according to his own preference. The structural descriptions will be given by output $->$ input mappings, while the behavioral (primitive level) descriptions will be specified by input $->$ output mappings.

The hierarchical description involves basically successive step-by-step refinements of the functional nodes. The power of such description may be exemplified by the available option of replacing a node by a linear array of nodes, thereby creating a new dimension of the array. As an illustrative example, an LU- decomposition array (cf. Fig. 11 in [27]) has in fact a (two-level) hierarchical representation: The first level being a horizontal one-dimensional array with N nodes and, at the second level, each of the nodes being further decomposed into N subnodes, resulting in the array as in Fig. 11 in [27]. The same scheme is also directly applicable to the computations for QR decomposition, and two- (or three-) dimensional image correlations, etc.

7.5.5 Transformations to VLSI Arrays with Pipelining, Partitioning, and Fault-Tolerance

In an earlier paper [27], it develops some powerful transformation procedures to systematically convert SFG's into pipelined structures, such as systolic or wavefront arrays, and eventually their VLSI hardware implementations. These include an automatic (cut-set) systolization procedure, optimal scheduling, and wavefront handshaking, etc. More importantly, the hierarchical and applicative approach offers the simplicity and flexibility necessary to deal with the issue of partitioning a larger size problem to fit into smaller arrays and/or imposing fault tolerance into arrays. These additional flexibilities will be critical to the effectiveness of the arrays.

The cut-set rules are also potentially very useful for designing fault-tolerant arrays. For systolic arrays without feedback, it has been shown in the literatures that a retiming along cut-sets allows a great degree of fault tolerance. The theoretical treatment in [27] (Section III-B) offers a theoretical basis for improving fault-tolerance of arrays with feedback via the cut-set retiming procedure. More interestingly, with a slight modification, the self-timed feature of wavefront arrays offers a way of achieving the same fault-tolerance efficiency without any need of retiming.

7.5.6 CAD Tools and Graphics-Based Design Systems

The main reason for having a graphics based approach lies in that it provides a simple communication language for humans (one picture is worth a thousand words). The human information processing capability makes it possible to identify at once the basic structure of an image. Together with this goes the fact that many typical constructs, like for instance 'array', have a natural graphical denotation. Especially, for the HIFI design, many array transformations are originally graphic based [24]. More importantly, graphics work stations are undergoing a rapid evolution, making them more available and more user-friendly. The development of graphics standards, like for instance GKS, has made the development of device independent software possible.

The graphics command language, mostly for the structural specifications, will be a simple applicative type language similar to the FP-type languages defined by Backus. In doing so the bidirectional mapping between graphics and textural-code becomes relatively simple. The textural-code is mainly meant for formal specification of the design. The other important roles of the design language, i.e., those of design tool and as a device for human communication, are taken care of by the graphics interface of the HIFI system.

When using the graphics work station, a HIFI designer will see the design object, say, an SFG network. He will select part of it, e.g., a node or a branch, and then apply an operation to the selected objects. This results in a modified design, whose formal specification is modified from the formal specification of the older design.

As an extension of the graphics command language there will be various editing functions. These editing functions allow the designer to create different versions, or alternatively change the current version of a computing array. The definition of the HIFI Data Base, containing all the design information, including design history, will be the starting point for the development of all software tools, including the graphics design tools just mentioned. In developing the graphic tool, our emphasis will be placed on the bi-directional mappability between the flow-graphic and the textural constructs. As an example, the name "array" will appear in both the graphic and textural constructs, facilitating the mappings. The definition of a Structural Description Language (SDL) for SFG's will be useful for the documentation of the design steps [29], as well as for interfacing the HIFI system with lower-level silicon compiler-like systems. The mapping from and to textual form should be one of the functions provided by the HIFI system.

Summary

In summary, for the hierarchical flow-graphic CAD pertaining to the design of VLSI arrays. From a software consideration, the proposed approach paves a way for applicative type programming techniques. These techniques, such as the applicative state transition (AST) systems, the data space or wavefront languages, which will have a major influence on VLSI array software. Most importantly, based on the above guidelines, a set of complete and integrated graphic based CAD hardware/ software synthesis packages, along with the simulation tools, may be developed. These would allow the user to simulate the design in different levels by the increasingly versatile and economic graphics design work stations.

7.6 CONCLUDING REMARKS

The advent of VLSI device components has precipitated the upgrading of signal processor technology, from the conventional transversal filtering to two-dimensional array processing. We have presented the evolution of VLSI architectures on an algorithmic analysis footing. Via a notion of wavefront processing, the data-driven computing scheme is shown to be naturally suitable for many signal processing algorithms. With respect to architectural design, this data-driven feature is especially noteworthy, as it avoids the need of global synchronization, a potential barrier in the design of ultralarge scale VLSI systems. There are already studies comparing synchronous versus asynchronous VLSI systems [14,15], and it appears that asynchronous processing will definitely become more favorable for extremely large systems. However, more study will be needed to determine the crossover threshold of array size. With the future wafer-scale integration technology, interconnections will often have to be reconfigured to permit rerouting around faulty PEs. This is bound to create more timing uncertainty, thus making the more flexible wavefront processing very appealing compared to the pure systolic processing. In a sense, the wavefront array blends the advantages of the globally synchronized systolic array and the general-purpose asynchronous data-flow organization.

Finally, the top-down-design methodology calls for closely coordinated design of algorithms, language, and architecture. In a wavefront array, the algorithm analysis dictates the language structure, which in turn determines the architecture of the computing network. A natural, although ambitious next stage could be the task of mapping the MDFL language structure into silicon compilers (see, e.g., [23]), further advancing modern CAD tools for VLSI systems.

In conclusion, the prospect of constructing high-speed (programmable) parallel signal processors using VLSI wavefront arrays appears to be very promising.

ACKNOWLEDGMENTS

The author wishes to acknowledge the major contributions to the development of the wavefront array processor by Drs. R. J. Gal-Ezer and Y. H. Hu, D. V. Bhaskar Rao, formerly of the University of Southern California, and K. S. Arun, and G. Sharma, of the University of Southern California.

Research supported in part by the Office of Naval Research under Contract N00014-81-K-0191; and by the National Science Foundation under Grant ECS-82-12479.

REFERENCES

[1] C. Mead and L. Conway, *Introduction to VLSI Systems*, Addison-Wesley, Reading, Mass., 1980.

[2] A. Oppenheim and R. Schafer, *Digital Signal Processing*, Prentice-Hall, Englewood Cliffs, N.J., 1975.

[3] L. R. Rabiner and B. Gold, *Theory and Application of Digital Signal Processing*, Prentice-Hall, Englewood Cliffs, N.J., 1975.

[4] T. Kailath, *Linear Systems*, Prentice-Hall, Englewood Cliffs, N.J., 1980.

[5] H. T. Kung, "Let's Design Algorithms for VLSI Systems," *Proc. Caltech Conf. VLSI*, Jan. 1979, pp. 65–90.

[6] H. T. Kung, "Why Systolic Architectures?" *IEEE Computer*, 15(1):37–46 (Jan. 1982).

[7] S. Y. Kung, K. S. Arun, R. J. Gal-Ezer, and D. V. Bhaskar Rao, "Wavefront Array Processor: Language, Architecture, and Applications," *IEEE Trans. Comput., Special Issue on Parallel and Distributed Computers*, C-31(11):1054–1066 (Nov. 1982).

[8] J. B. Dennis, "Data Flow Supercomputers," *IEEE Comput.*, Nov. 1980, pp. 48–56.

[9] K. P. Gostelow and R. E. Thomas, "Performance of a Simulated Data Flow Computer," *IEEE Trans. Comput.*, C-29(10):905–919 (Oct. 1980).

[10] S. Y. Kung, "VLSI Array Processor for Signal Processing," *Conf. Adv. Res. Integrated Circuits*, MIT, Cambridge, Mass., Jan. 28–30, 1980.

[11] J. M. Speiser and H. J. Whitehouse, "Architectures for Real Time Matrix Operations," *Proc. GOMAC*, Nov. 1980.

[12] R. J. Gal-Ezer, "The Wavefront Array Processor and its Applications," Ph.D. dissertation, University of Southern California, Dec. 1982.

[13] S. Y. Kung, and R. J. Gal-Ezer, "Eigenvalue, Singular Value and Least Square Solvers via the Wavefront Array Processor," In L. Snyder et al., eds., *Algorithmically Specialized Computer Organizations*, Academic Press, New York, 1983.

[14] S. Y. Kung and Y. H. Hu, "A Highly Concurrent Algorithm and Pipelined Architecture for Solving Toeplitz Systems," *IEEE Trans. Acoust. Speech Signal Process.*, 31(1) (Feb. 1983).

[15] Y. H. Hu, "New Algorithms and Parallel Architectures for Toeplitz System: With Applications to Spectrum Estimations," Ph.D. dissertation, University of Southern California, Dec. 1982.

[16] S. Y. Kung and R. J. Gal-Ezer, "Hardware Architectures of the Wavefront Array Processor," *Proc. Int. Comput. Symp.*, Taiwan, Dec. 1982.

[17] M. Franklin and D. Wann, "Asynchronous and Clocked Control Structures for VLSI Based Interconnection Networks," *Proc. 9th Annu. Symp. Comput. Architecture*, Austin, Tex., Apr. 1982.

[18] S. Y. Kung and R. J. Gal-Ezer, "Synchronous vs. Asynchronous Computation in VLSI Array Processors," *Proc. SPIE Conf.*, Arlington, Va., May 1982.

[19] A. L. Fisher and H. T. Kung, "Synchronizing Large Systolic Arrays," *Proc. SPIE Conf.*, Arlington, Va., May 1982.

[20] N. L. Owsley, "High Resolution Spectrum Analysis by Dominant Mode Enhancement," in S. Y. Kung, T. Kailath, and H. J. Whitehouse, eds., *Modern Signal Processing and VLSI Architectures*, Prentice-Hall, Englewood Cliffs, N.J., 1983.

[21] S. Y. Kung, ed., "VLSI and Modern Signal Processing," *Proc. USC Workshop on VLSI Mod. Signal Process.*, Los Angeles, Nov. 1–3, 1982.

[22] B. N. Parlett, *The Symmetric Eigenvalue Problem*, Series in Computational Mathematics, Cleve Moler (Adv.), Prentice-Hall, Englewood Cliffs, N.J., 1980.

[23] P. B. Denyer and D. Renshaw, "Case Studies in VLSI Signal Processing Using A Silicon Compiler," *Proc. IEEE ICASSP*, Boston, 1983, pp. 939–942.

[24] S. Y. Kung, J. Annevelink, and P. Dewilde, "Hierarchical Iterative Flow-graph Design for VLSI Array Processors." *IEEE International Conference on Computer-Aided Design*, Santa Clara, Calif., Nov. 1984.

[25] J. Backus, "Can Programming Be Liberated from the Von Neumann Style? A Functional Style and Its Algebra of Programs." *Communications of the ACM 21* (1978), 613–641.

[26] A. B. Cremers, and T. N. Hibbard, "Problem-Oriented Specification of Concurrent Algorithms." *Proceedings of U.S.C. Workshop on VLSI and Modern Signal Processing*, Los Angeles, Calif., Nov. 1982, pp. 64–68.

[27] S. Y. Kung, "On Supercomputing with Sytolic/Wavefront Array Processors." *Invited paper, Proceedings of the IEEE Vol. 72*, 7 (July 1984).

[28] A. B. Cremers, and S. Y. Kung, "On Programming VLSI Concurrent Array Processors." *Proc. IEEE Workshop on Languages for Automation*, Chicago, 1983, pp. 205–210, also in *INTEGRATIONS*, the VLSI Journal, Vol. 2, No. 1, March, 1984.

[29] W. Y-P. Lim, "HISDL-A Structure Description Language." *Communications of ACM 25*, 11 (Nov. 1982).

[30] W. Y-P. Lim, and C. K. C. Leung, "Computer Hardware Description Language and their Applications." In, T. Uehara and M. Barbacci, Eds., North Holland Pub., 1983, pp. 233–242.

8

Special-Purpose VLSI Architectures: General Discussions and a Case Study

ALLAN L. FISHER AND H. T. KUNG

Carnegie-Mellon University
Pittsburgh, Pennsylvania

8.1 INTRODUCTION

> It becomes apparent to one involved in pursuing research on new computer organizations that we are opening up a new era in digital data processing. It also becomes apparent that we may appear to be retrogressing to the concept of special-purpose computing equipment—and we believe this is going to be the case. In the future, we will see several lines of special-purpose but programmable computing devices for military and business. As the application of machine computation becomes even more widespread than today, it will become economical to market and to utilize special-purpose devices for performing operations such as file searching rather than utilizing time on a general-purpose computer.
>
> —excerpted from the preface of *Computer Organization*, 1963 [2]

A fundamental factor in any tool-building enterprise is the trade-off between generality and performance. In almost all cases, a general-purpose tool is less cost-effective in a particular case than is a tool designed especially for that case. In practice, this trade-off is usually decided by economic factors: the extra (per application) design cost of a special-purpose device is balanced against the lower efficiency of a general-purpose device.

In the area of signal processing, this trade-off often comes down on the side of special-purpose devices because of stringent real-time requirements. Computations requiring hundreds of millions of multiplications per second are common, and general-purpose systems with this speed cost millions of dollars apiece. Luckily, many signal processing tasks can be constructed from a relatively small set of primitives, so a special-prupose *building-block* approach can present a happy medium along the general purpose–special purpose spectrum.

In an equally important way, the appeal of the special-purpose approach is affected by implementation technology. VLSI has special advantages in this regard—transistors are cheap, and the availability of large functionality with small package count can dramatically lower the implementation and operating costs of many systems relative to SSI or MSI implementation. In terms of performance, VLSI technology promotes the use of massive homogeneous parallelism, since the incremental cost of replicating an existing processing element design is very low.

In practicality, however, the special-purpose approach has been more difficult than many people realize. Special-purpose computing devices have been proposed since the early days of digital computers, as evinced by the quotation in the beginning of this section, but only recently has technology begun to make special-purpose computers broadly practical. This chapter takes a two-pronged approach to the issues involved in the design and application of such devices. First, we discuss in a general way the factors that are critical to the success of the special-purpose approach in computation. Second, we illustrate the impact of these factors in the context of the development of a particular special-purpose architecture—the CMU *programmable systolic chip*.

8.2 FACTORS FOR SUCCESS

The special-purpose hardware approach can be successful only if certain criteria are met. We identify some of the criteria that we believe are among the most important.

The basis of our interest in the special-purpose approach is that of the dramatic *advances in hardware technology* in the last 20 years; without LSI or VLSI technology we would essentially have no alternative but to rely on general-purpose machines. It is the new technology that opened up the possibility of implementing many large systems directly in hardware in the first place. It is also the rapidly changing nature of technology that makes the task of mapping systems onto hardware always a challenging one. Although the semiconductor industry has, by and large, been driven by the needs of low-cost memory and processors for general-purpose systems [47], the technology developed can be applied to the implementation of special-purpose systems as well. Thus special-purpose system designers are often in a somewhat fortunate position of exploring the use of some existing technology rather than being involved in the invention of a new technology.

Availability of a technology establishes the basic feasibility for implementing special-purpose systems, but quick implementation in the technology must be easily accessible to system designers. In general, the development cycle for implementing a special-purpose system and using it to produce significant results must be short; otherwise, there is a risk that the system will never be used. First, its components may become obsolete before the system is operational, and thus its performance advantage over general-purpose systems may no longer exist. Second, application requirements for a special-purpose system may change during a long development cycle. *Fast-turnaround implementation* is therefore important for the special-purpose approach. Various CAD tools developed in recent years are essential in this regard. Moreover, for VLSI implementation, the development of simplified design rules and structured design methodology [45] have allowed system designers who are relatively naive in IC design to design their own chips quickly and confidently. Furthermore, structured approaches to testing and debugging [24] can facilitate the design validation and evaluation task.

Given that technology and CAD tools are available for quick, low-cost implementation, we next face the question of deciding what to implement in hardware for any given application area. The area must be mature enough so that there is enough experience to identify critical computation routines whose special-purpose implementation will greatly enhance total system performance. We must be convinced that there are no algorithmic or software solutions which could be more easily applied than a special-purpose hardware approach. We must make sure that the inner loops can be easily decoupled from the rest of the computation so that the interface between the host and the future special-purpose device will not be a bottleneck. Moreover, the inner loops must be "stable," in the sense that they will be used for a long period of time, to warrant the effort of hardware implementation. *Well-understood applications* are therefore important for providing the necessary motivation for special-purpose designs. One means of gauging the degree of our understanding of the computational needs of a particular area is to examine the level of hardware building blocks that have been proposed or built. Availability of building blocks implementing high-level system functions is an indication of the maturity of an application area. By this measure, signal processing is a relatively well-understood area—commercially available special-purpose chips, such as the NEC digital signal processor chip [46,51,60], AMI FFT chips, TRW multiplier chips, and TRW A/D converter chips, implement functions at a wide range of system levels.

After identifying tasks to be implemented in VLSI, *new algorithms* with high degrees of parallelism and regularity, with low communication overheads, and adaptable to various I/O constraints imposed by the outside environment have to be developed [33]. For VLSI implementation it is no longer important to minimize the number of multiplications, as people used to do; appropriate complexity measures should reflect trade-offs among chip area, time, and power [56]. The silicon area needed for an algorithm is heavily influenced by its degree of regularity [43], and the time required is equally heavily influenced by its degree of parallelism. Many classical problems will thus require a new set of algorithms. Examples are Gaussian elimination [25,36], GCD computation [7], and the Wiener–Levinson recurrence [9,40].

To implement algorithms in VLSI, appropriate architectures must be developed. *Architectural optimization* for the implementation of a narrow set of algorithms is usually not very difficult, but designing an architecture that can efficiently support a wide class of algorithms is a nontrivial task. The problem is to achieve a balance among many conflicting goals, such as the generality of the system versus ease of programming, flexibility versus efficiency, and performance of the system versus its design and implementation costs. "Optimal architectures" are functions of the technology that system implementation will be based on and the objectives that the system intends to achieve; one should avoid deluding oneself about the "universality" of a particular architectural approach. Plenty of interesting architectural ideas have been proposed or implemented; the challenge is to understand precisely the strengths and drawbacks of each approach so that a suitable architecture can be selected for a given environment.

The ultimate goal of any special-purpose system must be its effective use in a complete target system. Before such a special-purpose system is built, its specification must be evaluated in the context of the target system, and after being built it must be integrated into the target in such a way as to be easily used. For example, the use of a sorting chip requires extra memories to buffer intermediate data, a general-purpose host computer with software to control the sorting process, and hardware and software to interface the chip to the rest of the system. Without proper hardware and software interface aids, incorporating a custom chip into a complete system can be a larger project than the design of the chip itself, and calls for a different set of skills than those required by chip design. The *system integration* problem—now mostly being handled by ad hoc methods—is probably among the least understood of the issues that are crucial to the special-purpose approach. Because of this problem, very few custom chips designed by universities or research laboratories have so far been integrated into complete working systems. The most common method of integrating custom devices into a host system is to build an interface between the device and a standard host bus such as the UNIBUS. The availability of standard interface cards for these buses solves only some of the interfacing problems. Additional interface facilities, including a large memory for buffering frequently used data and interface processors for generating memory addresses for data to and from custom VLSI chips and their controlling signals, are typically needed.

8.3 PROGRAMMABLE SYSTOLIC CHIPS—A CASE STUDY

This section illustrates some of the discussions of the preceding section by considering a CMU design project in which the authors have participated. The project comprises the design and implementation of a *programmable systolic chip* (PSC) [20,21]—a single-chip special-purpose microprocessor designed for the implementation of a variety of *systolic architectures*. The PSC is both a special-purpose chip in the sense that it is designed for the implementation of special-purpose systolic systems and a general-purpose chip in the sense that it is a programmable microprocessor. The PSC example illustrates some of the breadth of scope of special-purpose designs. We first summarize some of the features and applications of systolic arrays, and then describe the PSC and some of the issues raised during its development.

8.3.1 An Overview of Systolic Architectures

Computational tasks can be conceptually classified into two families: compute-bound computations and I/O-bound computations. If the total number of operations in a computation is larger than the total number of input and output elements, the computation is compute-bound; otherwise, it is I/O-bound. For example, the ordinary matrix-matrix multiplication algorithm represents a compute-bound task, since every entry in a matrix is multiplied by all entries in some row or column of the other matrix. Adding two matrices, on the other hand, is I/O-bound, since the total number of adds is not larger than the total number of entries in the two matrices. It should be clear that any attempt to speed up an I/O-bound computation must rely on an increase in memory bandwidth. Memory bandwidth can be increased by the use of either fast components (which could be expensive) or interleaved memories (which could create complicated memory management problems). Speeding up a compute-bound computation, however, may often be accomplished in a relatively simple and inexpensive manner: that is, by the systolic approach.

A systolic system consists of a set of interconnected *cells*, each capable of performing some simple operation. Because simple, regular communication and control structures have substantial advantages over complicated ones in design and implementation, cells in a systolic system are typically interconnected to form a systolic array or a systolic tree. Information in a systolic system flows between cells in a pipelined fashion, and communication with the outside world occurs only at "boundary cells." For example, in a systolic array, only those cells at array edges may be I/O ports for the system.

The basic principle of a systolic architecture, a systolic array in particular, is illustrated in Figure 8.1. By replacing a single processing element (PE) with an array

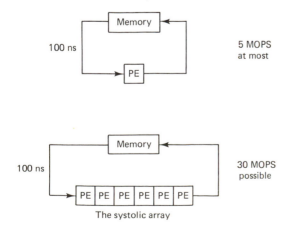

The systolic array

Figure 8.1 Basic principle of a systolic system.

of PEs, or cells in the terminology of this article, a higher computation throughput can be achieved without increasing memory bandwidth. The function of the memory in the diagram is analogous to that of the heart; it "pulses" data (instead of blood) through the array of cells. The crux of this approach is to ensure that once a data item is brought out from the memory it can be used effectively at each cell it passes while being "pumped" from cell to cell along the array. This is possible for a wide class of compute-bound computations where multiple operations are performed on each data item in a repetitive manner.

Although the pulsing of data is synchronous at the algorithm level [30], a machine implementing a systolic algorithm may operate either synchronously or asynchronously. Where feasible, synchronous design is usually preferred for reasons of simplicity. Given current technology, global synchronization is generally feasible for all but extremely large two-dimensional systolic arrays; in some cases, the structure of systolic architectures, such as linear arrays, may be exploited to extend the utility of synchronous design even when wire delays are large [19].

Being able to use each input data item a number of times (and thus achieving high computation throughput with only modest memory bandwidth) is just one of many advantages of the systolic approach. Other advantages, such as modular expansibility, simple and regular data and control flows, use of simple and uniform cells, elimination of global broadcasting and fan-in, and (possibly) fast response time, are also characteristic [33].

The following is a list of problems amenable to systolic solutions. The purpose of this catalog is twofold: to provide references for readers interested in exploring a systolic solution to a particular computational problem they might face, and to show the breadth of the approach.

1. Signal and image processing
   a. FIR and IIR filtering and one-dimensional convolution [13,17,29,31]
   b. Two-dimensional convolution and correlation [5,34,37,38,61]
   c. Discrete Fourier transform [29,31]
   d. Interpolation [37]
   e. One- and two-dimensional median filtering [18]
   f. Geometric warping [37]
2. Matrix arithmetic
   a. Matrix-vector multiplication [14,36]
   b. Matrix-matrix multiplication [36,58]
   c. Matrix triangularization (solution of linear systems, matrix inversion) [25,36]
   d. Solution of triangular linear systems [36]
   e. Solution of Toeplitz linear systems [9,40]
   f. QR decomposition (least-squares computations) [6,25,28]
   g. Singular-value decomposition [10]
   h. Eigenvalue problems [11,52]

3. Nonnumeric applications
   a. Data structures—stack and queue [26], searching [4,48,53], priority queue [42], and sorting [42,53]
   b. Graph algorithms—transitive closure [27], minimum spanning trees [3], and connected components [50]
   c. Geometric algorithms—convex hull generation [15]
   d. Language recognition—string matching [22] and regular expression [23]
   e. Dynamic programming [27]
   f. Polynomial algorithms—polynomial multiplication and division [32], and polynomial greatest common divisors [7]
   g. Integer greatest common divisors [8]
   h. Relational database operations [35,41]
   i. Monte Carlo simulation [59]

8.3.2 The Programmable Systolic Chip

Because of their regularity of structure and simplicity of basic components, systolic arrays should be inexpensive to design and implement, particularly in comparison to other structures of equal performance. This advantage is offset, however, by the fact that a given systolic algorithm is usually tailored to a particular application, and hence development costs cannot be amortized over the large number of units typical of a general-purpose processor. It is therefore appropriate to consider a range of points along a generality–performance trade-off curve. Applications with stringent throughput demands, such as many signal processing tasks, will often justify full custom design, while other applications may be handled by an existing, more flexible design. Existing systolic array implementations span this spectrum. Early test implementations [22,38] were full custom single-purpose devices, as is the GEC correlator chip [17]. In an intermediate range of flexibility are the ESL systolic processor [5,61] and a forthcoming ESL systolic chip set for floating-point matrix computations, both of which are programmable for a range of signal processing tasks. At the very general end of the spectrum is the NOSC systolic array test bed [12,54,55], which is assembled from general-purpose microprocessors and provides both one- and two-dimensional array communication structures. Another highly flexible approach, used in the PSC project, is the design of widely applicable building blocks.

Pending the development of design automation systems that can create efficient low-level designs from high-level mathematical specifications, it is useful to have some flexible means of assembling many different types and sizes of systolic arrays from a small number of building blocks. Such a tool, in order to support a wide variety of such algorithms, must provide programmability both within individual cells and at the interconnection level. Note that this configurability need only represent "design-time" rather than "run-time" flexibility; the structure of a systolic array and the behavior of its cells do not vary greatly during a computation.

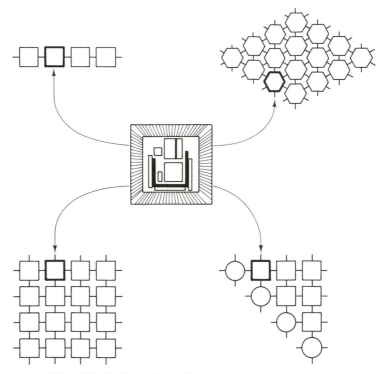

Figure 8.2   Building-block chip for a variety of systolic arrays.

One way of providing this flexibility is with a programmable single-chip processor, as depicted in Figure 8.2. If each chip constitutes one individually programmable cell, then many chips can be connected at the board level (or, in the future, with wafer-scale integration techniques) to build a systolic array. This solution is well suited to current technology, since chips are both big enough and small enough to hold one processor comfortably: big enough that an interesting processor can fit on one chip, thereby avoiding the costs of packaging and off-chip communication inherent to a chip-set design; and small enough that, even given a few factors of 2 increase in circuit density, chip area can profitably be devoted to enhancement of processor functionality rather than to multiple processors with their accompanying problems of configurability and large pinout. Note that two levels of design are involved here: systolic algorithms are designed for an application (e.g., signal processing), and a processor is designed to serve the application of implementing systolic algorithms.

This view was the starting point, in October 1981, of the programmable systolic chip (PSC) project at Carnegie-Mellon University. The goal of the project has been the design and implementation, in 4-μm nMOS, of a prototype PSC which would demonstrate the benefits of the PSC approach, and in addition would itself constitute a cost-effective means for implementation of many systolic arrays. At this writing, in March 1983, the chip design has been fabricated, tested, and debugged, and we expect to perform system demonstrations in the summer of 1983.

As we have noted above, the identification and understanding of the applications to be supported are crucial to the success of a special-purpose computing effort. In the case of a building-block design such as the PSC, a *range* of *target applications* must be chosen. The field of signal and image processing was an obvious candidate; many systolic algorithms for such problems have already been designed [31,36]. Another choice was error detection and correction coding, Reed–Solomon coding [44,49] in particular. In addition, disk sort/merge was chosen as a representative of data processing applications. In practice, the requirements of data processing applications are different enough from the others (in particular, memory is more valuable than multipliers for most such tasks) that a commercially oriented project would probably have concentrated on either numeric or nonnumeric computations; the difference in requirements was not enough, however, to force a research project to deal with one type of application and exclude the other.

During the early stages of the design, high-level CAD tools aided in the synergistic development of algorithms and architecture. The first strawman PSC design was described in ISPS [1], a register-transfer level hardware description language. Using a table-driven microassembler and the ISPS simulation facility, application microcode was written and simulated, providing quick feedback to the design process. The developing architecture, in turn, provided a concrete framework within which realistic systolic algorithms could be formulated; this aided in the development of algorithms for the Reed–Solomon decoding task [7].

As mentioned in Section 8.2, architecture is dependent on technology and on system goals. In the case of the PSC, the design of the architecture interacted in many ways with the implementation technology and the characteristics of the systolic algorithms to be implemented. Some of the issues involved were the following, all of which are important in many special-purpose designs.

1. *Locality and flexibility of control.* For some systolic arrays, including many signal processing and matrix arithmetic algorithms, simple global control would be sufficient. However, many other algorithms require different actions in different phases (e.g., loading of coefficients), as well as data-dependent actions within each systolic cycle. VLSI technology makes it economical to equip each processor with on-chip control store, achieving great flexibility at reasonable cost.

2. *Primitive operations.* The primitive arithmetic, logical, and control operations that a processor can perform are critical to its efficiency. In particular, fast multiplication is needed for the effective implementation of most signal and image processing algorithms. Provisions for multiple-precision arithmetic can extend a processor's utility.

3. *Intercell communication.* A principal feature of systolic arrays is the continuous flow of data between cells. Efficient implementation of such arrays requires wide I/O ports and data paths. This leads to an I/O structure quite different

from that of conventional microprocessors, where several primitive cycles are needed to pass a single word of data onto or off from a chip, and where addressing is a critical issue. Provisions must also be made for the transmission of pipelined *systolic control* signals, which are used to control variant computation phases such as loading of coefficients.

4. *Internal parallelism*. Partition of a processor's function into units that can operate in parallel enhances performance. The circuit density of VLSI allows a multiplicity of separate data paths to fit on a single chip.

5. *Word size*. Individually programmed processors are subject to a trade-off in word size: small word sizes lead to an imbalance between the hardware devoted to control and that devoted to data paths, and large words lead to large chips with large pinout. The solution to this trade-off is highly dependent on technology and application demands.

In brief, these considerations led to the adoption of the following features.

1. Three eight-bit data input ports and three 8-bit data output ports
2. Three one-bit control input ports and three 1-bit control output ports
3. Eight-bit ALU with support for multiple-precision and modulo 257 arithmetic (for Reed–Solomon coding)
4. Multiplier-accumulator with 8-bit operands and 16-bit accumulator
5. 64-word × 60-bit writable control store
6. 64-word × 9-bit register file
7. Three 9-bit on-chip buses
8. Stack-based microsequencer

The importance of fast-turnaround implementation of special-purpose designs is especially great in university environments. Chip fabrication for the project has been supplied by the DARPA MOSIS facility [16]. Another factor in turnaround is the scheduling of the design process. For the PSC, detailed design of some components of the chip was begun concurrently with the refinement of the high-level design. This scheduling had two advantages: it allowed one or two extra people to become involved in the project, thereby shortening the total design time; and it allowed the production of test chips for parts of the design in advance of the whole, thereby simplifying the testing task. It should be pointed out that the resulting "bottom-up" layout is probably somewhat less efficient than a "top-down" layout would have been; in a rapid development environment, a fraction of the chip area may be a reasonable price to pay for a fast design cycle.

We mentioned earlier the utility of CAD tools in architecture design; as useful as such tools are at a high level, they are especially important at the layout level. The layout of the PSC was performed with an interactive graphics editor. Layout design rules were checked by machine, and circuits were automatically extracted from the layout and simulated. Large circuits were simulated at the switch level, and small circuits in the chip's critical delay path were simulated at the circuit level for timing estimation. This approach has been quite successful: the only design errors found in test chips have been small logic bugs that went uncaught because simulation was not extensive enough. In view of the fact that most chip fabrication services take at least several weeks, extensive layout simulation seems definitely worthwhile as a way to avoid extra iterations. An alternative approach to layout validation is comparison of an extracted circuit with a previously validated schematic. Despite the extensive use of CAD tools in the layout of the chip, we often wished we had even more such tools on hand. The lack of good wire-routing programs and some powerful layout validation tools made the task more difficult than it needed to be.

The final phase in the development of a system is integration into an application environment. Since the PSC is a building block usable in many types of systems, no single answer will suffice. Note, however, that devices built from PSCs will not, in general, require complicated control signals or command interfaces, since each chip is internally programmable. As discussed in the preceding section, system integration of high-performance systolic arrays made of PSCs or any other circuits requires certain interface facilities. Figure 8.3 depicts a prototype interface system currently being developed at CMU, with the following features.

1. *High-bandwidth bus.* The I/O bandwidth of systolic arrays or custom VLSI chips is usually much higher than host buses such as the UNIBUS can support. To see the need for high-I/O bandwidth, we observe that a typical one-dimensional systolic array requires about 15 million words per second of I/O bandwidth—each of two input data streams and one output data stream requires 5 million words per second. The assumed 200-ns cycle time is typical for systolic cells that perform multiplications, such as those implemented with the PSC. For solving a complete problem, one can easily use several systolic arrays requiring an aggregate bandwidth of 100 million words per second or more. One or more special high-bandwidth buses are thus needed in the interface system.

2. *Buffer memory.* These memory units are used as buffers between the low-bandwidth host bus and the special high-bandwidth buses in the interface system. By holding data that are to be used repeatedly by the systolic arrays in this memory, the arrays can proceed with high speed, without consuming much host-bus bandwidth. Note that repeated use of data is common whenever a problem is decomposed into subproblems and a given systolic array then processes these subproblems one at a time [39,61].

3. *Systolic processor handler.* This programmable processor serves as the interface between the special high-bandwidth bus and a systolic array. In addition to handling bus protocols, it generates addresses for buffer memory accesses and run-time control signals for the systolic array.

4. *Interface processor.* This processor controls the interface between the host and the interface system. It controls the loading of the buffer memory, and schedules and monitors the computations carried out by the systolic arrays.

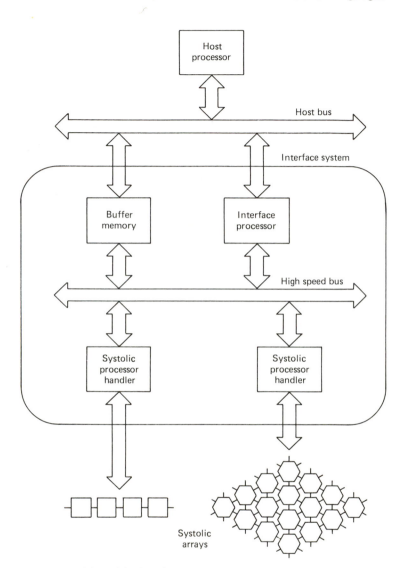

Figure 8.3   Interface system for custom VLSI chips.

8.4 CONCLUDING REMARKS

In summary, we suggest the following as a list of critical elements in the success or failure of a special-purpose design project:

1. *Technology*. Implementation technology has a pervasive effect on special-purpose design. Digital VLSI offers many advantages for signal processing, in

particular the feasibility of massive parallelism. Other technologies, such as integrated optoelectronics [14,57], offer different opportunities and trade-offs.

2. *Well-understood applications.* In order to justify the investment in special-purpose design, an application area must have a known set of widely encountered computational bottlenecks. Signal processing offers an ample supply of such tasks, which encourages the design of system building blocks. We expect such building blocks to play an important part in future special-purpose system design.

3. *Algorithms.* The algorithms that are most effective in a general-purpose environment may not be so in a special-purpose system. Algorithm efficiency is heavily dependent on technology; for VLSI, geometric considerations are as important as operation counts.

4. *Architectures.* Architectures must be chosen to strike the appropriate balance between generality and performance, taking into account the demands of the application and the strengths and weaknesses of the implementation technology.

5. *Fast-turnaround implementation.* Because of changing technology, changing applications needs, and the cost of design time, it is important to keep the design cycle for special-purpose systems as short as possible. Improved CAD tools, at all levels of abstraction, can help to reduce design cost. Silicon foundries can provide quick prototyping. Design for testability can also shorten the design cycle by easing the problem of testing and debugging.

6. *System integration.* The design of a special-purpose device is only part of the job. System integration of high-performance devices typically requires high-speed I/O facilities and powerful buffer memories. We anticipate the development of host systems which can provide these facilities in a flexible way to many special-purpose functional modules.

ACKNOWLEDGMENTS

This research was supported in part by the Defense Advanced Research Projects Agency (DoD), ARPA Order 3597, monitored by the Air Force Avionics Laboratory under Contract F33615-81-K-1539 and in part by the Office of Naval Research under Contracts N00014-76-C-0370, NR 044-422 and N00014-80-C-0236, NR 048-659. A. L. Fisher was supported in part by an IBM graduate fellowship.

REFERENCES

[1] M. R. Barbacci, "Instruction Set Processor Specifications (ISPS): The Notation and Its Application," *IEEE Trans. Comput.*, *C-30*(1):24–40 (Jan. 1981).

[2] A. A. Barnum and M. A. Knapp, eds., *Proceedings of the 1962 Workshop on Computer Organization*, Spartan Books, New York, 1963.

[3] J. L. Bentley, "A Parallel Algorithm for Constructing Minimum Spanning Trees," *J. Algorithms*, *1*:51–59 (1980).

[4] J. L. Bentley and H. T. Kung, "A Tree Machine for Searching Problems," *Proc. 1979 Int. Conf. Parallel Process.*, IEEE, Aug. 1979, pp. 257–266.

[5] J. Blackmer, P. Kuekes, and G. Frank, "A 200 MOPS Systolic Processor," *Proc. SPIE Symp., Vol. 298: Real-Time Signal Processing IV*, Society of Photo-optical Instrumentation Engineers, Aug. 1981.

[6] A. Bojanczyk, R. P. Brent, and H. T. Kung, *Numerically Stable Solution of Dense Systems of Linear Equations Using Mesh-Connected Processors*, Tech. Rep., Computer Science Department, Carnegie-Mellon University, May 1981.

[7] R. P. Brent and H. T. Kung, *Systolic VLSI Arrays for Polynomial GCD Computation*, Tech. Rep., Computer Science Department, Carnegie-Mellon University, May 1982.

[8] R. P. Brent and H. T. Kung, *Systolic VLSI Arrays for Integer GCD Computation*, Tech. Rep. TR-CS-82-11, Department of Computer Science, The Australian National University, 1982.

[9] R. P. Brent and F. T. Luk, "A Systolic Array for the Linear-Time Solution of Toeplitz Systems of Equations," *J. VLSI Comput. Syst.*, *Vol. 1, No. 1*, pp. 1–22, 1983.

[10] R. P. Brent and F. T. Luk, *A Systolic Architecture for the Singular Value Decomposition*, Tech. Rep. TR-CS-82-09, Department of Computer Science, The Australian National University, Aug. 1982.

[11] R. P. Brent and F. T. Luk, *A Systolic Architecture for Almost Linear-Time Solution of the Symmetric Eigenvalue Problem*, Tech. Rep. TR-CS-82-10, Department of Computer Science, The Australian National University, Aug. 1982.

[12] K. Bromley, J. J. Symanski, J. M. Speiser, and H. J. Whitehouse, "Systolic Array Processor Developments," in H. T. Kung, R. F. Sproull, and G. L. Steele, Jr., eds., *VLSI Systems and Computations*, Computer Science Department, Carnegie-Mellon University, Computer Science Press, Rockville, Md., Oct. 1981, pp. 273–284.

[13] P. R. Cappello and K. Steiglitz, "Digital Signal Processing Applications of Systolic Algorithms," in H. T. Kung, R. F. Sproull, and G. L. Steele, Jr., eds., *VLSI Systems and Computations*, Computer Science Department, Carnegie-Mellon University, Computer Science Press, Rockville, Md., Oct. 1981, pp. 245–254.

[14] H. J. Caulfield, W. T. Rhodes, M. J. Foster, and S. Horvitz, "Optical Implementation of Systolic Array Processing," *Opt. Commun.*, *40*(2):86–90 (Dec. 1981).

[15] B. Chazelle, *Computational Geometry on a Systolic Chip*, Tech. Rep., Computer Science Department, Carnegie-Mellon University, May 1982.

[16] D. Cohen and G. Lewicki, "MOSIS—The ARPA Silicon Broker," *Proc. 2nd Caltech Conf. VLSI*, California Institute of Technology, Jan. 1981.

[17] A. Corry and K. Patel, "A CMOS/SOS VLSI Correlator," *Proc. 1983 Int. Symp. VLSI Technol. Syst. Appl.*, 1983, pp. 134–137.

[18] A. L. Fisher, "Systolic Algorithms for Running Order Statistics in Signal and Image Processing," *J. Digital Syst.*, *VI*(2/3):251–264 (Summer/Fall 1982). A preliminary version appears in H. T. Kung, R. F. Sproull, and G. L. Steele, Jr., eds., *VLSI Systems and Computations*, Computer Science Press, Rockville, Md., 1981.

[19] A. L. Fisher and H. T. Kung, "Synchronizing Large Systolic Arrays," *Proc. SPIE Symp., Vol. 341: Real-Time Signal Processing V*, Society of Photo-optical Instrumentation

Engineers, May 1982, pp. 44–52. A revised version appears in *Proc. 10th Int. Symp. Comput. Architecture*, June 1983.

[20] A. L. Fisher, H. T. Kung, L. M. Monier, and Y. Dohi, "Architecture of the PSC: A Programmable Systolic Chip," *Proc. 10th Int. Symp. Comput. Architecture*, June 1983.

[21] A. L. Fisher, H. T. Kung, L. M. Monier, H. Walker, and Y. Dohi, "Design of the PSC: A Programmable Systolic Chip," in R. Bryant, ed., *Proceedings of the Third Caltech Conference on Very Large Scale Integration*, California Institute of Technology, Computer Science Press, Rockville, Md., Mar. 1983, pp. 287–302.

[22] M. J. Foster and H. T. Kung, "The Design of Special-Purpose VLSI Chips," *Computer 13*(1):26–40 (Jan. 1980). A reprint of the paper appears in *Digital MOS Integrated Circuits*, ed. M. I. Elmasry, IEEE Press Selected Reprint Series, IEEE Press, New York, 1981, pp. 204–217. A preliminary version of the paper entitled "Design of Special-Purpose VLSI Chips: Example and Opinions" also appears in *Proc. 7th Int. Symp. Comput. Architecture*, La Baule, France, May 1980, pp. 300–307.

[23] M. J. Foster and H. T. Kung, "Recognize Regular Languages with Programmable Building-Blocks," *J. Digital Syst.*, *6*(4):323–332 (1983). A preliminary version of the paper also appears in *VLSI 81*, ed. J. P. Gray, Academic Press, London, 1981, pp. 75–84.

[24] E. H. Frank and R. F. Sproull, "Testing and Debugging Custom Integrated Circuits," *Comput. Surv. 13*(4):425–451 (Dec. 1981).

[25] W. M. Gentleman and H. T. Kung, "Matrix Triangularization by Systolic Arrays," *Proc. SPIE Symp., Vol. 298: Real-Time Signal Processing IV*, Society of Photo-optical Instrumentation Engineers, Aug. 1981, pp. 19–26.

[26] L. J. Guibas and F. M. Liang, "Systolic Stacks, Queues, and Counters," *Proc. Conf. Adv. Res. VLSI*, Massachusetts Institute of Technology, Cambridge, Mass., Jan. 1982, pp. 155–164.

[27] L. J. Guibas, H. T. Kung, and C. D. Thompson, "Direct VLSI Implementation of Combinatorial Algorithms," *Proc. Conf. Very Large Scale Integration: Architecture, Design, Fabrication*, California Institute of Technology, Jan. 1979, pp. 509–525.

[28] D. E. Heller and I. C. F. Ipsen, "Systolic Networks for Orthogonal Equivalence Transformations and Their Applications," *Proc. Conf. Adv. Res. VLSI*, Massachusetts Institute of Technology, Cambridge, Mass., Jan. 1982, pp. 113–122.

[29] H. T. Kung, "Let's Design Algorithms for VLSI Systems," *Proc. Conf. Very Large Scale Integration: Architecture, Design, Fabrication*, California Institute of Technology, Jan. 1979, pp. 65–90. Also available as a Carnegie-Mellon University Computer Science Department technical report, Sept. 1979.

[30] H. T. Kung, "The Structure of Parallel Algorithms," in M. C. Yovits, ed., *Advances in Computers*, Vol. 19, Academic Press, New York, 1980, pp. 65–112.

[31] H. T. Kung, "Special-Purpose Devices for Signal and Image Processing: An Opportunity in VLSI," *Proc. SPIE, Vol. 241: Real-Time Signal Processing III*, Society of Photo-optical Instrumentation Engineers, July 1980, pp. 76–84.

[32] H. T. Kung, "Use of VLSI in Algebraic Computation: Some Suggestions," in P. S. Wang, ed., *Proc. 1981 ACM Symp. Symbolic Algebraic Comp.*, ACM SIGSAM, Aug. 1981, pp. 218–222.

[33] H. T. Kung, "Why Systolic Architectures?" IEEE *Comput. Mag. 15*(1):37–46 (Jan. 1982).

[34] H. T. Kung, L. M. Ruane, and D. W. L. Yen, "Two-Level Pipelined Systolic Array for Multidimensional Convolution," *Image Vision Comput.*, *1*(1):30–36 (Feb. 1983). A preliminary version appears in *VLSI Systems and Computations*, ed. H. T. Kung, G. L. Steele, Jr., and R. F. Sproull, Computer Science Press, Rockville, Md., 1981, pp. 255–264.

[35] H. T. Kung and P. L. Lehman, "Systolic (VLSI) Arrays for Relational Database Operations," *Proc. ACM-SIGMOD 1980 Int. Conf. Manag. Data*, ACM, May 1980, pp. 105–116.

[36] H. T. Kung and C. E. Leiserson, "Systolic Arrays (for VLSI)," in I. S. Duff, and G. W. Stewart, eds., *Sparse Matrix Proceedings 1978*, SIAM, Philadelphia, 1979, pp. 256–282. A slightly different version appears in C. A. Mead and L. A. Conway, *Introduction to VLSI Systems*, Addison-Wesley, Reading, Mass., 1980, Sec. 8.3.

[37] H. T. Kung and R. L. Picard, "Hardware Pipelines for Multi-dimensional Convolution and Resampling," *Proceedings of the 1981 IEEE Computer Society Workshop on Computer Architecture for Pattern Analysis and Image Database Management*, IEEE Computer Society Press, Nov. 1981, pp. 273–278.

[38] H. T. Kung and S. W. Song, "A Systolic 2-D Convolution Chip," in K. Preston, Jr. and L. Uhr, eds., *Multicomputers and Image Processing: Algorithms and Programs*, Academic Press, New York, 1982, pp. 373–384. An extended abstract appears in *Proceedings of 1981 IEEE Computer Society Workshop on Computer Architecture for Pattern Analysis and Image Database Management*, Nov. 11–13, 1981, pp. 159–160.

[39] H. T. Kung and S. Q. Yu, "Integrating High-Performance Special-Purpose Devices into a System," *Proc. SPIE Symp., Vol. 341: Real-Time Signal Processing V*, Society of Photo-optical Instrumentation Engineers, May 1982, pp. 17–22.

[40] S. Y. Kung and Y. H. Hu, "Fast and Parallel Algorithms for Solving Toeplitz Systems," *Proc. Int. Symp. Mini and Microcomputers in Control and Measurement*, San Francisco, May 1981. Also in *IEEE Trans. Acoust. Speech Signal Process. Vol. Assp-31, No. 1*, Feb. 1983, pp. 66–76.

[41] P. L. Lehman, "A Systolic (VLSI) Array for Processing Simple Relational Queries," in H. T. Kung, R. F. Sproull, and G. L. Steele, Jr., eds., *VLSI Systems and Computations*, Computer Science Department, Carnegie-Mellon University, Computer Science Press, Rockville, Md., Oct. 1981, pp. 285–295.

[42] C. E. Leiserson, "Systolic Priority Queues," *Proc. Conf. Very Large Scale Integration: Architecture, Design, Fabrication*, California Institute of Technology, Jan. 1979, pp. 199–214. Also available as a Carnegie-Mellon University Computer Science Department technical report, Apr. 1979.

[43] C. E. Leiserson, "Area-Efficient Graph Layouts (for VLSI)," *Proc. 21st Annu. Symp. Found. Comput. Sci.*, Oct. 1980, pp. 270–281.

[44] F. J. MacWilliams and N. J. A. Sloane, *The Theory of Error-Correcting Codes*, North-Holland, Amsterdam, 1977.

[45] C. A. Mead and L. A. Conway, *Introduction to VLSI Systems*, Addison-Wesley, Reading, Mass., 1980.

[46] T. Nishitani, Y. Kawakami, R. Maruta, and A. Sawai, "LSI Signal Processing Development for Communications Equipment," *Proc. ICASSP 80*, IEEE Acoustics, Speech and Signal Processing Society, Apr. 1980, pp. 386–389.

[47] R. N. Noyce, "Hardware Prospects and Limitations," in M. L. Dertouzos and J. Moses,

eds., *The Computer Age: A Twenty-Year View*, IEEE Press, New York, 1979, pp. 321–337.

[48] T. Ottmann, A. L. Rosenberg, and L. J. Stockmeyer, *A Dictionary Machine for VLSI*, Tech. Rep. RC 9060 (No. 39615), IBM Thomas J. Watson Research Center, Yorktown Heights, N.Y., 1981.

[49] W. W. Peterson and E. J. Weldon, Jr., *Error-Correcting Codes*, MIT Press, Cambridge, Mass., 1972.

[50] C. Savage, "A Systolic Data Structure Chip for Connectivity Problems," in H. T. Kung, R. F. Sproull, and G. L. Steele, Jr., eds., *VLSI Systems and Computations*, Computer Science Department, Carnegie-Mellon University, Computer Science Press, Rockville, Md., Oct. 1981, pp. 296–300.

[51] A. Sawai, "Programmable LSI Digital Signal Processor Development," in H. T. Kung, R. F. Sproull, and G. L. Steele, Jr., eds., *VLSI Systems and Computations*, Computer Science Department, Carnegie-Mellon University, Computer Science Press, Rockville, Md., Oct. 1981, pp. 29–40.

[52] R. Schreiber, "Systolic Arrays for Eigenvalue Computation," *Proc. SPIE Symp., Vol. 341: Real-Time Signal Processing V*, Society of Photo-optical Instrumentation Engineers, May 1982, pp. 27–34.

[53] S. W. Song, "On a High-Performance VLSI Solution to Database Problems," Ph.D. thesis, Computer Science Department, Carnegie-Mellon University, July 1981. Also available as a CMU Computer Science Department technical report, Aug. 1981.

[54] J. J. Symanski, "A Systolic Array Processor Implementation," *Proc. SPIE Symp., Vol. 298: Real-Time Signal Processing IV*, Society of Photo-optical Instrumentation, Aug. 1981.

[55] J. J. Symanski, "Progress on a Systolic Processor Implementation," *Proc. SPIE Symp., Vol. 341: Real-Time Signal Processing V*, Society of Photo-optical Instrumentation, May 1982, pp. 2–7.

[56] C. D. Thompson, "A Complexity Theory for VLSI," Ph.D. thesis, Carnegie-Mellon University, Computer Science Department, 1980.

[57] M. Tur, J. W. Goodman, B. Moslehi, J. E. Bowers, and H. J. Shaw, "Fiber-Optic Signal Processor with Applications to Matrix-Vector Multiplication and Lattice Filtering," *Opt. Lett.*, 7(9):463–465 (Sept. 1982).

[58] U. Weiser and A. Davis, "A Wavefront Notation Tool for VLSI Array Design," in H. T. Kung, R. F. Sproull, and G. L. Steele, Jr., eds., *VLSI Systems and Computations*, Computer Science Department, Carnegie-Mellon University, Computer Science Press, Rockville, Md., Oct. 1981, pp. 226–234.

[59] R. A. Whiteside, P. G. Hibbard, and N. S. Ostlund, Systolic Algorithms for Monte Carlo Simulations. Draft, Computer Science Department, Carnegie-Mellon University, June 1982.

[60] M. Yano, K. Inoue, and T. Senba, "A LSI Digital Signal Processor," *Proc. ICASSP 82*, IEEE Acoustics, Speech and Signal Processing Society, May 1982, pp. 1073–1076.

[61] D. W. L. Yen and A. V. Kulkarni, "Systolic Processing and an Implementation for Signal and Image Processing," *IEEE Trans. Comput.*, C-31(10):1000–1009 (Oct. 1982).

9

The Role of the CHiP Computer in Signal Processing

LAWRENCE SNYDER

Purdue University
West Lafayette, Indiana†

9.1 INTRODUCTION

Although a microcomputer is a general-purpose device, it is often used with a ROM in lieu of a special-purpose sequential circuit. The rationale is that the microcomputer is already designed, tested, and in production, and personalizing the ROM simply requires programming, an apparently easier task than the more demanding activity of circuit design.

The Configurable, Highly Parallel (CHiP) computer can serve the same purpose in the context of highly parallel signal processing systems. As a generic signal processor, CHiP devices can be produced independent of the application, developed and changed using the flexibility of programming, and because of their architectural characteristics, are fault tolerant and suitable for wafer-scale integration. Finally, should performance considerations mandate a direct VLSI implementation of the signal processor, the CHiP signal processor can serve as the first step to an orderly system design [1].

9.2 THE CHiP COMPUTER

The CHiP architecture family has been described elsewhere [2,3], but we recall its main features in preparation for an example. The main component of a CHiP computer is the *switch lattice*: a collection of processing elements (PEs) connected at regular intervals to a mesh of programmable switches (see Figure 9.1). The PEs are microprocessors with local program and data memory, and their own program

†Present address: Department of Computer Science, FR-35, University of Washington, Seattle, Washington 98195.

170

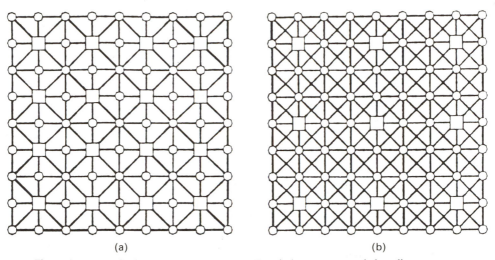

(a)   (b)

Figure 9.1. Two lattices: squares represent PEs; circles represent switches; lines represent data paths. Note that the scale is distorted—PEs are much larger than switches.

counters; there is no shared primary memory; the PEs are driven by a common clock. Rather than being connected to each other directly, the PEs are connected to a small number (typically eight) of adjacent switches. The number of switches connecting to a PE, the width of the data path, and so on, are parameters that characterize various CHiP family members.

The switches, each with a small amount of local memory, are programmable. A switch instruction, called a *configuration setting*, causes the switch to connect two or more of its incident data paths. By properly programming a sequence of switches between two PEs, a direct, static connection is established between them. This is circuit switching rather than packet switching, and it is established globally by a controlling computer, so the PEs do not have to know with whom they are communicating; they simply read or write to their local ports and let the processor interconnection structure programmed in the switches determine the data flow. Crossover and fan-out are possible at the switches. Switches on the perimeter are connected to the external environment.

The two lattices of Figure 9.1 differ in a characteristic called *corridor width*, the number of switches separating adjacent PEs. Corridor width is important in that it influences the convenience of programming processor interconnection structures: the wider the corridors, the more distinct data paths there are available for avoiding congestion when programming complicated graphs [3].

An example.   We view parallel algorithms generally and signal processing algorithms specifically as being defined by a graph and a set of processes with one assigned one per vertex. The graph describes the communication structure of the processes and the processes define the computing activity at each position in the structure. As an example, consider an eight-point fast Fourier transform (FFT)

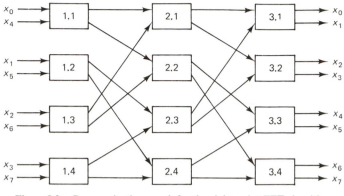

Figure 9.2   Communication graph for the eight-point FFT algorithm.

algorithm. Figure 9.2 gives the graph essentially as described in the literature [4]. The processes for the vertices are all the same;

$$\text{read } B, B';$$
$$C \leftarrow B + QB'; C' \leftarrow B - QB';$$
$$\text{write } C, C';$$

where Q is a constant. We will make the algorithm pipelined by embedding the foregoing process code in a loop.

Since the parallel algorithm is given in two parts, a graph and a set of associated processes, programming the CHiP computer will be a two-part activity: embedding the graph into the switches, and programming the activity of the processor elements. Figure 9.3 shows a direct embedding of the FFT graph into the lattice of Figure 9.1(b). Notice that the double corridor was required in order to handle the congestion across the center of the layout. Having embedded the communication graph, there is an automatic assignment of processes to processors. [This is not very important here since all PEs execute the same code, but for other algorithms (e.g., tree algorithms), it is.] To accomplish the other half of the programming process, we simply code the process in essentially the form given above and specify that all PEs receive the same code.

In preparation for running the program, the CHiP computer controller loads the lattice with programs (using a "skeleton" that is not shown). The FFT configuration settings are assigned to the switches (at a common memory location). The PEs are each assigned the object code for the inner product computation. Then the controller broadcasts a command to the switches causing them to "execute" the configuration setting in the specified location. This has the effect of interconnecting the processors in the manner illustrated in Figure 9.3. It also causes the PEs to begin executing their instructions, and the pipelined FFT begins. The interconnection remains static until it is changed by the controller, say to execute another algorithm.

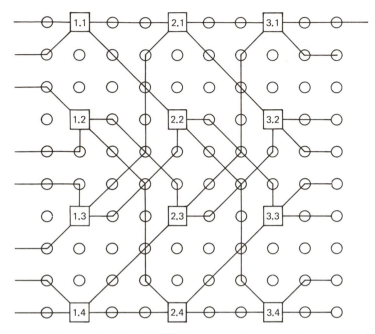

Figure 9.3   Direct embedding of the FFT processor interconnection graph. Notice that only active data paths are shown.

9.3 THE GENERIC CHiP SIGNAL PROCESSOR

The development of the FFT algorithm gives us a context in which to discuss the role of the CHiP computer as a generic signal processor. First, notice that although the algorithm we developed was pipelined, it is *not* a systolic array. Processor-to-processor communication is not limited to nearest-neighbor connections. The switches enable us to implement communication structures that include the fixed systolic meshes as well as more general structures. This permits designers of signal processing systems to consider a larger class of algorithms than they might consider were they to use a progammable systolic array [5,6].

This flexibility of programming different processor interconnection structures causes one to seek "better" solutions and to try to optimize the design. Whether a solution is the "best" solution depends on how costs are assigned. Since there is a tiny delay in passing through a switch, one criterion for "better" might be reduced length of data paths. Careful analysis indicates that for this simple FFT problem, it is possible to reduce path length by foregoing the unidirectional data flow. Figure 9.4 illustrates a processor interconnection structure that reduces data path length to nearest-neighbor connections and reduces the corridor width requirement to unity. Notice that this is simply a different embedding of the graph of Figure 9.2, so except for changes in the directions in which data are sent and received, the PE process

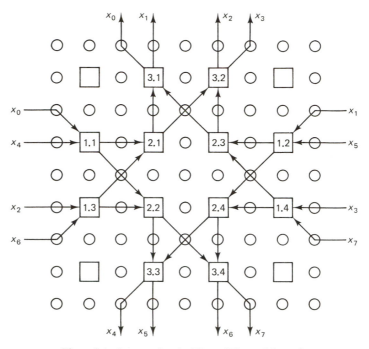

Figure 9.4   Improved embedding of Figure 9.2 graph.

codes are unchanged. (A compiler assigns port directions, so there is no burden on the designer.)

The problem of achieving a nearest-neighbor or at least a "local" communications graph for larger FFT problems is an interesting one. Thompson's results [7] imply that, asymptotically, the shuffle-exchange graph cannot even be embedded in a lattice with full processor utilization unless the corridor width is at least $\Omega(\sqrt{n}/\log n)$ for an n processor lattice. Local communication would seem unlikely even for such a well-endowed lattice. But asymptotic results can be misleading. Leighton and Miller [8] have shown surprisingly efficient embeddings for small shuffle-exchange graphs, and Morrissett has embedded the 64-node shuffle-exchange graph into a one-corridor lattice [9]. This latter result, although very nonlocal, is illustrated in Figure 9.5. The point is that having factored the problem of parallel programming into two parts, we can concentrate on each one in a largely independent manner.

The FFT is one of the most thoroughly studied signal processing problems and many algorithms are available. Since we are not constrained to use only systolic algorithms for the CHiP computer, we can explore the benefits of using other computational methods. Thompson [10] has analyzed eight FFT algorithms in terms of their area and time-delay properties in a VLSI implementation. Several of these would be appropriate for use with the CHiP computer.

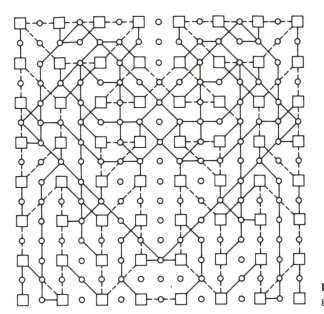

Figure 9.5   64-node shuffle-exchange graph.

9.4 IMPLEMENTATIONAL CONSIDERATIONS

We have postulated a situation in which the CHiP machine has been used to solve an arbitrary signal processing task. Although the signal processing domain is a good one for the CHiP computer because the algorithms tend to exhibit much regularity and parallelism, other characteristics of the domain make it less attractive for general-purpose solution, either CHiP or programmable systolic array solution. Two notable properties are the frequent need for very high performance and the fact that signal processors are often embedded systems solving only one task. We consider specializing the CHiP solution to improve performance.

One obvious specialization for an embedded system is to build a "special-purpose CHiP" machine. We could replace the RAMs with ROMs in both the PEs and switches. But we can do better with the switches; we can eliminate them altogether if the signal processor uses only one interconnection pattern. If the CHiP signal processor executes several algorithms or if it executes one algorithm that requires several interconnection structures, we may still be able to remove a substantial number of switches. In either case there will be some small area savings (after all the switches are not large), a reduction in propagation delay (which may be negligible except for very long data paths), a simplification or removal of the controller, and a complete loss in flexibility—the burned-in algorithm cannot be corrected or improved. Whether these benefits are worthwhile must be assessed on a case-by-case basis.

If the algorithm happens to be SIMD, the program memory can be eliminated by retooling the controller. Speed can be gained by replacing the processor/memory with a direct circuit implementation of the primitive operations. We have described how the CHiP architecture can be used as a design methodology for the development of the circuit implementation [1]. The process is essentially one of stepwise refinement in which the PE's computation is regarded as a problem to be solved on a CHiP machine with more primitive PEs. In the limit, the complexity of the PEs is reduced to simple gate functions that can be directly implemented in VLSI.

With current technology only a small portion of a CHiP lattice fits on a silicon chip. Although we expect improvements, a more fundamental improvement in integration can be realized through wafer-scale integration. We have described [11] how an entire wafer can be patterned with a lattice and how the switches can be used to route around faulty components. The result is a single, dense lattice of hundreds of PEs of the kind required for signal processing.

9.5 CONCLUDING REMARKS

The CHiP computer can be used as a generic signal processor to give the flexibility of convenient algorithm development and experimentation. Moreover, it represents the first step on a continuum of further specializations that reduce complexity, improve speed, and lower costs.

ACKNOWLEDGMENTS

It is a pleasure to thank Kye Hedlund, Stephan Bechtolsheim, and Paul Morrissett for their contributions to the ideas presented here.

The research reported herein is part of the Blue CHiP Project and is funded in part by the Office of Naval Research under Contracts N00014-80-K-0816 and N00014-81-K-0360; the latter is Special Research Opportunities Task SRO-100.

REFERENCES

[1] L. Snyder, "Configurable, Highly Parallel (CHiP) Approach to Signal Processing Applications," *Proc. Tech. Symp. East '82*, SPIE, 1982.

[2] L. Snyder, "Introduction to the Configurable, Highly Parallel Computer," *Computer*, *15*(1):47–56 (Jan. 1982).

[3] L. Snyder, "Overview of the CHiP Computer," in J. P. Gray, ed., *VLSI 81*, Academic Press, New York, 1981, pp. 237–246.

[4] H. S. Stone, "Parallel Processing with the Perfect Shuffle," *IEEE Trans. Comput.*, C-*20*(2):153–161 (1971).

[5] K. Bromley, J. J. Symanski, J. M. Speiser, and H. J. Whitehouse, "Systolic Array Processor Development," in H. T. Kung, B. Sproull, and G. Steele, eds., *VLSI Systems and Computations*, Computer Science Press, Rockville, Md., 1981, pp. 273–284.

[6] Y. Dohi, A. L. Fisher, H. T. Kung, and L. Monier, "The Programmable Systolic Chip: Project Overview," in L. Snyder, L. J. Siegel, H. J. Siegel, and D. B. Gannon, eds., *Algorithmically Specialized Computer Organizations*, Academic Press, New York, 1983.

[7] C. D. Thompson, "A Complexity Theory for VLSI," Ph.D. thesis, Carnegie-Mellon University, 1980.

[8] F. T. Leighton and G. L. Miller, "Optimal Layouts for Small Shuffle-Exchange Graphs," in J. P. Gray, ed., *VLSI 81*, Academic Press, New York, pp. 289–300, 1981.

[9] P. Morrissett, "Observations on Graph Embeddings," Blue CHiP Project Notes, Nov. 1981.

[10] C. D. Thompson, "Fourier Transform in VLSI," Tech. Rep. UCB/CSD 83/105, University of California at Berkeley, 1982.

[11] K. S. Hedlund and L. Snyder, "Wafer Scale Integration of Configurable, Highly Parallel Processors," *Proc. Int. Conf. Parallel Process.*, IEEE, pp. 262–264, 1982.

10

Yield Enhancement by Fault-Tolerant Systolic Arrays

Robert H. Kuhn

Northwestern University
Evanston, Illinois

10.1 INTRODUCTION

About the design of systolic arrays for fault tolerance, we know very little. This is especially true in the area of yield-enhancement techniques. Increases in the scale of integration result from two factors: decreasing minimum geometries and increasing chip size. One can delay fabrication of a device too large for current technology to take advantage of the former, but the other alternative is very attractive because it can be attempted immediately, providing an increased number of defects can be tolerated. Wafer-scale integration (WSI) techniques [1,3,4,6] attempt the latter. Most approaches to WSI have been concerned with connecting the functioning units after fabrication. This results in a virtual one-dimensional array which snakes its way around the wafer in an irregular but nearest-neighbor interconnection, avoiding defective units [see Figure 10.3(a)]. This technique is not suitable for two-dimensional arrays, however. The interstitial fault-tolerance technique permits the fabrication of one- or two-dimensional arrays in the presence of degraded yield in a straightforward manner.

In general, incorporating fault tolerance into a design implies redundancy. Redundancy may have two forms, increased hardware or increased computation time. Some WSI techniques [2], require increases in hardware and require overriding some of the efficiency of a fast local interconnection. The WSI techniques mentioned above maintain local interconnections, but they implicitly use hardware redundancy in the form of unreachable functional units. The interstitial fault tolerance technique however uses time redundancy. Because implementing multiple processing element systems on a chip has become feasible only with WSI technology and we seek to use this technology to implement highly parallel systems first and fault-tolerant systems second, few previous results hold.

178

Consider the conventional Price yield law [8]

$$Y = \left(1 + \frac{DA}{m}\right)^{-m}$$

where Y is the yield, D the defect density, A the effective chip area, and m the number of defect-inducing mechanisms, mainly the number of fabrication steps. Price's yield law may not be realistic in this case, for two reasons. First, it must be observed that processing elements are connected by only one or two layers of material in most systolic arrays, so that the system can be modeled by a *two-level model*.

1. On the lower level, fabricating each unit or processing element does follow Price's law, yielding a certain *unit defect density*, f, of nonfunctioning processors.
2. On the higher level, the distribution of failed processors based on independent events obeying the unit defect density determines the *system yield, Y*.

Second, in conventional non-fault-tolerant systems, yield is actually the probability that zero defects occur in area A. For a parallel processor fault-tolerant system such as the one proposed here, this definition does not carry over. A large number of failures may be tolerated, up to $O(N)$ out of N processing elements for interstitial fault tolerance, if the distribution is optimal. (In the WSI models mentioned above the yield degration is entirely graceful.) As determined by the two-level model, the proper measure of yield for an interstitial fault-tolerant system depends on the local fault distribution. If the system will function with *k or fewer adjacent faults*, the system yield of the fabrication process is the probability that the chip nowhere has more than k adjacent faults:

$$Y = \text{Pr(defects occur in no more than } k \text{ adjacent processors)}$$

10.2 INTERSTITIAL FAULT TOLERANCE

Any systolic algorithm can be represented by a skewed iteration space in which each operation can be assigned to a processor in space and a particular step in time. (A *skewing* is a linear transformation from the iteration space to a processor-time space which exposes systolic parallelism.) Assume now that one processor in the array is faulty. As the systolic algorithm sweeps through the iteration space as a sequence of parallel wavefronts, one sequence or line of computations will not be performed. This line is called the *fault line*. It will be shown that by proper skewing of the iteration space the fault line can be mapped to another line of idle processor cycles which occurs on an adjacent processor. This line of idle cycles is called the *repair line*, and the processor possessing the repair line is called the *repair processor* for this fault.

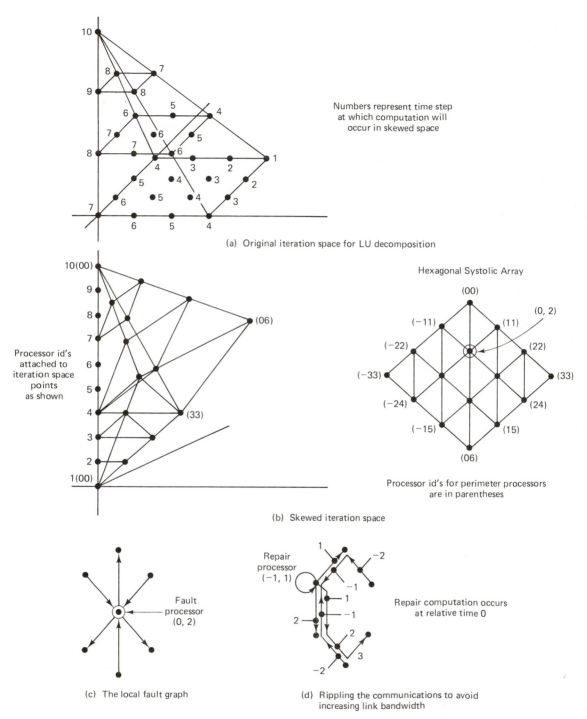

(a) Original iteration space for LU decomposition

Numbers represent time step
at which computation will
occur in skewed space

Processor id's
attached to
iteration space
points
as shown

(b) Skewed iteration space

Hexagonal Systolic Array

Processor id's for perimeter processors
are in parentheses

(c) The local fault graph

Fault
processor
(0, 2)

Repair
processor
(−1, 1)

Repair computation occurs
at relative time 0

(d) Rippling the communications to avoid
increasing link bandwidth

Figure 10.1 Skewing the hexagonal L-U decomposition algorithm for interstitial fault tolerance: (a) original iteration space for L-U decomposition; (b) skewed iteration space; (c) local
fault graph; (d) rippling the communications to avoid increasing link bandwidth.

180

In a properly skewed iteration space both fault and repair lines contain periodic compute and idle cycles. By design, at least one adjacent processor has idle cycles which mesh with the compute cycles of the faulty processor. Thus the method is called *interstitial fault tolerance*. That is, the time redundancy needed for fault tolerance is hidden in the interstices of the prospective repair processor's computation. Interstitial fault-tolerant arrays can be classified by a basic property of the skewing applied to the iteration space. If there exist n idle cycles per one compute cycle, the algorithm is called *level n interstitial fault tolerant*. This figure is closely related to the system yield. A level n interstitial fault-tolerant array will function with at least n adjacent faulty processing units.

Example: L-U decomposition.   Figure 10.1 illustrates the application of this technique to Kung and Leiserson's L-U decomposition algorithm. Figure 10.1(a) shows the original iteration space for the standard L-U decomposition. Figure 10.1(b) shows how the iteration space is skewed to realize the algorithm on a hexogonal mesh. If processor (0, 2) is faulty, the fault line of computations that would have been performed by processor (0, 2) can be computed by processor $(-1, 1)$, the repair processor. The *repair line* is an *interstitial repair line*. That is, no other computation is being performed at the times when a *fault-line* computation must occur. L-U decomposition is level 2 interstitial fault tolerant. Figure 10.1(c) and (d) show the local fault graph and the local remapping of communication to effect the repair if the interconnection network at the faulty processor has failed. Three routing steps per computation cycle are required to mask the rerouting time in this case.

10.3 YIELD ANALYSIS

Yield analysis for interstitial fault tolerance has been conducted analytically and by simulation. Melzak [7] and others have studied a mathematical model based on forbidden configurations and coincidences that can be adapted to evaluate the yield for interstitial fault-tolerant arrays. A statement of the model in terms of our yield model is given by the following parameters:

1. d: the systolic array dimension
2. n: the number of faulty processors in the system
3. f: a unit defect density function which holds for all defects
4. k: The maximum number of adjacent faulty processors

Using the inclusion–exclusion principle, with this model it can be shown that

$$Y_k = \sum_{r=1}^{\binom{n}{k}} \sum_{i=1}^{t(r,k)} (-1)^{r+1} N_{rik}(m) \int_{G_{ri}} \cdots \int f \, dx_1 \cdots dx_{v(G_{ri})}$$

where the G_{ri} are coincidence graphs with $v(G_{ri})$ vertices such as the ways in which triples of pairs of adjacent failed processors can share a failed processor, $t(r, k)$ is the number of such graphs for a fixed term r in the inclusion–exclusion series, and $N_{rik}(m)$ is a combinatorial coefficient expressing the number of occurrences of the graph G_{ri} in the d times n-dimension problem domain. However, evaluating this yield function is extremely cumbersome. Melzak points out that due to the oscillating and diminishing nature of the inclusion–exclusion formula, the leading terms can be computed and an error bound proportional to the last term computed can be found.

Figure 10.2 shows the results of simulating three interstitial fault-tolerant networks. The three networks simulated are:

1. A linear systolic array with level 1 interstitial fault tolerance.
2. A linear systolic array with level 2 interstitial fault tolerance.
3. A linear systolic array with level 2 interstitial fault tolerance laid out in a snakelike manner on a mesh. In the linear snakelike network we allowed a processor in one row to repair a processor in an adjacent row. See Figure 10.3(b) for an example.

The method of simulation was to inject uniformly distributed random faults into the network. Each fault failed exactly one working processor. An adjacent processor that could execute the processing needed by the failed processor during its idle cycles was selected as the interstitial repair processor. The adjacent processors were checked in the same order for each failed processor. If no adjacent processor

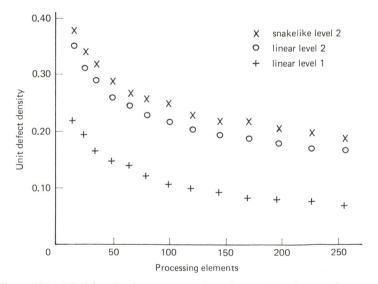

Figure 10.2   PE defect density versus number of processing elements for system yield of 0.5.

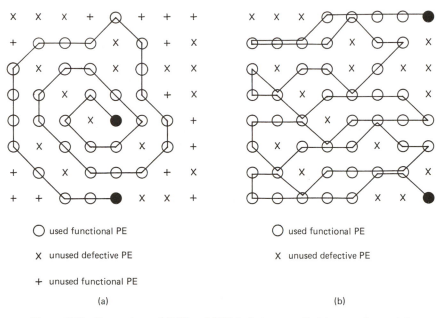

used functional PE

unused defective PE

unused functional PE

(a)   (b)

Figure 10.3 Comparison of WSI and IFT techniques applied to a random-point-defect distribution: (a) simple WSI technique; (b) IFT technique.

with interstitial idle periods could be found, the entire network was considered failed and the trial was terminated recording the number of faults it survived. No attempt was made to reassign repair processors in an optimal way. This model models a reliable computation model closely, but it is more pessimistic than a yield model should be in that multiple defects per processor should be allowed.

The results of the simulation for networks ranging from 16 to 256 processors are shown in Figure 10.2. Figure 10.2 shows the unit defect density that can be tolerated for a system yield of 0.5. (System yield as a function of unit defect density curves is contained in the report.) The interstitial fault-tolerance technique compares favorably with the WSI techniques cited in the references, especially for the snake-like linear layout. For example, Figure 10.3 shows an 8×8 processor array with random unit defect with a density of 29% corresponding to a unit yield of 71%. Using a simple WSI algorithm, a usable yield of 50% is achieved by the connection shown in Figure 10.3(a). A level 1 interstitial fault-tolerant network laid out in a snakelike manner with the repair processor assignment shown in Figure 10.3(b) would yield a functioning system using all 71% of the unit yield. The WSI approach has inherently used hardware redundancy, whereas the interstitial fault-tolerant approach has used time redundancy because the array will run at half the speed. (More sophisticated WSI algorithms have been developed, such as [4], but they also have used some time redundancy.)

This discussion has centered around one type of defect, independent point defects; other types of defects have been observed on ICs, such as line defects, point cluster defects, and area defects. In addition, researchers have observed considerable radial variation in defect density [5]. That is, defects are more frequent on the perimeter of the wafer than at its center due to temperature gradients during fabrication, mask warping, and handling. Interstitial fault tolerance may be well suited to such distributions if the original iteration space is skewed so that the computation density decreases linearly toward the perimeter of the wafer. The full paper contains a two-dimensional array that has been skewed to minimize the effect of radially increasing defect density. The ability to adapt systolic arrays in this manner is due to our abstract representation of systolic arrays as skewed computations on an iteration space.

10.4 CONCLUDING REMARKS

Highly parallel systems must provide some form of fault tolerance to be useful, especially in the VLSI area, where defects are not uncommon. Interstitial fault tolerance is a natural technique for incorporating fault tolerance into systolic arrays and it compares well with currently known WSI techniques. The technique that we propose permits the linear speed-up found in systolic arrays. And, using the analysis methods mentioned above, it can easily be used by an IC designer even after an initial layout of the systolic array without fault tolerance. Since interstitial fault tolerance uses time redundancy, it does not require large amounts of hardware redundancy and it is especially well suited for configurable and programmable systolic arrays which already incorporate some form of local switches or routing tables. It also handles two-dimensional arrays where WSI techniques do not.

REFERENCES

[1] R. C. Aubusson and I. Catt, "Wafer Scale Integration—A Fault Tolerant Procedure," *IEEE J. Solid-State Circuits, SC-13*(3):339–344 (June 1978).

[2] Y. Egawa, T. Wada, Y. Ohmori, N. Tsuda, and K. Masuda, "A 1-Mbit Full-Wafer MOS RAM," *IEEE J. Solid State Circuits, CS-15*(4):677–686 (Aug. 1980).

[3] D. Fussell and P. Varman, "Fault-Tolerant Wafer-Scale Integration for VLSI," *9th Symp. Comput. Architecture*, 1982, pp. 190–198.

[4] D. Gajski and A. H. Sameh, "A WSI-Multiprocessor for Iterative Algorithms," *Proc. GOMAC—82*, Nov. 1982.

[5] A. Gupta, W. A. Porter, and J. W. Lathrop, "Defect Analysis and Yield Degradation of Integration Circuits," *IEEE J. Solid-State Circuits, SC-9*:96–103 (June 1974).

[6] R. M. Lea and M. Sreetharan, "VLSI Distributed Logic Memories," *Proc. Caltech Conf. Very Large Scale Integration*, Caltech Computer Science Department, Jan. 1979.

[7] Z. A. Melzak, *Companion to Concrete Mathematics*. Wiley, New York, 1973.

[8] J. E. Price, "A New Look at Yield of Integrated Circuits," *Proc. IEEE, 58*:1290–1291 (Aug. 1970).

11

Partitioning Big Matrices for Small Systolic Arrays

DON HELLER

Pennsylvania State University
University Park, Pennsylvania†

11.1 INTRODUCTION

The object of this chapter is to show how systolic array hardware designs and data structures may be adapted for multipass matrix computations. Systolic arrays are usually designed for one-pass computation; we cosider how to partition the data to fit the device and route the data in a pattern consistent with the problem at hand. We assume that the reader is familiar with the basic literature on systolic arrays, in particular [3,8,11,12,19], and has some familiarity with vector pipeline computation (e.g., [5,9]). Other special-purpose devices for VLSI implementation might also be considered along similar lines.

Our basic premise is a form of Murphy's law:

No matter what special-purpose device is available, there is a problem too large for it. The problem will manifest itself only after the device is acquired and can no longer be modified. The problem cannot be ignored.

It is therefore necessary to engage in defensive computer design:

Only construct special-purpose devices that can be used to solve problems of arbitrary size (up to the limits placed by the overall system).

and a special brand of defensive algorithm design for special-purpose devices:

Don't make it all worse than software on a general-purpose computer.

†Permanent address: Physics and Computer Science Department, Shell Development Company, Bellaire Research Center, P.O. Box 481, Houston TX 77001.

A natural conclusion is that no one systolic array design will be sufficient for any given class of problems over the entire range of problem sizes. It is therefore desirable to incorporate more than one design into a given device, to avoid multiple devices for one problem. Some kind of programmability, perhaps on-the-fly replacement of programs, may then be useful. Among the set of devices and drivers for a variety of problems, it is desirable to have common I/O patterns so that devices may be chained and memory suport simplified. There may need to be devices whose sole use is to reformat data between incompatible computing devices. We shall demonstrate some of these ideas here.

Systolic array sizes are usually specified in terms of the matrix dimensions (order n, bandwidth w, etc.), so we use barred symbols for the actual array dimensions ($\bar{n}$, $\bar{w}$, etc.). We usually assume that $\bar{n} < n$, and so on. Several passes over the data will be necessary in this case, each pass operating on a subset of the data determined by some partitioning method. The underlying principle is that larger arrays may be simulated with extra memory in each processor or by recycling data through the array.

One goal is to minimize or eliminate delay time between passes. For a non-systolic processor array this may be done by storing a large portion of the matrix within the array; Johnsson [16] fully develops this idea for L-U factorizations of banded matrices. Two relevant questions are: "What data format is best for partitioning?" and "How should the data be laid out on a disk?" Note that these layouts may be modified through on-the-fly filters (see Chapter 22 of this volume).

The usual methods of partitioning follow the row and column structure of the matrix. Partitioning by diagonals or into nonrectangular blocks are some nontraditional techniques that are needed for systolic arrays. A major theme is to develop long, skinny matrices from short, squat ones. This might be expected since systolic arrays are pipelines that work best with long, steady input streams.

It is most natural to think of partitioning methods in terms of a "window" over the matrix. The window consists of a submatrix with which some computation is to be performed; as the window scans the matrix we approach the desired result. In a similar fashion there are major windows and minor windows, the latter being a window over the major window. Related concepts include the "computational window," that portion of the matrix residing in the systolic array at a given time [16,17], and the "wavefront," the leading edge of the window.

The matrix representation may sometimes be manipulated to improve performance of multipass systolic computations. We consider two examples, using additive splittings and a tensor product representation. Memory in the processor array often trades off against the number of processors, allowing some variation in data storage. Fewer processors generally means more passes over the data, but more memory may simplify the simulation of a larger array. A less elegant but nevertheless useful trick is to "double up" a computation, using the idle processor time inherent in some systolic array designs.

In [10] we discussed the concept of a hardware library, where functional units are in relation to the host computer as subroutines from a software library are to a production code. The original view of systolic arrays as external functional units attached to a host through various controllers fits into this model. We also want to argue that systolic arrays will make good internal functional units, taking the role for simple matrix computations that vector pipelines now have for simple vector computations. The partitioning methods discussed here and elsewhere (e.g., [20] and Chapter 22 of this volume) will contribute to the programming of systolic arrays and the design of entire systems for their support.

11.2 ADDITIVE SPLITTINGS

Additive splittings such as $A = A_1 + A_2$ are used primarily with matrix multiplication devices. The goal may be to decrease bandwidth, to increase density within a band, or to use a symmetry property to reduce resource requirements. Perhaps the simplest example of bandwidth reduction is to write a symmetric matrix M as $L + L^T$, where L is lower triangular, and to compute $\mathbf{x}^T M \mathbf{x}$ as $2\mathbf{x}^T L \mathbf{x}$.

Priester et al. [22] addressed the problem of bandwidth reduction for use with the Kung–Leiserson band matrix-vector product linear systolic array [19]. The following generalizes their method. The diagonals of an $n \times n$ matrix A are indexed from $-n + 1$ to $n - 1$, so matrix element a_{ij} lies in diagonal $j - i$. The bandwidth of A is w if there are integers r and s, $-n + 1 \leq r \leq s \leq n - 1$, such that $a_{ij} = 0$ if $j - i < r$ or $j - i > s$, and $w = s - r + 1$. Thus all nonzero elements of A lie within w diagonals of A. For example, a strictly lower-triangular matrix would have $r = -n + 1$, $s = -1$, and a tridiagonal matrix would have $r = -1, s = 1$.

Given any matrix A and integer $\bar{w} \geq 1$, we can write

$$A = \sum_{i=1}^{k} A_i$$

where the bandwidth of A_i is at most $\bar{w}$. Matrix vector products $\mathbf{y} := A\mathbf{x}$ may then be accomplished by

$$\mathbf{y} := 0 \qquad \text{for } i := 1 \text{ to } k \text{ do } \mathbf{y} := \mathbf{y} + A_i \mathbf{x}$$

using a $\bar{w}$-cell Kung–Leiserson linear systolic array for the inner loop. The window here selects $\mathbf{x}$ and a diagonal band of A and moves across the matrix in a pattern determined by further details of the matrix.

If $\bar{w} = 1$, this method reduces to the "matrix multiplication by diagonals" scheme of Madsen et al. [21]. The PRT (partial row translation) technique of Priester et al. [22] is the case where A is dense, $\bar{w} = n$, $k = 2$, $A_1 =$ upper-triangular part of A, $A_2 =$ strictly lower-triangular part of A. The PCT (partial column translation) technique [23] is similar, but $A_1 =$ strictly lower triangular, $A_2 =$ upper triangular. For very small values of $\bar{w}$ one should probably use instead a linearly connected array with broadcasting, which can increase throughput by a factor of 2 [13]. The matrices A_i and the vector $\mathbf{x}$ may be trimmed around the edges to avoid computing zero products, and the time between passes through the loop may be reduced by alternating the A_i's between upper- and lower-triangular matrices (see Figs. 4 and 5 of [22]).

The additive splitting method can be extended to matrix products by

$$AB = \sum_{i=1}^{k} \sum_{j=1}^{l} A_i B_j$$

where neither A_i nor B_j has bandwidth greater than some specified integers. Huang and Abraham [13] discuss a variant of this for $k = l = 3$ using an $n \times n$ hexagonally connected dense matrix multiplier, where the goal is to improve processor utilization as well as bandwidth reduction.

Another extension is given by Priester et al. [22]: by adding one more cell to the linear array and by recycling data properly, the PRT method can implement the Gauss–Seidel or SOR iterations. Similarly, the PCT method may be adapted to compute $A^i \mathbf{x}$ with suitable recycling [23].

The methods above are designed to decrease bandwidth in the data when too few cells are available in a systolic array. There is a decrease in speed corresponding to the decrease in cells from w to $\bar{w}$, perhaps with an additional penalty for maintaining idle cells in one or two passes. In some cases, however, it may be advantageous to increase density within the band in return for an increase in speed. A common technique is to manipulate a succinct representation of the matrix in place of the full matrix itself.

For example, when solving certain elliptic partial differential equations by finite-difference methods, we might obtain a matrix of the form

$$A = I \otimes B + C \otimes I$$

where B and C have small bandwidth, usually tridiagonal. ($\otimes$ is the Kronecker product; B and C represent second differences in orthogonal space dimensions.) If B and C are $n \times n$ and tridiagonal (which we now suppose), then A is $n^2 \times n^2$ with bandwidth $2n + 1$, but A is quite sparse within the band, having only five nonzero diagonals, indexed $-n$, -1, 0, 1, n. The approximate solution to the differential equation may be represented either as an n^2-long vector $\mathbf{u}$ or an $n \times n$ matrix U, Fortran-style. The vector $A\mathbf{u}$ may then be represented as the matrix $BU + UC^T$. The matrix-vector product $A\mathbf{u}$ may be treated as above, taking advantage of sparsity within the band, using

> 1 cell, 2 input lines, time $= 5n^2 + O(n)$, 5 passes, or
> 3 cells, 4 input lines, time $= 4n^2 + O(n)$, 3 passes, or
> 5 cells plus $2n - 4$ buffers, 6 input lines, time $= 2n^2 + O(n)$, 1 pass

To compute the matrix $BU + UC^T$ by combining two Kung–Leiserson band matrix-product systolic arrays and a linear systolic array for matrix addition, we could use $14n + 7$ cells, $4n + 8$ input lines, and time $3n + 4$. Of course, the matrix U may be split along its diagonals, rows, or columns if it is necessary to reduce the cell count after having increased it. The point is that by changing representation we may go from an effective bandwidth of five to any bandwidth between 1 and $2n + 1$, thereby fitting the data to the available device. Of related interest is recent work by Adams [1] on extensions of the red-black ordering for iterative solution of linear systems on parallel computers.

11.3 QR FACTORIZATIONS

We next consider the orthogonal reduction of a matrix to triangular form by use of Givens rotations. The basic operation forms a linear combination of two matrix rows and replaces both of them in the matrix. The rotation is an ideal candidate for systolic execution since more than one update operation may be applied to the same pair of rows as long as the updates propagate in the correct sequence. If each processor is equipped with a local memory capable of storing one column of the matrix, the Householder triangularization may be attractive; for banded matrices the storage requirement is not onerous [17].

Givens rotation systolic arrays may be divided into three categories: fixed rotations and moving elements (hold Q in the array); moving rotations and fixed elements (hold R in the array); and moving rotations and moving elements. Partitioning methods for the Heller–Ipsen array [11,12], in the third group, will be discussed in further detail. Arrays in the second group are described by Gentleman and Kung [8] and Ciminiera et al. [4], the latter using on-line arithmetic to speed up the processors. Schreiber and Kuekes (Chapter 22 of this volume) present a complete analysis of partitioning methods based on a $\bar{p} \times \bar{q}$ trapezoidal systolic array (a truncated $\bar{q} \times \bar{q}$ Gentleman–Kung triangle) and the following scheme:

1. Partition $A = (A_1 \quad A_2 \quad \cdots)$, $A_1 = n \times \bar{p}$, $A_i = n \times (\bar{q} - \bar{p})$, $2 \le i$.
2. Reduce A_1 to $\binom{R}{0}$, $R = Q_1^T A_1$, upper triangular, $\bar{p} \times \bar{p}$.
3. Recycle the rotations, obtaining $Q_1^T A_i$, $2 \le i$.
4. Omit the first $\bar{p}$ rows and columns of the updated A and repeat.

This method is standard and will be put to further use shortly.

In the first group of methods, the processor cell computes an annihilating rotation when first presented with an (x, y) pair and a control signal: it then outputs $((x^2 + y^2)^{1/2}, 0)$ and saves the rotation. Upon further receipt of (x, y) pairs, these are rotated and the results passed on. It is essential that data entry occurs row per port, to preserve the algorithm's integrity. Two basic designs exist, with the order of elimination from bottom to top (Gannon [7], Bojanczyk et al. [3]) or top to bottom (Ahmed et al. [2]). The execution time for each is the same, $n + 2\bar{p} - 1$ steps to reduce a $\bar{p} \times n$ matrix to upper-trapezoidal form using a $\bar{p} \times \bar{p}$ triangular array ($\bar{p}(\bar{p} + 1)/2$ cells).

Gannon's partitioned elimination method for dense matrices [7] augments the $\bar{p} \times \bar{p}$ triangular array with half of a $(\sqrt{2\bar{p}}) \times (\sqrt{2\bar{p}})$ triangular array (another

$\bar{p}(\bar{p} + 1)/2$ cells) and a perfect shuffle permutation network. The augmented array is simply half of a $(2\bar{p}) \times (2\bar{p})$ triangle. Let A be $n \times m, n \le m$, partitioned as

$$A = \begin{bmatrix} A_1 \\ A_2 \\ \vdots \end{bmatrix}$$

where A_i is $\bar{p} \times m$. In the first $k = \lceil n/\bar{p} \rceil$ passes through the $\bar{p} \times \bar{p}$ triangle, we reduce A_i to upper-trapezoidal form, giving a sawtooth edge to the matrix. In succeeding passes the window size increases to $2\bar{p}$ rows as we execute

> for j = 2 to k by 1
> > for i = k to j by −1
> >
> > shuffle the rows of $\begin{pmatrix} A_{i-1} \\ A_i \end{pmatrix}$
> >
> > reduce the result to upper-trapezoidal form using the augmented array, in time $(m - (j - 2)\bar{p}) + \bar{p} + 1$
> >
> > replace the rows in A

This finally brings A to triangular form in time $O(k^2 m + k^3 \bar{p})$. If $n = m$, the time is $O(n^3/\bar{p}^2)$. The $(2\bar{p}) \times m$ window moves from bottom to top (controlled by i) within a larger window, which moves from top to bottom (controlled by j).

We now describe partitioning methods for the QR decomposition of banded matrices using the Heller–Ipsen systolic array [11,12]. This device consists of $\bar{q} \cdot \bar{w}$ processors orthogonally connected as $\bar{q} \cdot \bar{w}$-wide linear meshes (Figure 11.1 without the triangles); each linear mesh is capable of annihilating one subdiagonal from a $\bar{w}$-banded matrix. Suppose that the matrix A has bandwidth w, with p superdiagonals and q subdiagonals ($w = p + q + 1$), and that $q \le p$. The design presented in [11,12] assumes that $w = \bar{w}, q = \bar{q}$, and reduces A to the upper-triangular matrix R in one pass by repeated Givens rotations across the rows. The order of elimination is bottom to top in succeeding columns, as for the fixed-rotation devices [3,7]. The leftmost processor of a linear mesh generates a Givens rotation, and the remaining processors propagate the rotations. The input to the systolic array is taken one antidiagonal per step starting with the upper left corner of the matrix; each input or output port receives or delivers elements of the same diagonal in alternate steps. When $w = \bar{w}, q = \bar{q}$, the time to produce R from A is $2(n + q - 1)$; the bandwidth of R is always that of A, due to fill-in.

If $w \le \bar{w}$ (no assumed relation between q and $\bar{q}$), we can pad the input with $\bar{w} - w$ zero superdiagonals and route the matrix through the processor array $\lceil q/\bar{q} \rceil$ times, clearing the array to zero before each pass. Each pass strips off $\bar{q}$ subdiagonals and propagates rotations across the rows without need for recycling; $\bar{q}$ superdiagonals fill in as a result. The factorization time is $\lceil q/\bar{q} \rceil(2n + 2\bar{q} - 2 + t_0)$, where t_0 is the time to clear the array between passes. The extra superdiagonals may be removed from the final output without harm, as they will remain zero. On the last pass it may occur that we eliminate all subdiagonals before the matrix exits from the

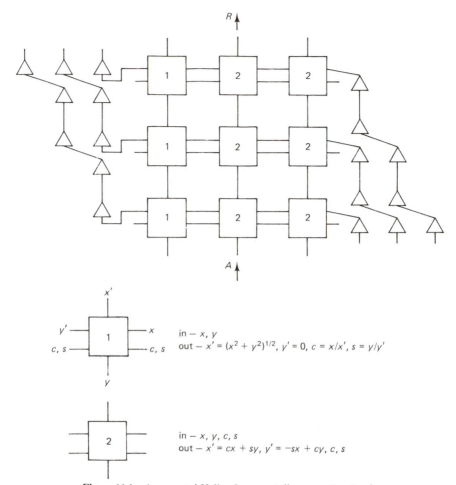

in — x, y
out — $x' = (x^2 + y^2)^{1/2}, y' = 0, c = x/x', s = y/y'$

in — x, y, c, s
out — $x' = cx + sy, y' = -sx + cy, c, s$

Figure 11.1   Augmented Heller–Ipsen systolic array, $\bar{w} = \bar{q} = 3$.

processor array (i.e., if $\bar{q}$ does not divide q evenly). Then the main diagonal enters the leftmost processor of some layer; since the datalines are initialized to zero, this case is treated as though the matrix were padded with a leading row of zeros and $q = 1$. The rotations now generated simply exchange the zero row with succeeding rows, and the matrix passes through unchanged.

Now consider the cases where the matrix is too wide for the processor array ($w > \bar{w}$). As before, it is necessary to process the matrix in several passes by submatrices. Some hardware modifications are also necessary, to recycle rotations that had not moved across a complete row. Since the submatrices may be dense (bandwidth ≈ dimension) and the processor array is designed for sparse matrices (bandwidth ≪ dimension), some degree of inefficiency should be expected; we will compensate for this by a simple hardware augmentation.

We consider two schemes. The first is similar to that of Chapter 22: annihilate all elements below the main diagonal in succeeding groups of k columns, using a total of $\lceil (n-1)/k \rceil$ passes; $k \geq 1$ is to be determined from $\bar{w}$ and $\bar{q}$. In each pass we need only consider a major window of $r \geq q + k$ rows and n columns, where r is to be determined. The major window contains all elements to be annihilated and moves down through the matrix, starting in rows 1, $k+1$, $2k+1$, and so on, in successive passes. On the last pass, we take nonexistent rows to be zero. The major window will always contain some number of leading- and trailing-zero columns, which we can safely ignore; there are at most $r + p$ nonzero columns.

Since the processor array is capable of annihilating only $\bar{q}$ subdiagonals in one pass, if $q > \bar{q}$ it will be necessary to process the r rows in $\lceil q/\bar{q} \rceil$ passes, from bottom up. For this we use a semimajor window consisting of $r_1 = \bar{q} + k$ rows and n columns and moving upward within the major window. Partition the r_1 rows into a leading block with s_1 columns and trailing blocks with s_2 columns each, s_i to be determined. There will be at most $m = \lceil (r + p - s_1)/s_2 \rceil$ nonzero trailing blocks. We choose to force $k \leq s_1$ so that all necessary rotations may be determined from the first block and recycled through the processor array as they are applied to the nonzero trailing blocks. This requires that the one cell in each layer of the array which normally generates rotations be switched to a receiving mode for recycling; the rotations themselves may be stored outside the array in shift registers, one register per layer.

The $r_1 \times s_1$ block taken from the last r_1 rows of the leading $r \times s_1$ block has at most $\bar{q}$ nonzero subdiagonals, so we eliminate these in one pass through the systolic array, apply the newly generated rotations to the m trailing $r_1 \times s_2$ blocks, move the semimajor window up by $\bar{q}$ rows, and repeat. The minor windows therefore consist of s_i columns and move right across the semimajor window. Figure 11.2 summarizes the window placements and elimination sequence.

Processor utilization may be improved if we augment the $\bar{q} \times \bar{w}$ array with $2\bar{q}^2$ delay elements in two triangles, as shown in Figure 11.1 for $\bar{q} = \bar{w} = 3$. This allows additional data entry to the upper layers of the array in a consistent pattern. Pairs of delay elements are equivalent to a standard cell with a fixed identity rotation, where the entering rotation is passed through unused. From consideration of the numbers of nonzero diagonals and input ports (one diagonal per port) we require that

$$\text{(leading block, } r_1 \times s_1) \quad \bar{q} + s_1 \leq \bar{w} + \bar{q}$$

$$\text{(trailing blocks, } r_1 \times s_2) \quad r_1 + s_2 - 1 \leq \bar{w} + \bar{q}$$

This establishes all the basic parameters and constraints.

A quick estimate of the factorization time is

$$(n/k)(q/\bar{q})(\text{leading block} + m \cdot \text{trailing block} + m \cdot t_0)$$

From the formula for m it is apparent that it will be minimized if we minimize r and

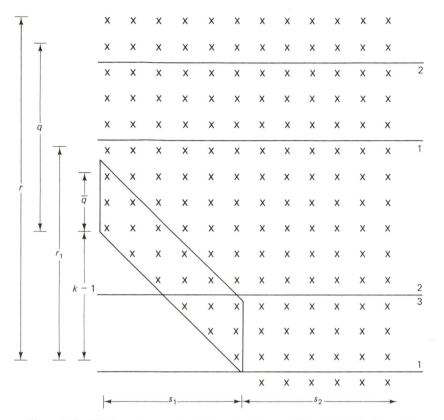

Figure 11.2 Window placement and elements to be annihilated: 1, beginning and end of first semimajor window placement; 2, beginning and end of second semimajor window placement; 3, beginning of second semimajor window placement, alternative method. Elements to be annihilated by the first pass are outlined.

maximize s_i, so we take

$$s_1 = \bar{w}$$
$$s_2 = \bar{w} + \bar{q} - r_1 + 1 = \bar{w} - k + 1$$
$$r = q + k$$
$$m = \lceil (r + p - s_1)/s_2 \rceil = \lceil w/s_2 \rceil - 1$$

With these choices the time to process the leading block is $\bar{w} + 3\bar{q} + k - 2$ and $\bar{w} + 3\bar{q} - 1$ for a trailing block. It now remains to choose k between 1 and $\bar{w}$ to minimize execution time; this will occur at $k \approx \bar{w}/2$.

Notice from Figure 11.2, however, that $\bar{q} \cdot \bar{w}$ rotations are generated from the leading block, and these are to be recycled across $\bar{q} + k$ rows. The delay time between trailing blocks may be eliminated if we take $k = \bar{w}$ and use a fixed rotation method instead of the original moving rotations. Figure 11.3 describes the new situation. Rotation (i, j) is the one generated to annihilate element (i, j) of the leading block, where we index relative to the semimajor window. Following the cell indexing of Figure 11.3, we fix rotation $(i + j, j)$ in cell (i, j), $1 \leq i \leq \bar{q}$, $1 \leq j \leq \bar{w}$. The delay elements require no alteration.

Loading the rotations for the second pass is easy if we add a control signal. In the first pass, a rotation is generated at the left and passed to the right; when it reaches "home," a copy is made before it is passed on. Home can be identified by a control signal passed to the left and up starting from the lower right cell, indexed

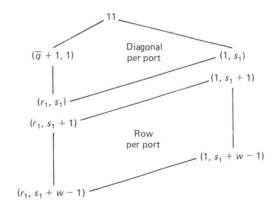

Cell indexing:

$(1, \bar{w} + \bar{q}) \ldots (1, \bar{w} + 1)$ | $(1, \bar{w})$ $\ldots$ $(1, 1)$ | $(1, 0)$

$(\bar{q}, \bar{w} + 1)$ | $(\bar{q}, \bar{w})$ $\ldots$ $(\bar{q}, 1)$ | $(\bar{q}, 0) \ldots (\bar{q}, -\bar{q} + 1)$

Fixed rotations for second pass:

I $\ldots$ I | $(\bar{w} + 1, \bar{w}) \ldots (2, 1)$ | I

I | $(\bar{w} + \bar{q}, \bar{w}) \ldots (\bar{q} + 1, 1)$ | I $\ldots$ I

Data entry:

element (i, j), $1 \leq i \leq j + \bar{q}$, $1 \leq j \leq s_1$, enters cell $(\bar{q}, \bar{w} - \bar{q} + i - j)$ at time $i + j - 1$, exits cell $(1, \bar{w} + i - j)$ at time $2(\bar{q} - 1) + i + j$

element $(i, s_1 + j)$, $1 \leq i \leq r_1$, $1 \leq j \leq w - 1$, enters cell $(\bar{q}, i - \bar{q})$ at time $\bar{w} + i + j - 1$, exits cell $(1, i)$ at time $2(\bar{q} - 1) + \bar{w} + i + j$

Figure 11.3   Fixed rotation pattern and data entry for augmented array.

$(\bar{q}, 1)$, when rotation $(\bar{q} + 1, 1)$ reaches it. By the end of the first pass, the rotations are all in place and we can start the second pass with no delay (Figure 11.4). The total time to process a semimajor window is thus $w + 3(\bar{q} + \bar{w} - 1)$, so the total time for the reduction is

$$\lceil (n - 1)/\bar{w} \rceil \lceil q/\bar{q} \rceil (w + 3(\bar{q} + \bar{w} - 1) + t_0)$$

This is essentially optimal.

An advantage of the moving-fixed rotations array versus the fixed rotations arrays is that we only need $\bar{q}$ cells which generate rotations, as opposed to $\bar{q} \cdot \bar{w}$. The control signal does not add much complexity to the cell design. There is also no need to store rotations outside the array, at least as far as the partitioning method is concerned. One disadvantage is the need for two methods of data entry, first diagonal per port and then row per port. This can be handled by sending the data row per port through a rearrangement filter whose size is proportional to $\bar{q} \cdot \bar{w}$.

There is a second elimination sequence, closer in spirit to the original scheme of [11,12]: annihilate the $\bar{q}$ outermost subdiagonals, and repeat using a total of $\lceil q/\bar{q} \rceil$ passes. Now the major window consists of rows $(\tilde{q} + 1 - \bar{q}) \cdots n$, where $\tilde{q}$ is the number of remaining nonzero subdiagonals; its top row moves up by $\bar{q}$ rows. The semimajor window has r_1 rows as before but moves down by k rows through the major window, starting at the top (Figure 11.2). As can be seen by counting the number of nonzeros eliminated, the total time will be the same as before.

y	x	Time	Rotation used		y	x	Time	Rotation used		y	x	Time	Rotation used
21	11	6	21		22	12	7	21		23	13	8	21
										24	14	9	21
32	22	8	32		33	23	9	32		25	15	10	21
					34	24	10	32		26	16	11	21
43	33	10	43		35	25	11	32		27	17	12	21
31	21	5	31		32	22	6	31		33	23	7	31
										34	24	8	31
42	32	7	42		43	33	8	42		35	25	9	31
					44	34	9	42		36	26	10	31
53	43	9	53		45	35	10	42		37	27	11	31
41	31	4	41		42	32	5	41		43	33	6	41
										44	34	7	41
52	42	6	52		53	43	7	52		45	35	8	41
					54	44	8	52		46	36	9	41
63	53	8	63		55	45	9	52		47	37	10	41

Figure 11.4 Timing for Figure 11.3, $\bar{q} = \bar{w} = 3$. Rotations and matrix elements identified by indices only. Values not mentioned are $x - y = 0$, rotation $= I$.

11.4 BLOCK PARTITIONS

The partitioning methods already discussed try to use submatrices extending across a complete row. Since it is often easier to develop algorithms treating small submatrices, we mention two cases where the competing methods are combined. As a design technique, the idea is to start with a small block partitioning and modify it to become a long block partitioning.

Consider first the dense matrix product

$$C = AB = \begin{bmatrix} A_1 \\ A_2 \\ \vdots \end{bmatrix} [B_1 \quad B_2 \quad \cdots] = \begin{bmatrix} C_{11} & C_{12} & \cdots \\ C_{21} & C_{22} & \cdots \\ \vdots & \vdots & \ddots \end{bmatrix}$$

where A is $n \times r$ and B is $r \times m$. The formula above displays a block inner product algorithm, which is most directly implemented using a systolic processor array with an accumulating inner product cell. Suppose that the processor array has $\bar{n} \cdot \bar{m}$ orthogonally connected processors, computes $(\bar{n} \times k) \times (k \times \bar{m})$ matrix products for arbitrary k, and stores the result in the array. This corresponds to the engagement processor of Speiser and Whitehouse [24] among others. We take A_i to be $\bar{n} \times r$ and B_j to be $r \times \bar{m}$, padding with zero rows and columns if necessary. Katona [18] describes a method based on control signals and data interweaving to read out the blocks of C while computing succeeding $A_i B_j$ products. When $\bar{n} = 1$, as considered by Kung and Yu [20], the unloading is immediate and no special tricks are required.

A major disadvantage of the block inner product algorithm is that A_i enters row per port while B_j enters column per port, which requires either dual storage mechanisms or a large but simple systolic array for on-the-fly transposition. An exhaustive analysis of alternative methods for vector pipeline machines is given by Dongarra et al. [5], and could be extended to systolic arrays.

The L-U factorization of a block partitioned matrix is discussed frequently as a technique for all manner of parallel processors (e.g., [9,14,15,16]). Partial pivoting may be achieved through additional hardware [14] or local storage [16] at the expense of time. The disadvantage from a numerical point of view is that pivot searches only within a diagonal block are not sufficient for numerical stability, in the absence of further assumptions about A.

The data movement into a systolic array can be enhanced if we combine a block Crout factorization with orthogonal reduction of block rows and a few other basic operations. Suppose that we have fixed-rotation hardware to reduce a $\bar{p} \times n$ matrix to upper trapezoidal form. Partition the $n \times n$ matrix A into $\bar{p} \times \bar{p}$ blocks, and set $m = \lceil n/\bar{p} \rceil$. Partition the factors L and U similarly. The block triangularization is thus

for $i = 1$ to $m - 1$ **do**

reduce

$$(\tilde{A}_{ii} \cdots \tilde{A}_{in}) = (A_{ii} \cdots A_{in}) - (L_{i1} \cdots L_{i,i-1}) \begin{bmatrix} U_{1i} & \cdots & U_{1n} \\ \vdots & & \vdots \\ U_{i-1,i} & \cdots & U_{i-1,n} \end{bmatrix}$$

to upper-trapezoidal form, obtaining $(U_{ii} \cdots U_{in})$ and rotations representing the orthogonal matrix L_{ii} (or L_{ii} itself). Compute

$$\begin{bmatrix} \tilde{A}_{ii} \\ \vdots \\ \tilde{A}_{ni} \end{bmatrix} = \begin{bmatrix} A_{ii} \\ \vdots \\ A_{ni} \end{bmatrix} - \begin{bmatrix} L_{i1} & \cdots & L_{i,i-1} \\ \vdots & & \vdots \\ L_{n1} & \cdots & L_{n,i-1} \end{bmatrix} \begin{bmatrix} U_{1i} \\ \vdots \\ U_{i-1,i} \end{bmatrix}$$

and

$$\begin{bmatrix} L_{ii} \\ \vdots \\ L_{ni} \end{bmatrix} = \begin{bmatrix} \tilde{A}_{ii} \\ \vdots \\ \tilde{A}_{ni} \end{bmatrix} U_{ii}^{-1}$$

The final step may be accomplished with block transposes and repeated lower-triangular system solutions. As a practical matter it is best not to compute L_{ii} directly in the last step, as it will almost surely not be orthogonal due to round-off error. The usual Crout algorithm obtains an L-U factorization of $\tilde{A}_{ii}$, but is otherwise identical.

11.5 SPARSE MATRICES

An interesting open problem is the treatment of sparse matrices with systolic arrays. Duff [6] observes that since the L-U factorization of a sparse matrix A tends to cause the lower left corner of A to fill in to a dense submatrix, it may be useful to change representations during the factorization. That is, we start with A completely stored as a sparse matrix (index, value, and pointer components in a linked list) and end with

$$L = \begin{bmatrix} L_{11} & 0 \\ L_{21} & L_{22} \end{bmatrix} \qquad U = \begin{bmatrix} U_{11} & U_{12} \\ 0 & U_{22} \end{bmatrix}$$

where L_{11}, L_{21}, U_{11}, and U_{12} are stored as sparse matrices and L_{22} and U_{22} are stored as dense matrices.

Similar ideas may be used with systolic arrays, and indeed may be necessary to avoid unstructured memory reference patterns. For example, the additive splittings

$$A = A_{\text{dense}} + A_{\text{sparse}} = A_{\text{structured}} + A_{\text{unstructured}}$$

might be used for matrix-vector products or with preconditioned conjugate gradient iterations (preconditioner $= A_{\text{dense}}$ or $A_{\text{structured}}$). The threshold between dense and sparse is a matter of experimentation, as is the design of systolic arrays for unstructured sparse matrices.

ACKNOWLEDGMENT

This work was supported in part by the National Science Foundation under Grant MCS-8202372.

REFERENCES

[1] L. M. Adams, "Iterative Algorithms for Large Sparse Linear Systems on Parallel Computers," Ph.D. thesis, Dept. of Appl. Math. and Comp. Sci., University of Virginia, Nov. 1982.

[2] H. M. Ahmed, J. M. Delosme, and M. Morf, "Highly Concurrent Computing Structures for Matrix Arithmetic and Signal Processing," *Computer*, Jan. 1982, pp. 65–82.

[3] A. Bojanczyk, R. P. Brent, and H. T. Kung, "Numerically Stable Solution of Dense Systems of Linear Equations Using Mesh-Connected Processors," Comp. Sci. Dept., Carnegie-Mellon University, May 1981. To appear in *SIAM J. Sci. Stat. Comput.*

[4] L. Ciminiera, A. Serra, and A. Valenzano, "Fast and Accurate Matrix Triangularization Using an Iterative Structure," *Proc. 5th Symp. Comput. Arith.*, May 1981, pp. 215–221.

[5] J. J. Dongarra, F. G. Gustavson, and A. Karp, "Implementing Linear Algebra Algorithms for Dense Matrices on a Vector Pipeline Machine," Math. and Comp. Sci. Div., Argonne Nat. Lab., Sept. 1982.

[6] I. S. Duff, "Full Matrix Techniques in Sparse Gaussian Elimination," in Lecture Notes in Mathematics, Vol. 912: *Numerical Analysis*, Springer-Verlag, New York, 1981, pp. 71–84.

[7] D. Gannon, "A Note on Pipelining a Mesh-Connected Multiprocessor for Finite Element Problems by Nested Dissection," *Proc. Int. Conf. Parallel Process.*, 1980, pp. 197–204.

[8] W. M. Gentleman and H. T. Kung, "Matrix Triangularization by Systolic Arrays," *SPIE Proc.*, Vol. 298: *Real-Time Signal Processing IV*, 1981, pp. 19–26.

[9] D. E. Heller, "A Survey of Parallel Algorithms in Numerical Linear Algebra," *SIAM Rev.*, 20:740–777 (1978).

[10] D. E. Heller, "Mathematical Hardware—Design Issues and Responsibilities," *Purdue Workshop on Algorithmically-Specialized Computer Organizations*, Sept. 1982.

[11] D. E. Heller and I. C. F. Ipsen, "Systolic Networks for Orthogonal Decomposition, with Applications," Comp. Sci. Dept., Pennsylvania State University, Aug. 1981. To appear in *SIAM J. Sci. Stat. Comput.*

[12] D. E. Heller and I. C. F. Ipsen, "Systolic Networks for Orthogonal Equivalence Transformations and Their Applications," *Proc. MIT Conf. Adv. Res. VLSI*, Jan. 1982, pp. 113–122.

[13] K. H. Huang and J. A. Abraham, "Efficient Parallel Algorithms for Processor Arrays," *Proc. Int. Conf. Parallel Process.*, 1982, pp. 271–279.

[14] K. Hwang and Y. H. Cheng, "VLSI Computing Structures for Solving Large-Scale Linear Systems of Equations," *Proc. Int. Conf. Parallel Process.*, 1980, pp. 217–227.

[15] K. Hwang and Y. H. Cheng, "Partitioned Algorithms and VLSI Structures for Large-Scale Matrix Computations," *Proc. 5th Symp. Comput. Arith.*, May 1981, pp. 222–232.

[16] L. Johnsson, "Computational Arrays for Band Matrix Equations," Comp. Sci. Dept., California Institute of Technology, May 1981.

[17] L. Johnsson, "A Computational Array for the QR Method," *Proc. MIT Conf. Adv. Res. VLSI*, Jan. 1982, pp. 123–129.

[18] E. Katona, "Cellular Algorithms for Binary Matrix Operations," *Conpar 81*, Lecture Notes in Computer Science, Vol. 111, Springer-Verlag, New York, 1981, pp. 203–216.

[19] H. T. Kung and C. E. Leiserson, "Systolic Arrays (for VLSI)," in I. S. Duff and G. W. Stewart, eds., *Sparse Matrix Proceedings 1978*, SIAM, Philadelphia, 1979, pp. 256–282; and in C. A. Mead and L. A. Conway, *Introduction to VLSI Systems*, Addison-Wesley, Reading, Mass., 1980, Sec. 8.3.

[20] H. T. Kung and S. Q. Yu, "Integrating High-Performance Special Purpose Devices into a System," *SPIE Proc.* Vol. 341: *Real-Time Signal Processing V*, 1982, pp. 17–22.

[21] N. K. Madsen, G. H. Rodrigue, and J. I. Karush, "Matrix Multiplication by Diagonals on Vector/Parallel Processors," *Inf. Proc. Lett.*, 5:41–45 (1976).

[22] R. W. Priester, H. J. Whitehouse, K. Bromley, and J. B. Clary, "Signal Processing with Systolic Arrays," *Proc. Int. Conf. Parallel Process.*, 1981, pp. 207–215.

[23] R. W. Priester, J. H. Whitehouse, K. Bromley, and J. B. Clary, "Problem Adaptation to Systolic Arrays," *SPIE Proc.*, Vol. 298: *Real-Time Signal Processing IV*, 1981.

[24] J. M. Speiser and H. J. Whitehouse, "Parallel Processing Algorithms and Architectures for Real-Time Signal Processing," *SPIE Proc.*, Vol. 298: *Real-Time Signal Processing IV*, 1981, pp. 2–9.

12

Executable Specification
of Concurrent Algorithms in Terms
of Applicative Data Space Notation

ARMIN B. CREMERS

University of Dortmund
Dortmund, Federal Republic of Germany

THOMAS N. HIBBARD

Jet Propulsion Laboratory
Pasadena, California

12.1 INTRODUCTION

VLSI challenges the programmer to take advantage of the concurrency hidden in computational problems. The ease with which a larger number of processing and memory elements can be combined in a circuit makes it desirable to look for algorithmic solutions that process large amounts of information concurrently. An increase of efficiency can be expected if the algorithm arranges for a balanced distribution of work load while observing the requirement of locality (i.e., short communication paths). These properties of load distribution and information flow serve as guidelines to the designer of algorithms for VLSI [8,13].

In the past several years, a great number of algorithms in rather disjoint application areas have been discovered adequate to the properties of VLSI. Many of these algorithms have sequential counterparts that had been known for decades but had been discarded for lack of efficiency. No widely applicable methodology has evolved yet to support a review of the art of computer programming for suitable VLSI solutions to a given computational problem.

In view of the high cost of reliable VLSI implementation, and in view of the rising complexity of application demands, it is important to exercise great care at the point of algorithmic specification. Without disputing the usefulness of diagrams, commented examples, and hints at known sequential versions, the definition of a VLSI algorithm doubtlessly requires a precise notation capable of reflecting the concurrent nature of the algorithm. The algorithmic definition serves as input to the VLSI implementation cycle that starts with chip design and ends with the tested chip. In addition, the definition serves as a basis for the functional description of the device to the environment. An important advantage of a precise and adequate notation for concurrent algorithms is that it may be executable and thus offer a potential for high-level testing and diagnosis.

The main purpose of this chapter is to suggest an executable algorithmic notation for VLSI solutions which, in many cases, may be more natural than traditional programming notation. The notation is based on the concept of a *data space*, introduced by the authors several years ago [2,3] as a formal model of abstract machines. The data-space notion is very close in nature to the *modules* of Parnas [14] and the *processes* of Horning and Randell [10]. A data space encapsulates control-oriented and data-oriented modeling, and thus may be a vehicle for combining the advantages of both approaches. Basically, a data space is a transition system whose states are given an explicit information structure. Each transition is computed as a function of the whole state. The model therefore includes the possibility of powerful changes to the state in a single step. The latter point has also been one of the central ideas of the applicative state-transition systems suggested by Backus [1].

Initially introduced as an operational semantic model of programming, data spaces have been investigated since 1979 from the point of view of software specification. A large class of finitely specifiable data spaces has been identified and given a computable syntax. These data spaces are called *syntactic* or *context-free*. Two versions of the syntax have been implemented in PL/I and Pascal, respectively, and have been successfully used as vehicles of specification in different application areas [7]. Our notation is *applicative* in the sense that during the computation of a state transition, there are no side effects on the state of the data space (Section 12.2).

For our purposes in this chapter, the syntax has been extended by some new concepts. Primarily, these are *subspaces*, intended as an encapsulation of computational substructures, and *synchronized types*, a very basic communication mechanism, somewhat similar to the one of Hoare's CSP [9].

We emphasize that we are not proposing a new programming language: our notation is not designed to produce efficient code directly. On the other hand, our notation is not intended to generate input directly for an automatic verification system. To keep the notation simple, yet problem oriented, we had to "separate concerns." However, it is quite conceivable that our proposal could be extended in either direction. (cf. [Cremers and Kung, 15].)

In Section 12.3 we convert two typical systolic array algorithms, convolution and matrix multiplication, to data spaces. In both we let synchronization be global. In the convolution example we include all the details of pipelining output and resetting for a new problem. In the second example we omit these details since they are of the same general nature as in the first example and not as intricate. A complete executable specification of a matrix multiplication is listed in the Appendix to this chapter. In Section 12.4 we turn to local synchronization and introduce some additional constructs for the purpose. We convert a bubble-sorting array to a data space, and show matrix multiplication again.

12.2 DATA SPACES AND ALGORITHMIC NOTATION

Before presenting the algorithmic notation, we review some fundamental aspects of the computational model on which the notation is based.

12.2.1 The Data-Space Model

As mentioned in Section 12.1, a data space $(X, \mathscr{F}, p)$ is made up of

1. A transition system (X, p), where X denotes a set of states and p a partial function from X to X

2. An information structure $\mathscr{F}$, which is a set of total functions defined on X

The functions in $\mathscr{F}$ may be used to obtain information from a given state; however, changes to the state are only made by p. In general, we expect $\mathscr{F}$ to give us a complete description of any given state; on the other hand, we ought to be able to identify a subset of $\mathscr{F}$ for the purpose of a nonredundant description of the states. In [4], properties of nonredundant information structures are formally investigated. In addition to the formal relationship between $\mathscr{F}$ and X, the data-space model includes various aspects of the relationship between $\mathscr{F}$ and p. These are made precise in, for example, [5,6].

In the present chapter we attempt to keep the mathematics of data spaces in the background and focus on the consequences of this work for the *specification* of concurrent algorithms. We propose to specify a concurrent algorithm formally in terms of a data space. The main advantages of this approach are:

1. History sensitivity
2. Encouragement of functional style
3. Simple, yet powerful communication mechanism
4. Executability
5. Transition-directed diagnostics
6. Problem orientation

The set of examples given in the following sections is intended to illustrate these properties. The set includes algorithms that are rather easy to specify precisely, together with some rather involved ones. Readers are encouraged to compare the specifications to their own favorite precise algorithmic notation in the light of properties 1 to 6 above.

12.2.2 Data-Space Syntax

For a restricted class of data spaces, which, however, would include the data spaces that arise in practice, we are able to give a computable syntax. In this syntax a data space can be defined completely and precisely. The syntax provides mechanisms for the definition of computational structures on problem-oriented levels of abstraction. The basic idea of the syntax presented here is the applicative specification of transitions; that is, every processor step is computed by means of a side-effect-free evaluation of expressions. The following is a brief and informal introduction into the elements of the syntax.

A (syntactic) data space is defined by

1. Some data-type declarations, some of which are declared to be names of cells that hold some data type.
2. Some function definitions, one of which specifies the state transitions for the data space.

DATA TYPES.   *Primitive types* are certain "universal" types such as Boolean, integer, real, (character) string, and assgts (set of name-value pairs). Of these types, which are to be considered built in, Boolean, integer, and real are self-explanatory. String will not be used in the examples presented in this paper, and assgts will be explained further in connection with cells and expressions.

Defined types are obtained by means of *aggregation* (concatenation), *generalization*, (discriminated union), and *recursion*. All three mechanisms may be used in the same definition. Thus, on the whole, the defined types make up an unambiguous, marked, context-free grammar with as many start symbols as there are data types. The productions of the grammar are of the form

$$\text{TYPE } T \rightarrow c \; s_1 \; T_1 \; s_2 \; T_2 \cdots s_n \; T_n \qquad (n \geq 0)$$

where c is the *constructor* name of the production for data type T, and the s_i are the *selector* names of the types T_i on the right-hand side. For notational brevity, we sometimes make the type names do double duty: If a constructor name c is not specified, we may use the data type name T of the left-hand side for construction, and if a selector name s_i is not specified, we may use the corresponding T_i for selecting the ith component of the right-hand side.

Example

A (fully marked) set of productions for type "binary tree of integer" is

$$\text{TYPE tree} \rightarrow \Delta \quad \text{root integer}$$
$$\text{left tree}$$
$$\text{right tree} \qquad |$$
$$\square \; \{\text{empty tree}\} \quad ||$$

As usual, "|" separates the productions for the same left-hand side, and "||" ends the list of alternatives.

CELLS.   Memory is specified in terms of a list of definitions of the form

$$\text{CELLS} \quad T_1 \quad \text{HOLD} \quad T_2$$

where T_1 and T_2 are data types: T_1 the cell name type, T_2 the contents type. For example, "CELLS tree HOLD integer" expresses that values of type "tree" are used to address values of type "integer." For the sake of brevity, in the case of defined types, we allow either T_1 or T_2 or both in the cell declaration scheme above to be replaced with their definitions. As mentioned already, there is a built-in data type "assgts" which has as values sets of (cell name, contents) pairs, where each set constitutes a function (i.e., multiple occurrences of cell names on the left are ruled out).

FUNCTIONS.   In addition to built-in functions belonging to primitive types (e.g., arithmetic), and in addition to the constructors and selectors mentioned above, there are *defined functions*:

$$\text{FUNCTION } F(t_1 \ T_1 \ t_2 \ T_2 \cdots t_n \ T_n) \ T = E$$

The argument types T_i are given argument names t_i which may be used in the defining expression E. T is the result type of the function. The function definition may be (directly or indirectly) recursive. The state is an implicit argument of every defined function.

EXPRESSIONS.   Each value of a primitive type as well as each argument name is an expression. Further, for the definition of expressions, there are the following mechanisms:

1. *Application:*

$$F \ E_1 \ E_2 \ \cdots \ E_n$$

Function F is applied to the values of the argument expressions $E_1, \ldots, E_n$. The values must be of the argument types specified in the definition of the function.

2. *Logical branching:*

$$\text{IF} \ E \ E_1 \ E_2$$

E must evaluate to Boolean; the values of E_1 and E_2 must be of the same type.

3. *Structural branching:*

$$\text{TEST} \ E \ c_1 \ E_1 \ c_2 \ E_2 \cdots c_n \ E_n$$

If the value of E decomposes by the production for constructor c_i, the expression evaluates to the value of the corresponding expression E_i.

4. *Memory fetch:*

$$\text{REF } E$$

E must evaluate to a value of a cell name type. The contents associated to the cell name is the value of the memory fetch expression.

5. *Memory store:*

$$\text{ø} E_1 \cdots E_n \text{ø}$$

Each E_i is either of the form

$$\text{LET } E = E'$$

where E evaluates to a cell name and E' to a value of the associated contents type (the value of E_i being of type assgts), or an expression that evaluates to data type assgts. The value of the whole expression delimited by ø (or any other delimiter) is the union of the values of the E_i (i.e., of data type assgts).

Example

$$\text{ø LET } x = \text{REF } y$$

$$F z \text{ ø}$$

The value of this store expression is $\{(x, \text{contents of } y)\} \cup \{\text{value of } F z\}$, and may be illustrated by a binary tree.

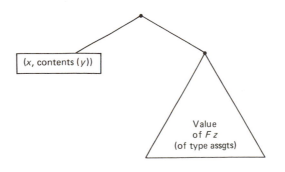

In the examples here we also use a do construction in memory store expressions:

$$\text{DO FOR } i = m \text{ TO } n \text{ BY } d \; E(i)$$

which for m, $d \leq n$ is equivalent to $E(m)$, $E(m + d)$, ..., $E(m + (\lfloor n/d \rfloor - 1)d)$. The NIL, or empty, memory store expression is convenient, for example, in

$$\text{IF} \quad B \quad E \quad \text{NIL}$$

OMEGA.   The processor (i.e., the transition function) has the standard name OMEGA and is a function with no explicit arguments (the state being an implicit argument as with every function), defined by an expression E. For each transition, E is evaluated to an element of type assgts. The transition then consists in the application of the set of (cell name, contents) pairs to the memory defined. Only after the application has been completed, E is evaluated anew. The processor halts when either the list of definitions computed is empty (no new state computed) or E becomes undefined (next state undefined).

LOAD.   By means of this special function, a list of (cell name, contents) pairs is computed, again in a side-effect-free way, whose application to memory puts the data space into an initial state.

Remark. In order to improve the readability of a syntactic specification, we shall make optional use of delimiters in the following sections. For instance, $F(x)$ instead of $F\ x$, IF E THEN E_1 ELSE E_2 for IF $E\ E_1\ E_2$. An expression IF B E NIL may be written IF B THEN E. We frequently will rely on indentation to delimit expressions, particularly in TEST and memory store expressions whose delimiters, although technically necessary for correct parsing, only raise the noise level in text meant just for human consumption.

12.3 GLOBAL SYNCHRONIZATION

To illustrate the specification of systolic algorithms in terms of data spaces, we have chosen two rather well-known examples: convolution on a linear systolic array [11] and matrix multiplication on a wavefront array [12], using a global clock.

12.3.1 Convolution

Given two sequences $w_1, ..., w_k$ and $x_1, ..., x_n$, with $n \geq k$ (probably substantially greater), to compute

$$y_i = w_1 x_i + w_2 x_{i+1} + \cdots + w_k x_{i+k-1}$$

for $i = 1, ..., n - k + 1$. Kung [11] gives the systolic diagram of Figure 12.1. The diagram assumes that $n - k = 2$, so that there are three y_i's to compute. We show the state two cycles after it has started. For each y_i there is a processing element that computes it.

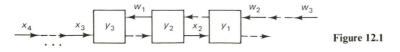

Figure 12.1

It is crucial that the inputs from both sequences are accepted only every two cycles. Thus, in the cycle about to occur in the figure, only the cells for y_1 and y_3 will be accumulating. Of course, it would not be satisfactory to ask the user of such a processor to put in the "blank" inputs; the array processor will do that. Nor would it be reasonable to ask the user to load the x sequence in its starting position; it should be necessary only for him to present the w's at the front port and the x's at the back port, and furthermore to be able to present a new problem at the main ports without having to do anything special to help the processor recover from having done the first problem.

The algorithm we are about to give takes care of the details of output and getting ready to do the next convolution. For this purpose both the w_i and x_i are accompanied by flags marking the end of each sequence. When a processor encounters the end of the w sequence, it then starts receiving outputs, through the y line, from the processors to its left. The x sequence keeps coming at the same time and the output phase ends when the last x arrives. Meanwhile, processor 1 holds up the new w sequence until the last x arrives, then (in the next transition) zeroes its y in preparation for the new problem.

We remark that the algorithm will work as long as the length of the x sequence does not exceed the length of the w sequence by more than the number of processors. Note that a global clock is implicit in the fact that each execution of OMEGA effects one cycle of each processor.

The processors are not identified as explicit syntactic elements, but are there implicitly as the collection of operations occurring over x_i, y_i, and so on, for a particular i. In the next section on local synchronization we shall introduce a syntactic construct for individual processors.

We reemphasize that the LETs do not take effect until OMEGA is finished, and that they constitute a set, rather than a sequence, of assignments. If two LETs with the same target were to be generated, the result would be undefined. (Nevertheless, at execution time, an arbitrary choice will be made.)

CONVOLUTION

```
TYPE value & flag →
    val REAL
    flag BOOLEAN   ||
CELLS w INTEGER HOLD value & flag          "initially flagged
                                            in all processors"
CELLS x INTEGER HOLD value & flag          "initially not
                                            flagged in all
                                            processors"

CELLS y INTEGER HOLD REAL
CELLS state INTEGER HOLD accumulate|output|wait    "initially wait
                                            except for proces-
                                            sor n, initially out-
                                            put"
```

```
TYPE in_out_sequence →
            empty
          |   first value & flag
              rest in_out_sequence    ||
CELLS x_in | w_in | y_out HOLD in_out_sequence
CELLS cycle HOLD even | odd                          "initially odd"
```

Note that, for example, w in the cell definition above serves as both constructor/selector and data type name. The definition is less cumbersome than the equivalent one: TYPE W → w w INTEGER together with CELLS W HOLD value & flag.

```
FUNCTION new_y_1 (   ) assgts =
  TEST REF state 1
      accumulate
          LET y 1 = REF y 1 + val REF x 1 * val REF w 1
          IF flag REF w 1 THEN
                IF flag REF x 1 THEN
                    LET state 1 = wait
                ELSE
                    LET state 1 = output
      output
          LET y 1 = REF y 2
      wait
          NIL


FUNCTION pass_data_1 (   ) assgts =
  LET x 1 = REF x 2
  TEST REF state 1
      accumulate
          LET w 1 = first REF w_in
          LET w_in = rest REF w_in
      wait
          IF flag REF x 1 THEN
              LET w 1 = first REF w_in
              LET w_in = rest REF w_in
              LET y 1 = 0
              LET state = accumulate
      output
          IF flag REF x 1 THEN
              LET y_out = in_out_sequence (REF y 1, TRUE) REF y_out
              TEST REF w_in
                empty
                    LET state 1 = wait
                in_out_sequence
```

```
                    LET w 1 = first REF w_in
                    LET w_in = rest REF w_in
                    LET state 1 = accumulate
                    LET y 1 = 0
            ELSE
                LET y_out = in_out_sequence (REF y 1, FALSE) REF y_out

FUNCTION new_y(i INTEGER "1 < i < n") assgts =
    TEST REF state i
        accumulate
            LET y i = REF y i + val REF w i * val REF x i
            IF flag REF w i THEN
                IF flag REF x i THEN
                    LET state i = wait
                ELSE
                    LET state i = output
        output
            LET y i = REF y i+1
            IF flag REF x i THEN
                LET state i = wait
        wait
            IF NOT flag REF w i THEN
                LET y i = val REF x i * val REF w i
                LET state i = accumulate

FUNCTION pass_data(i : integer "1 < i < n") assgts =
        LET w i = REF w i−1
        LET x i = REF x i+1

FUNCTION new_y_n(   ) assgts =
    TEST REF state n
        accumulate
            LET y n = REF y n + val REF x n * val REF w n
            IF flag REF w n THEN LET state n = wait
        wait
            IF NOT flag REF w n THEN
                LET y n = val REF w n * val REF x n
                LET state n = accumulate
        output
            NIL

FUNCTION pass_data_n(   ) assgts =
    TEST REF state n
        output
```

```
        TEST REF x_in
            empty
                NIL
            in_out sequence
                LET x n = (0, TRUE)
                LET state n = wait
wait
    LET w n = REF w n−1
    TEST REF x_in
        empty
            LET x n = (0, FALSE)
            LET state n = output
        in_out_sequence
            LET x n = first REF x_in
            LET x_in = rest REF x_in
accumulate
    same as wait
```

Remark. State "output" of processing element n serves the special purpose of accepting a new x sequence.

```
FUNCTION OMEGA =
  TEST REF cycle
    odd
        new_y_1
        DO FOR i = 3 TO n BY 2
            new_y(i)
        DO FOR i = 2 TO n − 1 BY 2
            pass_data(i)
        pass_data_n            "assuming that n is even"
        LET cycle = even
    even
        pass_data_1
        DO FOR i = 3 TO n BY 2
            pass_data(i)
        DO FOR i = 2 TO n − 1 BY 2
            new_y(i)
        new_y_n                "assuming that n is even"
        LET cycle = odd
```

12.3.2 Matrix Multiplication by Wavefront Technique (cf. [12])

Two $n \times n$ matrices A and B are presented at the input ports; each row of A is presented at each of n row ports and each column of B at each of n column ports. There are n^2 processing elements arranged in a square array. The rows of A propagate across and the columns of B propagate down. Element a_{ik} meets element b_{kj} at (i, j)th processing element, where the product $a_{ik} b_{kj}$ is accumulated. The processing elements are not present specifically as syntactic elements, but are there implicitly as all of the operations involving a_{ij}, b_{ij}, and c_{ij} for a given i, j. The data space follows.

MATRIX MULTIPLICATION

```
CELLS a INTEGER INTEGER HOLD REAL
CELLS b INTEGER INTEGER HOLD REAL
CELLS c INTEGER INTEGER HOLD REAL                  "initially 0"
TYPE input_sequence →
             empty
            | first REAL rest input_sequence ||
CELLS a_in INTEGER | b_in INTEGER HOLD          input_sequence
CELLS step HOLD fetch | execute | stop          "initially fetch"
CELLS k HOLD INTEGER                            "initially 1"
FUNCTION OMEGA =
   TEST REF step
      fetch
          DO FOR ALL (i,j) SUCH THAT    1 ≤i≤n & 1≤j≤n &
                                        i+j ≤ REF
                                        k + 1

              IF j = 1 THEN
                  IF i>k−n THEN
                      LET a i j = first REF a_in i
                      LET a_in i = rest REF a_in i
                  ELSE NIL
              ELSE
                  IF i+j>k−n+1 THEN
                      LET a i j = REF a i j−1
                  ELSE NIL
              IF i = 1 THEN
                  IF j>k−n THEN
                      LET b i j = first REF b_in j
                      LET b_in j = rest REF b_in j
                  ELSE NIL
              ELSE
                  IF i+j>k−n+1 THEN
                      LET b i j = REF b i−1 j
                  ELSE NIL
          LET step = execute
      execute
          DO FOR ALL (i,j) SUCH THAT 1≤i≤n & 1≤j≤n &
                                  i+j ≤ REF k + 1
```

```
            IF i+j>k−n+1 THEN
                LET c i j = REF c i j + REF a i j * REF b i j
            ELSE NIL
        IF REF k = 3n−2 THEN
            LET step = stop
        ELSE
            LET k = REF k + 1
            LET step = fetch
    stop
        NIL
```

Remark. DO FOR ALL ... SUCH THAT ... is not a legal notation in our implemened syntax. We use it for clarity, translation to legal notation being straightforward (see the Appendix for a complete specification).

12.4 LOCAL SYNCHRONIZATION

In the previous examples the processing elements were not explicit, but only implicitly identifiable as the operations occurring on a certain group of cells. We now introduce the constructs *subspace* and *synchronized* cells. The processing elements will be identifiable with the subspace. This will allow us to express local synchronization.

12.4.1 Subspace Concept

We first define the notion of a subspace theoretically.

Definition. A data space with a nondeterministic transition system (X, p) (i.e., p a relation on X) is termed *nondeterministic*.

For a nondeterministic data space $(X, \mathscr{F}, p)$, x a member of X, and a subset $\mathscr{F}'$ of $\mathscr{F}$, let $[x, \mathscr{F}']$ denote the subset of X

$$\{y \mid f(y) = f(x) \text{ for all } f \in \mathscr{F}'\}$$

[For $z = [x, \mathscr{F}']$ and $f \in \mathscr{F}'$ we will assume $f(z)$ to mean $f(x)$ even though technically we are using the same name for functions with different domains.]
A *subspace* of $(X, \mathscr{F}, p)$ is a data space $(X', \mathscr{F}', p')$ where $\mathscr{F}' \subseteq \mathscr{F}$, $X' = [x, \mathscr{F}']$, $x \in X$, and such that if $y' \in p'(x')$, $x \in x'$, $y \in y'$, and $f(y) = f(x)$ for all $f \in \mathscr{F} - \mathscr{F}'$, then $y \in p(x)$. The definition says that every transition of the subspace can be made in the main space without regard to, or disturbing, the memory outside of $\mathscr{F}'$.

In case $(X, \mathscr{F}, p)$ is deterministic, the subspace, once started, must continue until it halts. But here we will be looking at nondeterministic main spaces with a number of concurrent subspaces. All of the subspaces will be deterministic and the nondeterminism of the main space will arise from the choice of which subspace to push next.

Remark. In our previous work, the subspace concept plays an important role in connection with the decomposition of large spaces into manageable parts. A more extensive theoretical paper is in preparation. Concurrent subspaces have been useful in our studies of "collective memory" algorithms.

12.4.2 Simple Communication Mechanism

The $\mathscr{F}'$ of one subspace may have a nonempty intersection with the $\mathscr{F}''$ of another: the cells in the intersection are shared.

The syntactic mechanism for building a main space out of several subspaces is the following:

SPACE ⟨name⟩ ⟨data space definition⟩

Any types, cell types or functions defined in the data-space definition, are, with certain exceptions involving new constructs presently to be explained, strictly local to the subspace. The subspace does, however, have access to type declarations, but not in general cell or function declarations, which are global to it.

The only declarations that are allowed to be nonlocal are cell declarations preceded by SHARED or SYNCHRONIZED, for example,

SYNCHRONIZED CELLS x HOLD REAL.

Then under certain conditions which we will not specify completely here, the cell x of the subspace S may be referred to outside the subspace by the expression $S.x$.

Each member of a synchronized cell type has two colors in addition to the data it is declared to hold:

1. Ready to read
2. Ready to write

If any OMEGA, subspace or main space, generates a REF to a cell that is not ready to read, the transition of that space is held up until the cell is made ready to read. Whenever the transition is completed, that is, as part of the transition, the cell is made ready to write and thus is no longer ready to read. Similarly, if the synchronized cell is the target of an assignment generated by some OMEGA, the transition of that space is held up until the cell is made ready to write by another space.

The only reference outside a subspace to a member of a type declared synchronized or shared inside is in the new construct

EQUIVALENT S1.a, S2.b

which declares that a of subspace S1 is the same cell as b of subspace S2. The two types to which a and b belong must have been declared to hold the same global type.

We reemphasize that our notation is not intended to support directly an accompanying proof of correctness. The correctness of a nondeterminisitc algorithm is much more difficult to show than that of a deterministic one. Therefore, executable specification, the theme of this chapter, becomes even more important for the case of locally synchronized algorithms.

12.4.3 Locally Synchronized Sorting Algorithm

As an example we define a "sorting array" in data-space notation. The sorting array is made up of subspaces S_i, $1 \leq i \leq n$, and is intended for sorting an input sequence of length up to n. The mode of operation is as follows: S_i is synchronized with S_{i+1} on a channel big enough to accommodate an input plus flag. In addition, they share an "acknowledgment bit." Every input is pipelined through the array on the basis of simple comparisons at each processing element: If the new input is bigger than the current contents, hand on the input; otherwise, keep the input and hand on the contents. Thus processing element S_i eventually becomes activated upon the ith input from the outside. The last input is flagged TRUE. The flag is handed on together with the input value, and thus eventually reaches the last processor element S_k, where k is the length of the input sequence. S_k prepares for output by putting its contents, flagged TRUE, into the channel to its left, together with an acknowledgment, then waits for the acknowledgment to get turned off, and finally, returns to idle. In fact, every S_i does all that except that the S_i, $i \neq k$, flag their values FALSE. Upon receiving an acknowledgment from the right, outputting starts by way of pipelining to the left. As soon as the flag TRUE is encountered, S_i turns off the acknowledgment bit to its right. S_1's acknowledgment TRUE signals to the outside that the output is ready; the last output in sorted order is flagged TRUE. Upon receiving it, the outside agent is supposed to turn off S_1's acknowledgment bit.

The formal definition follows:

```
SPACE S_i
    TYPE value & flag → value REAL flag BOOLEAN ||
    CELLS contents HOLD REAL
    SYNCHRONIZED CELLS left | right HOLD value & flag
    SYNCHRONIZED CELLS ack-left | ack-right HOLD BOOLEAN
    CELLS state HOLD idle | active | send-contents | wait-for-ack-left |
                        wait-for-ack-right | outputting | send-last-output
```

Global to all subspace definitions we declare for $i = 1, \ldots, n - 1$:

$$\text{EQUIVALENT } S_i.\text{right, } S_{i+1}.\text{left}$$
$$\text{EQUIVALENT } S_i.\text{ack-right, } S_{i+1}.\text{ack-left}$$

Initially, ack-left is FALSE and state is idle for every S_i. Inputs are delivered into S_1's left; the last input is flagged TRUE. Initially, all synchronized cells are ready to write.

The processor of each SPACE S_i is defined as follows:

```
FUNCTION OMEGA =
  TEST state
    idle
        LET contents = value REF left
        IF flag REF left THEN
            LET ack-left = TRUE
            LET state = send-last-output
        ELSE
            LET state = active
    active
        IF REF contents > value REF left THEN
            LET right = (contents, flag REF left)
            LET contents = value REF left
        ELSE
            LET right = REF left
        IF flag REF left THEN
            LET state = send-contents
        ELSE NIL
    send-contents
        LET ack-left = TRUE
        LET left = (REF contents, FALSE)
        LET state = wait-for-ack-right
    wait-for-ack-right
        IF REF ack-right THEN
            LET state = outputting
        ELSE NIL
    outputting
        LET left = REF right
        IF flag REF right THEN
            LET ack-right = FALSE
            LET state = wait-for-ack-left
        ELSE NIL
    send-last-output
        LET left = (REF contents, TRUE)
```

```
        LET state = wait-for-ack-left
wait-for-ack-left
        IF NOT REF ack-left THEN
            LET state = idle
        ELSE NIL
```

This gives the array

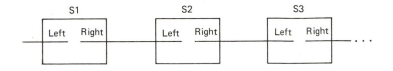

Such a construction not only has simplicity of notation but carries the additional advantage that we may equip it with a global synchronized value, say input_output, equivalenced to S1.left, and then it will be complete in the sense that it may be "plugged in" to any data space as a subspace. The larger space would equivalence one of its cells to sorter.input_output.

12.4.4 Locally Synchronized Wavefront Matrix

The wavefront matrix of Kung [12] can now be expressed as follows (the use of local instead of global synchronization will allow the wavefront to be crooked): For each i, j, $1 \leq i, j \leq n$, define

```
SPACE S_{ij}
    SYNCHRONIZED  CELLS a_left | a_right | b_up | b_down
                    HOLD REAL
    CELLS c HOLD REAL "initially 0"
    FUNCTION OMEGA =
        LET c = REF c + REF a_left * REF b_up
        LET a_right = REF a_left "omitted for S_in"
        LET b_down = REF b_up "omitted for S_ni"
```

The equivalences are

$$\text{EQUIVALENT } S_{ij}.\text{a\_right, } S_{i\,j+1}.\text{a\_left,} \quad (j < n)$$
$$S_{ij}.\text{b\_down, } S_{i+1\,j}.\text{b\_up} \quad (i < n)$$

The processor is not really complete because we have made no provision for outputting and resetting for the next problem. This would be done in a way similar to that used in the sorter.

ACKNOWLEDGMENTS

Thanks are due to Dick Lau and David Mizell for pointing out the subject area to us, as well to H. Heckhoff and G. Krüger for testing the algorithms on the data-space machine.

Work supported in part by ONR Contract N00014-77-C-0536 through the University of Southern California.

REFERENCES

[1] J. Backus, "Can Programming Be Liberated from the Von Neumann Style? A Functional Style and Its Algebra of Programs," *Commun. ACM*, *21*:613–641 (Aug. 1978).

[2] A. B. Cremers and T. N. Hibbard, "Formal Modeling of Virtual Machines," *IEEE Trans. Software Eng.*, *4*:426–436 (1978).

[3] A. B. Cremers and T. N. Hibbard, "Functional Behavior in Data Spaces," *Inf.-Forschungsber.*, No. 37/77, Universität Dortmund; *Acta Inf.*, *9*:293–307, 1978.

[4] A. B. Cremers and T. N. Hibbard, "Orthogonality of Information Structures," *Inf.-Forschungsber.*, No. 31/76, Universität Dortmund; *Acta Inf.*, *9*:243–261 (1978).

[5] A. B. Cremers and T. N. Hibbard, "On the Formal Definition of Dependencies between the Control and Information Structure of a Data Space," *Theor. Comput. Sci.*, *5*:113–128 (1977).

[6] A. B. Cremers and T. N. Hibbard, "Data Spaces with Indirect Addressing," *Inf.-Forschungsber.*, No. 43/77, Universität Dortmund; *Math. Syst. Theory*, *12*:151–173 (1978).

[7] A. B. Cremers and T. N. Hibbard, "Specification of Data Spaces by Means of Context-Free Grammar-Controlled Primitive Recursion," *Inf.-Forschungsber.*, No. 107/80, Universität Dortmund (1980).

[8] L. S. Haynes, R. L. Lau, D. P. Siewiorek, and D. W. Mizell, "A Survey of Highly Parallel Computing," *IEEE Comput.*, Jan. 1982, pp. 9–24.

[9] C. A. R. Hoare, "Communicating Sequential Processes," *Commun. ACM*, *21*:666–677 (1978).

[10] J. J. Horning and B. Randell, "Process Structuring," *ACM Comput. Surv.*, *5*:5–30 (1973).

[11] H. T. Kung, "Why Systolic Architectures?" *IEEE Comput.*, *15*(1):37–46 (Jan. 1982).

[12] S. Y. Kung, K. S. Arun, R. J. Gal-Ezer, and D. V. Bhaskar Rao, "Wavefront Array Processor: Language, Architecture, and Applications," *IEEE Trans. Comput.*, C-*31*(11):1054–1066 (Nov. 1982).

[13] C. Mead and L. Conway, *Introduction to VLSI Systems*, Addison-Wesley, Reading, Mass., 1980.

[14] D. L. Parnas, "A Technique for Software Module Specification with Examples," *Commun. ACM*, *15*:330–336 (1972).

APPENDIX

For the matrix multiplication algorithm in Section 12.3.2 we include the listing of the data-space definition together with the complete executed transition-directed steps of the successful execution trace.

For simplicity, the definition of OMEGA has been decomposed into auxiliary functions. On the data-space machine used for execution, some of the syntactic shortcuts mentioned in Section 12.2.2 have not yet been implemented. However, the difference in notation is minimal.

```
TYPE AIJ  →  A  A1  INTEGER
                 A2  INTEGER : :
TYPE BIJ  →  B  B1  INTEGER
                 B2 INTEGER : :
TYPE CIJ  →  C  C1  INTEGER
                 C2 INTEGER : :
TYPE INPUT_SEQUENCE  →  , FIRST REAL
                          REST  INPUT_SEQUENCE :

                           . : :

TYPE STEP  →  STEP : :
TYPE OPERATION  →  FETCH : EXECUTE : STOP : :
TYPE INT_CELL  →  K : N : :
TYPE A_B_IN  →  A_IN A_IN_S INTEGER :
                B_IN B_IN_S INTEGER : :


CELLS AIJ HOLD REAL
CELLS BIJ HOLD REAL
CELLS CIJ HOLD REAL
CELLS A_B_IN HOLD INPUT_SEQUENCE
CELLS STEP HOLD OPERATION
CELLS INT_CELL HOLD INTEGER

FUNCTION A_FUNC (K INTEGER N INTEGER I INTEGER
J INTEGER) ASSGTS  =
    IF < I  + N 1
      IF AND  < J  + N 1  < + I J  + K 2
        : IF = J 1
            IF > I  − K N
              : LET A I J =  FIRST REF A_IN I
                LET A_IN I =  REST REF A_IN I :
              SKIP
          IF > + I J + − K N 1
              LET A I J =  REF A I  − J 1
```

```
              SKIP
          A_FUNC K N I + J 1 :
          A_FUNC K N + I 1 1
      SKIP

FUNCTION B_FUNC (K INTEGER N INTEGER I INTEGER
J INTEGER) ASSGTS =
    IF < I + N 1
      IF AND < J + N 1  < + I J  + K 2
        : IF = I 1
            IF > J  − K N
              : LET B I J = FIRST REF B_IN J
                LET B_IN J = REST REF B_IN J :
              SKIP
            IF > + I J + − K N 1
              LET B I J = REF B − I 1 J
              SKIP
          B_FUNC K N I + J 1 :
          B_FUNC K N + I 1 1
      SKIP

FUNCTION C_FUNC (K INTEGER N INTEGER I INTEGER
J INTEGER) ASSGTS =
    IF < 1 + N 1
      IF AND < J + N 1  < + I J  + K 2
        : IF > + I J + − K N 1
            LET C I J = + REF C I J  * REF A I J  REF B I J
            SKIP
          C_FUNC K N I + J 1 :
          C_FUNC K N + I 1 1
      SKIP

FUNCTION C_INIT_HELP (N INTEGER IND INTEGER) ASSGTS =
    IF > IND 0
      :  LET C IND N = 0.0
         LET C N IND = 0.0
         C_INIT_HELP N − IND 1 :
      SKIP

FUNCTION C_INIT (N INTEGER) ASSGTS =
    : LET C N N = 0.0
      C_INIT_HELP N − N 1
      IF > N 1
        C_INIT − N 1
        SKIP :
```

```
OMEGA
    TEST REF STEP :
          FETCH : A_FUNC REF K   REF N   1   1
                  B_FUNC REF K   REF N   1   1
                  LET STEP = EXECUTE :
          EXECUTE : C_FUNC REF K   REF N   1   1
                    IF = REF K − * 3 REF N 2
                    LET STEP = STOP
                    : LET K = + REF K 1
                      LET STEP = FETCH : :
          STOP SKIP :

LOAD
    : LET K = 1
      LET N = 3
      C_INIT 3
      LET A_IN 1 = , 1.0 , 2.0 , 1.0 .
      LET A_IN 2 = , 2.0 , 2.0 , 1.0 .
      LET A_IN 3 = , 2.0 , 1.0 , 2.0 .
      LET B_IN 1 = , 3.0 , 4.0 , 5.0 .
      LET B_IN 2 = , 4.0 , 5.0 , 3.0 .
      LET B_IN 3 = , 5.0 , 3.0 , 4.0 .
      LET STEP = FETCH :
```

(.K: = 1.)
(.N: = 3.)
(.C33: = + 0.00000E + 00.)
(.C23: = + 0.00000E + 00.)
(.C32: = + 0.00000E + 00.)
(.C13: = + 0.00000E + 00.)
(.C31: = + 0.00000E + 00.)
(.C22: = + 0.00000E + 00.)
(.C12: = + 0.00000E + 00.)
(.C21: = + 0.00000E + 00.)
(.C11: = + 0.00000E + 00.)
(.A_IN1: = , + 1.00000E + 00, + 2.00000E + 00, + 1.00000E + 00..)
(.A_IN2: = , + 2.00000E + 00, + 2.00000E + 00, + 1.00000E + 00..)
(.A_IN3: = , + 2.00000E + 00, + 1.00000E + 00, + 2.00000E + 00..)
(.B_IN1: = , + 3.00000E + 00, + 4.00000E + 00, + 5.00000E + 00..)
(.B_IN2: = , + 4.00000E + 00, + 5.00000E + 00, + 3.00000E + 00..)
(.B_IN3: = , + 5.00000E + 00, + 3.00000E + 00, + 4.00000E + 00..)
(.STEP: = FETCH.)
→ → →
(.A11: = + 1.00000E + 00.)

(.A_IN1: =, +2.00000E +00, +1.00000E +00..)
(.B11: = +3.00000E +00.)
(.B_IN1: =, +4.00000E +00, +5.00000E +00..)
(.STEP: = EXECUTE.)
→ → →
(.C11: = +3.00000E +00.)
(.K: = 2.)
(.STEP: = FETCH.)
→ → →
(.A11: = +2.00000E +00.)
(.A_IN1: =, +1.00000E +00..)
(.A12: = +1.00000E +00.)
(.A21: = +2.00000E +00.)
(.A_IN2: =, +2.00000E +00, +1.00000E +00..)
(.B11: = +4.00000E +00.)
(.B_IN1: =, +5.00000E +00..)
(.B12: = +4.00000E +00.)
(.B_IN2: =, +5.00000E +00, +3.00000E +00..)
(.B21: = +3.00000E +00.)
(.STEP: = EXECUTE.)
→ → →
(.C11: = +1.09999E +01.)
(.C12: = +4.00000E +00.)
(.C21: = +6.00000E +00.)
(.K: = 3.)
(.STEP: = FETCH.)
→ → →
(.A11: = +1.00000E +00.)
(.A_IN1: =..)
(.A12: = +2.00000E +00.)
(.A13: = +1.00000E +00.)
(.A21: = +2.00000E +00.)
(.A_IN2: =, +1.00000E +00..)
(.A22: = +2.00000E +00.)
(.A31: = +2.00000E +00.)
(.A_IN3: =, +1.00000E +00, +2.00000E +00..)
(.B11: = +5.00000E +00.)
(.B_IN1: =..)
(.B12: = +5.00000E +00.)
(.B_IN2: =, +3.00000E +00..)
(.B13: = +5.00000E +00.)
(.B_IN3: =, +3.00000E +00, +4.00000E +00..)
(.B21: = +4.00000E +00.)
(.B22: = +4.00000E +00.)

```
(.B31: = +3.00000E+00.)
(.STEP: = EXECUTE.)
→  →  →
(.C11: = +1.59999E+01.)
(.C12: = +1.39999E+01.)
(.C13: = +5.00000E+00.)
(.C21: = +1.39999E+01.)
(.C22: = +8.00000E+00.)
(.C31: = +6.00000E+00.)
(.K: = 4.)
(.STEP: = FETCH.)
→  →  →
(.A12: = +1.00000E+00.)
(.A13: = +2.00000E+00.)
(.A21: = +1.00000E+00.)
(.A_IN2: = ..)
(.A22: = +2.00000E+00.)
(.A23: = +2.00000E+00.)
(.A31: = +1.00000E+00.)
(.A_IN3: = , +2.00000E+00..)
(.A32: = +2.00000E+00.)
(.B12: = +3.00000E+00.)
(.B_IN2: = ..)
(.B13: = +3.00000E+00.)
(.B_IN3: = , +4.00000E+00..)
(.B21: = +5.00000E+00.)
(.B22: = +5.00000E+00.)
(.B23: = +5.00000E+00.)
(.B31: = +4.00000E+00.)
(.B32: = +4.00000E+00.)
(.STEP: = EXECUTE.)
→  →  →
(.C12: = +1.69999E+01.)
(.C13: = +1.09999E+01.)
(.C21: = +1.89999E+01.)
(.C22: = +1.79999E+01.)
(.C23: = +1.00000E+01.)
(.C31: = +1.00000E+01.)
(.C32: = +8.00000E+00.)
(.K: = 5.)
(.STEP: = FETCH.)
→  →  →
(.A13: = +1.00000E+00.)
(.A22: = +1.00000E+00.)
(.A23: = +2.00000E+00.)
```

```
(.A31: = +2.00000E +00.)
(.A_IN3: =..)
(.A32: = +1.00000E +00.)
(.A33: = +2.00000E +00.)
(.B13: = +4.00000E +00.)
(.B_IN3: =..)
(.B22: = +3.00000E +00.)
(.B23: = +3.00000E +00.)
(.B31: = +5.00000E +00.)
(.B32: = +5.00000E +00.)
(.B33: = +5.00000E +00.)
(.STEP: = EXECUTE.)
→  →  →
(.C13: = +1.50000E +01.)
(.C22: = +2.09999E +01.)
(.C23: = +1.59999E +01.)
(.C31: = +2.00000E +01.)
(.C32: = +1.29999E +01.)
(.C33: = +1.00000E +01.)
(.K: =6.)
(.STEP: = FETCH.)
→  →  →
(.A23: = +1.00000E +00.)
(.A32: = +2.00000E +00.)
(.A33: = +1.00000E +00.)
(.B23: = +4.00000E +00.)
(.B32: = +3.00000E +00.)
(.B33: = +3.00000E +00.)
(.STEP: = EXECUTE.)
→  →  →
(.C23: = +2.00000E +00.)
(.C32: = +1.89999E +01.)
(.C33: = +1.29999E +01.)
(.K: =7.)
(.STEP: = FETCH.)
→  →  →
(.A33: = +2.00000E +00.)
(.B33: = +4.00000E +00.)
(.STEP: = EXECUTE.)
→  →  →
(.C33: = +2.09999E +01.)
(.STEP: = STOP.)
→  →  →
→  →  →
```

13

Concurrent Algorithms as Space-Time Recursion Equations

Marina C. Chen and Carver A. Mead

California Institute of Technology
Pasadena, California

13.1 INTRODUCTION

Recent developments in the technology of fabricating large-scale integrated circuits have made it possible to implement computing systems that use many hundred thousands of transistors to achieve a given task. An interesting design will have high computational complexity rather than merely vast numbers of identical simple components such as memory elements. Such a design can be represented as a fully instantiated implementation of objects of the implementation medium (e.g., transistors in VLSI technology) or as successive hierarchical levels of implementations where each level is constructed of objects which are abstract models of the implementation at the level below it. The former allows implementation details at the bottom level to penetrate throughout the whole design. Such representation may be suited for machine execution but is hard to deal with from the designer's point of view, and verifying both its functionality and physical layout is costly. As the complexity of the design grows, the limitation of this approach becomes more apparent. The second approach is aimed at managing the complexity of a design. One breaks the design into successive levels of subsystems until each is of a manageable complexity—the hierarchical design method [11].

13.1.1 A Hierarchical Design

Imagine that one wants to carry out the computation of a matrix multiplication; there are many possible ways to do it. One possible implementation of this function is by the algorithm [8] shown in Figure 13.1. The elementary building block has three inputs and three outputs each of which is an integer with a bound. Each element performs $a_{out} = m_{in} \times b_{in} + a_{in}$, $m_{out} = m_{in}$ and $b_{out} = b_{in}$. How each ele-

224

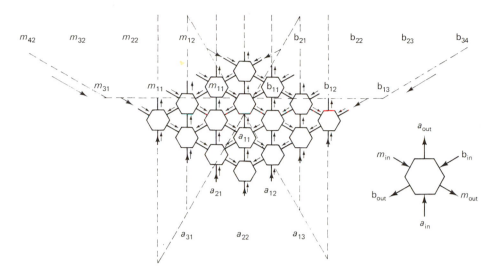

Figure 13.1   Systolic algorithm.

ment is implemented is not of concern here, only the behavioral description of each of the elements used in designing and reasoning about the systolic algorithm. The complexity or performance of this algorithm are also discussed using that of each element as a unit without knowing the actual value for each unit. Once the design of this algorithm is completed, namely, it is verified to be correct and to satisfy the requirements in performance, one can move on and focus on the design of each individual element.

Such an element can be implemented by serial operation on each bit of a binary number or by operations on a parallel word that stores the binary number. A possible bit-serial implementation [9,18] is shown in Figure 13.2; three sequences of input bits are shifted into the pipeline and the result sequences come out the other end. The elementary building block is now a half-adder with some shift registers. Effort may be spent on how to minimize the total delay so the maximum throughput is possible. Boolean algebra is used in verifying the correctness of the algorithm.

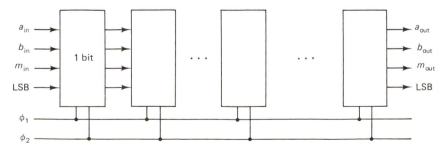

Figure 13.2   Pipelined algorithm.

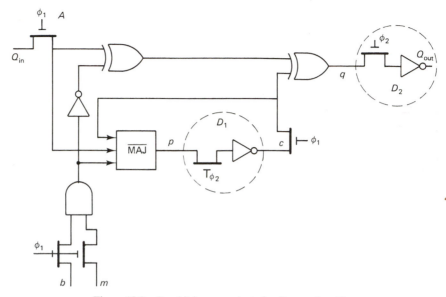

Figure 13.3   One-bit inner product circuit—an algorithm.

When the design at this level is completed, one moves on to a detailed implementation of each element.

Again a particular implementation is proposed, the logic circuit shown in Figure 13.3 is an "algorithm" describing the design. At this level, switching logic elements are used. In turn, these logic elements are implemented in a certain technology. Transistors, capacitors, and so on, are used to implement these logic functions as shown in Figure 13.4. At this level, algebra of signals [2] will be used.

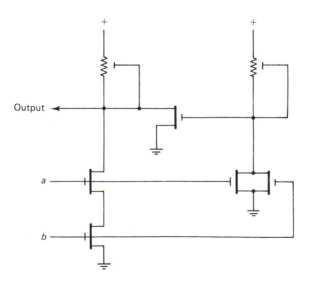

Figure 13.4   Transistor network—an algorithm.

Eventually, when the design is completed, it will be realized in a physical domain. By partitioning design into these levels, the designer need only concentrate on one level at a time. The details of lower levels are completely hidden.

13.1.2 Interplay of Space and Time

In each of the algorithms chosen above, the exact location and time step where the data arrive at a processing element are important. The interplay of space and time does not occur in a program in a conventional "high-level" language since only one thing happens at a time, and one cannot care less where each item of data is physically stored. This is not the case in the design of VLSI systems. Data that are far apart cost more energy and take more time to access, since a longer wire means larger capacitance and resistance. What is considered as a good algorithm in a traditional language is not necessarily good when physical cost is taken into account. For example, the algorithm Quicksort has performance advantage $n \log n$ over n^2 of the Interchange sort algorithm. Yet Quicksort involves swapping of data that are arbitrarily far apart, each unit of cost is as expensive as accessing data that are farthest away. On the other hand, interchange sort only swaps data that are next to each other; each unit of cost is that involved in accessing only the neighboring elements [17]. This locality of communication is the key to an algorithm that is amenable to VLSI implementation. Hence VLSI is a richer environment in which both space and time need to be considered for design, verification, and analysis of algorithms.

Notation for describing VLSI systems.   So far the examples of algorithms for VLSI above are described by pictures. It is possible to reason about an algorithm informally using pictures when an algorithm is relatively simple. A formal notation will not only be helpful in reasoning about the design, but can also be used to generate simulation and drive the compilation or assembly of physical layouts. Moreover, it is the first step toward an automatic verification tool (theorem prover) for VLSI systems. With the advent of computer-aided-design tools for VLSI circuits, specifying VLSI designs in an *executable* form has become well accepted. However, the specification must be *manipulable* as well for the following reason. The hierarchical design method demands proper interfaces among pieces of design at each level, and consistency between the abstract model at one level and the implementation at the level below. At a very low level, where the inputs and internal states of a subsystem are only a few bits, consistency checking can be done by exhaustively verifying all possible cases. For any larger subsystem, the consistency check amounts to verification of a program, where the program corresponds to the specification of the subsystem. Therefore, the specification language must have a formal semantics so that the specification of a given design can be formally verified.

Another important property of a language for VLSI systems is its applicability to objects of all levels. It must be able to describe a network of processes as well as a transistor network. It must be capable of describing a realistic system in a simple way and its semantics can be obtained easily. A language is of no use if either the description or the semantics is insurmountably complicated.

In [3], a framework for specifying and verifying VLSI systems is presented. The same language is used for systems ranging from the level of transistors up to the level of communicating processes. Fixed-point semantics [10,16] is used for abstracting the behavior of a system from its implementation. The consistency between an implementation and its description can therefore be verified. The framework can be viewed as a concurrent programming notation when describing communicating processes, a hardware description notation when specifying integrated circuits.

This chapter contains an introduction to the framework. The model of computation for concurrent systems is first presented. It differs from "communicating sequential processes" [6] in two respects.

1. It separates the deterministic concurrent model from a more general nondeterministic one. The former is presented here and the extension to the latter in [3].
2. An applicative state transition in the sense of [1] is introduced so that a functional style of programming is possible. The description of a system is called a space-time algorithm, since space and time are explicit in the description.

A "pipelined" algorithm is described and the proofs of its correctness are given. The power of this framework cannot be illustrated by this single example; interested readers should refer to [3].

13.2 MODEL OF COMPUTATION

The model consists of a collection of *processes* [5–7,18]. Each process has

1. A *control state* register for determining the communication of the process with other processes.
2. A data store.
3. The machinery for computing a *state transition function* [1].
4. Input ports and output ports. A port is *filled* in the sense that a place in a Petri net [13,14] is filled by a token and *emptied* as a token is removed from a place when the corresponding transition is fired.

To describe the relationship among processes, coordinate systems are used.

5. Each process is located in a space coordinate system where each coordinate is taken from a countable set.
6. It is often convenient to use a global time coordinate to index the operations occurring in the ensemble when there exists a total ordering of these oper-

ations. In general, operations occurring in the ensemble of processes are concurrent and cannot be totally ordered. Therefore, each process has a local time coordinate, taken from a countable well-ordered set, that is used to index the operations that each process has performed. A set of relations among these local time coordinates of the processes is derived.

7. The relationship among processes is established by identifying input port of process s_1 at its t_{s_1}th operation with the output port of process s_2 at its t_{s_2}th operation. The counting of the operations is defined below.

A process operates according to the following procedure:

8. The state of the control state register is used to select a set of input ports. The state of this register is used only in this way and is not used as an argument to a state transition function.

9. If all of the selected input ports are filled, we say that communication is established. The process now starts an operation consisting of (a) emptying the selected input ports and (b) evaluating the state transition function (firing a transition as in Petri net), called the tth *invocation* of the process, where t is the time coordinate of the process. If some input ports are not filled, the process waits until they are.

10. Data in the selected input ports and the state in the data store are arguments to the state transition function. After the function is evaluated, the result is used to update the data store and control state register and to fill the designated output ports. If some of the designated output ports are not emptied, the process waits until they are emptied and then fills them. After all designated output ports are filled, the tth invocation is completed and the process starts step 8 again for the $(t + 1)$th invocation, where $t + 1$ denotes the next element in the well-ordered set for the time coordinate.

13.3 THE LANGUAGE CRYSTAL†

The language and its semantics are based on the typed λ-calculus version [12] of Scott's theory of computable functions [15]. Each of the parts or operations of the computation model above is described by either a constant value, a variable, a function, or a function of functions.

Data types.   There are various data types for state and inputs/outputs for a wide range of systems of interests. Examples of data types are:

1. The set of analog voltage values $\mathscr{A} \equiv \{a : 0 \leq a \leq V\}$: a subset of the set of real numbers

†CRYSTAL stands for "Concurrent Representation of Your Space-Time Algorithm."

2. The set of Boolean values $\mathscr{B} \equiv \{\hat{0}, \hat{1}\}$

3. The set of n-bit words in two's-complement representation $\{i : -2^{(n-1)} \le i < 2^{(n-1)}\}$: a subset of the set of integers

State transition functions.   According to the model of computation, a state transition function is the basic unit used in constructing a system. Depending on the systems of interest, different functions are used as the primitive state transition functions. In designing a VLSI system, a primitive can be an analog model of a transistor when detailed electrical characteristics of the system are desired. A switch-level model of a transistor is used if the logic values computed by the circuit are desired. An adder is the primitive, for example, when a multiplier is built. A state transition function can also be implemented by a collection of existing systems rather than given as a primitive, in which case it is called a composite state transition function. Its meaning will be precisely defined below.

In describing a primitive state transition function, say, a model of a transistor, functions such as addition, minimum, *and* maximum on subsets of integers are used. These functions are primitives of the language and must not be confused with those of the system under construction. We may well use the plus function (which models parts of the behavior of a transistor) to construct a piece of machinery for computing the plus function (which models an n-bit adder, for example).

Let $\mathbf{x} \equiv (x_1, x_2, \ldots, x_m)$ denote the arguments of a state transition function f where $x_i \in \mathscr{D}_i$ (the data type of x_i), $i = 1, 2, \ldots, m$.

$$f : \mathscr{D}^m \to \mathscr{D}^n$$
$$f = (\lambda \mathbf{x} \cdot f_1, \lambda \mathbf{x} \cdot f_2, \ldots, \lambda \mathbf{x} \cdot f_n)$$

(13.1)

Each component, $\lambda \mathbf{x} \cdot f_i$, of such a function is an element of $[\mathscr{D}^m \to \mathscr{D}]$; in the syntax below, we call each component a state transition function.

Coordinates.   To express the relationship among invocations of processes in the space-time coordinate system, or to express different state transition functions (or different relationship of invocations) at different points in the coordinate system, it is necessary to use functions on these coordinates. These functions are primitives of the language and like the plus function, must not be confused with elements of the system being described. Although expressions of coordinates are not part of the computation of the system being designed, they are part of how the system is going to be constructed (space coordinates) and used (time coordinates), that is, what timing discipline is imposed. Examples of such data types:

1. The discrete time domain $\mathscr{T} = \{0, 1, 2, \ldots\}$ (the set of nonnegative integers).
2. The discrete space domain $\mathscr{S} \equiv \{(x, y) : 0 \le x < n, \ 0 \le y < n, \ x, \ y \ \text{are integers}\}$.
3. The set of k-tuples $\mathscr{E} \equiv \{(e_1, e_2, \ldots, e_{k-1}, e_k)\}$, where $e_1, e_2, \ldots, e_k$ are space and time coordinates. This class of data types is called the space-time domain.

Data streams.   Each process is a point in a space coordinate system, and each invocation of a process is a point in a space-time coordinate system where the space coordinates are the same as the corresponding space coordinates of the process and the time coordinate is local within the process. Control state, data, inputs, and outputs of all processes as used in the invocation have the same coordinates as the invocation. The state and input/output values are defined in the space-time coordinates as the computation proceeds. They are expressed as unknown functions from the space-time domain to a certain value domain. In the beginning of the computation, only the initial state and initial inputs are defined. As computation goes on, more state and inputs/outputs are defined in the space-time domain; computation ends when no more of them are defined. Each of these unknowns is called a data stream.

Space-time algorithm.   An algorithm describing a system consists of the description of the computational part (applying state and input/output to state transition functions) and the communicative part (equating or identifying an input with an output at another point in the space-time domain). The result is a system of recursion equations in the space-time coordinates.

13.4 THE BEHAVIOR OF AN ALGORITHM

The semantics of a space-time algorithm is the solution of the system of equations above. It is a description of values computed by the algorithm as a function of space and time. If an algorithm is to be treated as a black box when used to construct some other algorithms, we must describe outputs at the end of a computation in terms of inputs at the beginning of the computation. From the solution, the behavioral description of an algorithm is derived. It is a system of state transition functions that maps inputs and current states $(X_1, \ldots, X_{m'})$ to outputs and next states $(Y_1, \ldots, Y_{n'})$. It therefore can serve as a primitive building block for constructing a more complex system.

13.5 SIMPLE EXAMPLE OF A SPACE-TIME ALGORITHM

The following is a very simple space-time algorithm which corresponds to the program that computes the factorial function in an assignment-based language. Given an input a, this program computes *fac* as the result.

```
count ← a, fac ← 1
while count > 0 do
begin fac ← fac × count ; count := count − 1 ; end ;
```

For any $t \geq a$, in the following corresponding space-time algorithm, **fac**$(t) = a!$. Since this algorithm is sequential, only one process is needed and the space domain degenerates to one point. Let $\mathcal{T} = 0, 1, 2, \ldots$ be the time domain.

$$
\mathbf{count}(t) = \begin{cases} t = 0 \rightarrow a \\ t > 0 \rightarrow \begin{cases} \mathbf{count}(t-1) > 0 \rightarrow \mathbf{count}(t-1) - 1 \\ \mathbf{count}(t-1) \leq 0 \rightarrow \mathbf{count}(t-1) \end{cases} \end{cases}
$$
$$
\hspace{10cm} (13.2a)
$$
$$
\mathbf{fac}(t) = \begin{cases} t = 0 \rightarrow 1 \\ t > 0 \rightarrow \begin{cases} \mathbf{count}(t-1) > 0 \rightarrow \mathbf{fac}(t-1) \times \mathbf{count}(t-1) \\ \mathbf{count}(t-1) \leq 0 \rightarrow \mathbf{fac}(t-1) \end{cases} \end{cases}
$$

Data streams in this example are **count** and **fac**. Both **count**(t) and **fac**(t) are of type $\mathcal{N} \equiv \{0, 1, 2, \ldots\}$. Three state transition functions in the algorithm are $f = (\lambda \mathbf{x} \cdot a, \lambda \mathbf{x} \cdot 1)$, where each component is a constant function, $g = (\lambda \mathbf{x} \cdot (x_1 - 1), \lambda \mathbf{x} \cdot (x_2 \times x_1))$, and $h = (\lambda \mathbf{x} \cdot x_1, \lambda \mathbf{x} \cdot x_2)$. Notice that in the algorithm above, **count** is used as a variable for keeping track of the number to be multiplied to the partial result. Why do we not write the algorithm as the following equation, where the time coordinate t is used in the computation of factorial a?

$$
\mathbf{fac}(t) = \begin{cases} t = 0 \rightarrow 1 \\ t > 0 \rightarrow \begin{cases} t \leq a \rightarrow \mathbf{fac}(t-1) \times t \\ t > a \rightarrow \mathbf{fac}(t-1) \end{cases} \end{cases}
$$

The reason is that t does not exist physically; it is only a reference frame for us to envision and reason about the computation. To see that the algorithm computes the factorial function, we claim that the following is the solution of equation (13.2a) [the least fixed point of the functional on the right-hand side of (13.2a)].

$$
\mathbf{count}^{\infty}(t) = \begin{cases} t > a \rightarrow 0 \\ t \leq a \rightarrow a - t \end{cases}
$$
$$
\hspace{10cm} (13.2b)
$$
$$
\mathbf{fac}^{\infty}(t) = \begin{cases} t > a \rightarrow a! \\ t \leq a \rightarrow \dfrac{a!}{(a-t)!} \end{cases}
$$

By a straightforward induction on t, (13.2b) can be shown to be the solution of (13.2a). Notice that for all $t > a$, both **count**$(t) = $ **count**$(t-1)$ and **fac**$(t) = $ **fac**$(t-1)$. Thus the time domain can be restricted to such that $t \leq a$. In general, a system reaches its *steady state* at t_{steady}, where t_{steady} is the maximum of all t_i, where t_i is the smallest t such that data stream i at t is the same as it is at $t - 1$, that is,

$$
t_{steady} = \max_{t_i} \{\mu t_i [\lambda s \cdot (Stream_i(s, t) - Stream_i(s, t-1)) = \lambda s \cdot 0] : i \in \{1, 2, \ldots, n\}\}
$$

where μ is the minimalization operator and i is the index for data streams.

The behavior of the algorithm is $\lambda a \cdot \mathbf{fac}(t_{steady}) = \lambda a \cdot a!$, and thus the algorithm implements the factorial function correctly.

13.6 SPACE-TIME ALGORITHM FOR A CONCURRENT SYSTEM

The pipelined architecture (Figure 13.2) is very similar to the systolic architecture (Figure 13.1) in that local communication is used to avoid long propagation delay. It is usually simpler to describe and analyze because of the one-dimensional structure in space. In the example presented below, each stage of the pipeline has internal state, which is not the case in the systolic array example. In CRYSTAL, since the time coordinate is explicitly used, internal state does not pose difficulty in describing a system.

The following is the space-time algorithm for an n-stage pipelined inner product element IPE given the behavioral description of a 1-bit inner product element IPB shown in Figure 13.5. The function this pipeline implements is $B \times M + A_{in}$, where B and M are n-bit nonnegative binary numbers and A_{in} is a $2n$-bit nonnegative binary number.

We show that the behavior of an IPE element (the composition of n IPB elements and several n-bit shift registers) is in fact the function above. The symbols " $\cdot$ " and " $+$ " denote the "and" and "or" operations on Boolean values "$\hat{0}$" and "$\hat{1}$". Let $\mathscr{S} \equiv \{0, 1, 2, \ldots, n\}$ be the space domain for indexing the IPB elements and shift registers and $\mathscr{T} \equiv \{0, 1, 2, \ldots\}$ be the set of nonnegative integers which indicates the steps of computation. Since the behavior of this pipeline inner product element is periodic in time, the value of the expression $t - s$ can be restricted to within 1 and $2n$. The space-time domain $\mathscr{E}$ for the pipeline is defined as

$$\mathscr{E} \equiv \{(s, t) : s \in \mathscr{S}, t \in \mathscr{T}, \text{ and } 1 \le t - s \le 2n\}$$

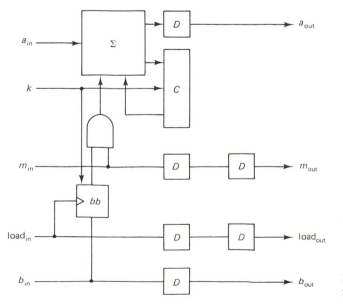

Figure 13.5 Single bit inner product element IPB.

The value of $\mathbf{a}(s, t)$ is the accumulated partial sum computed by IPB element s at step t. $\mathbf{c}(s, t)$ and $\mathbf{bb}(s, t)$ are the carry of an adder and the sth bit of multiplier B, respectively; both are internal states of IPB element s. Signal $\mathbf{load}(s, t)$ is for loading $\mathbf{bb}(s, t)$ from $\mathbf{b}(s, t)$, a bit of the multiplier. Signal $\mathbf{k}(s, t)$ is used to clear the internal state of the carry and multiplier of the previous word before a new word starts. The value of $\mathbf{k}(s, t)$ computed at the $(k - 1)$th cycle (word) is used as initial conditions for the kth word. In this case, t and s such that $t - s = 0$, which is not in the specified space-time domain, is used in place of t and s such that $t - s = 2n$ in the description of $\mathbf{k}$.

$$\mathbf{a} \in [\mathscr{E} \to \mathscr{B}]$$

$$\mathbf{a}(s, t) = \begin{cases} s > 0 \to \mathbf{a}(s - 1, t - 1) \oplus (\mathbf{bb}(s, t) \cdot \mathbf{m}(s - 1, t - 2)) \oplus \mathbf{c}(s, t - 1) \\ s = 0 \to A_0(t) \end{cases} \tag{13.3}$$

$$\mathbf{c} \in [\mathscr{E} \to \mathscr{B}]$$

$$\mathbf{c}(s, t) = \begin{cases} s > 0 \to \\ \quad \text{MAJ}(\mathbf{a}(s - 1, t - 1), (\mathbf{bb}(s, t) \cdot \mathbf{m}(s - 1, t - 2)), \mathbf{c}(s, t - 1)) \\ \quad \cdot \bar{\mathbf{k}}(s - 1, t - 1) + \hat{0} \cdot \mathbf{k}(s - 1, t - 1) \\ s = 0 \to \bot \end{cases} \tag{13.4}$$

$$\mathbf{bb} = \in [\mathscr{E} \to \mathscr{B}]$$

$$\mathbf{bb}(s, t) \begin{cases} s > 0 \to (\mathbf{bb}(s, t - 1) \cdot \overline{\mathbf{load}}(s - 1, t - 2) \\ \quad + \mathbf{b}(s - 1, t - 1) \cdot \mathbf{load}(s - 1, t - 2)) \cdot \bar{\mathbf{k}}(s - 1, t - 1) \\ \quad + \hat{0} \cdot \mathbf{k}(s - 1, t - 1) \\ s = 0 \to \bot \end{cases} \tag{13.5}$$

The following are shift registers for propagating the operands; their functional behavior is derived.

$$\mathbf{load} \in [\mathscr{E} \to \mathscr{B}]$$

$$\mathbf{load}(s, t) = \begin{cases} (s > 0) \land (t \geq 2) \to \mathbf{load}(s - 1, t - 2) \\ s = 0 \to \begin{cases} t = 0 \to \hat{1} \\ t > 0 \to \hat{0} \end{cases} \end{cases} \tag{13.6}$$

therefore

$$\mathbf{load}(s, t) = \begin{cases} t = 2s \to \hat{1} \\ s < t < 2s \to \bot \\ 2s < t \leq 2n + s \to \hat{0} \end{cases}$$

$$\mathbf{k} \in [\mathscr{E} \to \mathscr{B}]$$

$$\mathbf{k}(s, t) = \begin{cases} (s > 0) \wedge (t \geq 1) \to \mathbf{k}(s - 1, t - 1) \\ s = 0 \to \begin{cases} t = 0 \to \hat{1} \\ 0 < t \leq 2n - 1 \to \hat{0} \end{cases} \end{cases} \tag{13.7}$$

therefore

$$\mathbf{k}(s, t) = \begin{cases} t = s \to \hat{1} \\ s < t \leq 2n + s - 1 \to \hat{0} \end{cases}$$

$$\mathbf{b} \in [\mathscr{E} \to \mathscr{B}]$$

$$\mathbf{b}(s, t) = \begin{cases} (s > 0) \wedge (t \geq 1) \to \mathbf{b}(s - 1, t - 1) \\ s = 0 \to \begin{cases} 0 < t \leq n \to B(t - 1) \\ n < t \leq 2n \to \perp \end{cases} \end{cases} \tag{13.8}$$

therefore

$$\mathbf{b}(s, t) = \begin{cases} s < t \leq n + s \to B(t - 1 - s) \\ n + s < t \leq 2n + s \to \perp \end{cases}$$

$$\mathbf{m} \in [\mathscr{E} \to \mathscr{B}]$$

$$\mathbf{m}(s, t) = \begin{cases} (s > 0) \wedge (t \geq 2) \to \mathbf{m}(s - 1, t - 2) \\ s = 0 \to \begin{cases} 0 \leq t < n \to M(t) \\ n \leq t < 2n \to \hat{0} \end{cases} \end{cases} \tag{13.9}$$

therefore

$$\mathbf{m}(s, t) = \begin{cases} s < t < 2s \to \perp \\ 2s \leq t < n + 2s \to M(t - 2s) \\ (n + 2s \leq t \leq 2n + s) \to \hat{0} \end{cases}$$

From the functional description of shift registers, (13.6), (13.7), and (13.8), the functional description of the internal state **bb** can be derived.

$$\mathbf{bb}(s, t) = \begin{cases} t = s \to \hat{0} \\ t = 2s \to \mathbf{b}(s - 1, t - 1) = B(t - 1 - s) = B(s - 1) \\ (s < t \leq 2s - 1) \vee (2s < t \leq 2n + s - 1) \to \mathbf{bb}(s, t - 1) \\ t = 2n + s \to \hat{0} \end{cases} \tag{13.10}$$

$$= \begin{cases} s < t \leq 2s - 1 \to \hat{0} \\ 2s \leq t \leq 2n + s - 1 \to B(s - 1) \\ t = 2n + s \to \hat{0} \end{cases}$$

To prove the correctness of this algorithm, the space-time structure is first mapped to another structure. The multiplication of two n-bit binary numbers can be represented by a recurrence of the partial product. In the following i is used to indicate the partial product at stage i and j is used to indicate bit j. Let

$$\mathcal{N}_1 \equiv \{0, 1, 2, \ldots, n\} \quad \text{and} \quad \mathcal{N}_2 \equiv \{0, 1, 2, \ldots, 2n - 1\}$$

and A and C be two functions in $[\mathcal{N}_1 \times \mathcal{N}_2 \to \mathcal{B}]$ where

$$A(i, j) \equiv \mathbf{a}(i, i + j + 1) \quad \text{and} \quad C(i, j) \equiv \mathbf{c}(i, i + j + 1)$$

Inversely,

$$\mathbf{a}(s, t) \equiv A(s, t - 1 - s) \quad \text{and} \quad \mathbf{c}(s, t) \equiv C(s, t - 1 - s) \tag{13.11}$$

Using definition (13.11) and substituting the result obtained in (13.9) and (13.10) into (13.3) and (13.4), we obtain the following relations among the new functions A, C, B, and M.

$$\mathbf{a}(s, t) = A(s, t - 1 - s)$$

$$= \begin{cases} s > 0 \to \begin{cases} 2s \le t \le n + 2s - 1 \to \\ \quad A(s - 1, (t - 1) - 1 - (s - 1)) \oplus (B(s - 1) \\ \quad \quad \cdot M((t - 2) - 2(s - 1))) \\ \quad \quad \oplus C(s, (t - 1) - s) \\ (s < t \le 2s - 1) \lor (n + 2s - 1 < t \le 2n + s) \to \\ \quad A(s - 1, (t - 1) - 1 - (s - 1)) \oplus C(s, (t - 1) - 1 - s) \end{cases} \\ s = 0 \to A_0(t) \end{cases} \tag{13.12}$$

$$\mathbf{c}(s, t) = C(s, t - 1 - s)$$

$$= \begin{cases} s > 0 \to \begin{cases} 2s \le t \le n + 2s - 1 \to \\ \quad \text{MAJ}(A(s - 1, (t - 1) - 1 - (s - 1)), (B(s - 1) \\ \quad \quad \cdot M((t - 2) - 2(s - 1))), \\ \quad C(s, (t - 1) - 1 - s)) \\ (s < t \le 2s - 1) \lor (n + 2s - 1 < t \le 2n + s) \to \\ \quad \text{MAJ}(A(s - 1, (t - 1) - 1 - (s - 1)), 0, \\ \quad C(s, (t - 1) - 1 - s)) \\ t = 2n + s - 1 \to \hat{0} \end{cases} \\ s = 0 \to \perp \end{cases} \tag{13.13}$$

Let $i = s$ and $j = t - 1 - s$; substituting them into (13.11) and (13.12), we obtain

$$A(i, j) = \begin{cases} i > 0 \rightarrow \begin{cases} i - 1 \leq j < n + i - 1 \rightarrow \\ \quad A(i - 1, j) \oplus (B(i - 1) \cdot M(j - (i - 1))) \oplus C(i, j - 1) \\ (0 \leq j < i - 1) \vee (n + i - 1 \leq j < 2n) \rightarrow \\ \quad A(i - 1, j) \oplus C(i, j - 1) \end{cases} \\ i = 0 \rightarrow A_0(j) \end{cases} \quad (13.14)$$

$$C(i, j) = \begin{cases} i > 0 \rightarrow \begin{cases} i - 1 \leq j < n + i - 1 \rightarrow \\ \quad MAJ(A(i - 1, j), (B(i - 1) \cdot M(j - (i - 1))), C(i, j - 1)) \\ (0 \leq j < i - 1) \vee (n + i - 1 \leq j < 2n - 1) \rightarrow \\ \quad MAJ(A(i - 1, j), \hat{0}, C(i, j - 1)) \\ j = 2n - 1 \rightarrow \hat{0} \end{cases} \\ i = 0 \rightarrow \perp \end{cases} \quad (13.15)$$

In the two equations above, if j is not in $\mathcal{N}_2$, it is understood as $2n + j$ of the previous word computed in the pipeline. We now proceed to show that (13.14) and (13.15) embody an algorithm for computing the inner product of positive numbers in the binary representation.

Let $x(k)$, $y(k)$, and $z(k)$ for $k = 0, 1, 2, \ldots, n - 1$ be the binary representation of nonnegative integers x, y, and z, respectively. A recurrence formula for addition of binary numbers $z = x + y$ is

$$z(k) = y(k) \oplus x(k) \oplus w(k - 1)$$

$$w(k) = MAJ(y(k), x(k), w(k - 1))$$

$$\text{for } k = 0, 1, \ldots, n - 1 \text{ and } z(n) = w(n - 1) \quad (13.16)$$

where $w(k)$'s are the carry bits and

$$w(-1) = \hat{0}$$

Notice that the Boolean *exclusive-or* function is actually modulo 2 addition of bits and the *majority* function is the floor ("integer" division) of sum of bits divided by 2, that is,

$$z(k) = (y(k) + x(k) + w(k - 1))(\text{mod } 2)$$

$$w(k) = \left\lfloor \frac{(y(k) + x(k) + w(k - 1))}{2} \right\rfloor \quad (13.17)$$

By a straightforward induction on n, it can be shown that

$$\sum_{k=0}^{n} 2^k z(k) = \sum_{k=0}^{n-1} 2^k x(k) + \sum_{k=0}^{n-1} 2^k y(k)$$

Due to the finiteness of machines, usually $z = x + y \pmod{2^n}$ is the operation performed. This operation is defined similarly by

$$\mathbf{z}(k) = \mathbf{y}(k) \oplus \mathbf{x}(k) \oplus \mathbf{w}(k-1) \qquad \text{for } k = 0, 1, \ldots, n-1$$

$$\mathbf{w}(k) = \mathrm{MAJ}(\mathbf{y}(k), \mathbf{x}(k), \mathbf{w}(k-1)) \qquad \text{for } k = 0, 1, \ldots, n-2 \text{ and } \mathbf{w}(-1) = \hat{0} \qquad (13.18)$$

and

$$\mathbf{z}(n) = \mathbf{w}(n-1) = \hat{0}$$

For the pipeline inner product element, we need to show that

$$\sum_{j=0}^{2n-1} 2^j \, A(n, j) = \left(\sum_{k=0}^{n-1} 2^k \, B(k) \times \sum_{k=0}^{n-1} 2^k \, M(k) + \sum_{j=0}^{2n-1} 2^j \, A_0(j) \right) \pmod{2^{2n}}$$

Since

$$\sum_{k=0}^{n-1} 2^k \, B(k) \times \sum_{k=0}^{n-1} 2^k \, M(k) + \sum_{j=0}^{2n-1} 2^j \, A_0(j)$$

$$= \sum_{i=0}^{n-1} \left(2^i \, B(i) \times \sum_{k=0}^{n-1} 2^k \, M(k) \right) + \sum_{j=0}^{2n-1} 2^j \, A_0(j)$$

$$= \sum_{i=0}^{n-1} d_i + \sum_{j=0}^{2n-1} 2^j \, A_0(j)$$

where

$$d_i = 2^i \, B(i) \times \sum_{k=0}^{n-1} 2^k \, M(k) = \sum_{j=i}^{n+i-1} 2^j \, M(j-i) B(i)$$

$$= \left(\left(\cdots \left(\left(\sum_{j=0}^{2n-1} 2^j \, A_0(j) + d_0 \right) + d_1 \right) + \cdots \right) + d_{n-1} \right)$$

$$= p_n \text{ (the nth partial product)}$$

where

$$p_i = p_{i-1} + d_{i-1} \qquad i = 1, 2, \ldots, n$$

$$p_0 = \sum_{j=0}^{2n-1} 2^j \, A_0(j)$$

By induction on i and (13.16), it can easily be shown that

$$p_i \pmod{2^{2n}} = \sum_{j=0}^{2n-1} 2^j \, A(i, j)$$

and therefore

$$\sum_{j=0}^{2n-1} 2^j \, A(n, j) = p_n \pmod{2^{2n}}$$

$$= \left(\sum_{k=0}^{n-1} 2^k \, B(k) \times \sum_{k=0}^{n-1} 2^k \, M(k) + \sum_{j=0}^{2n-1} 2^j \, A_0(j) \right) \pmod{2^{2n}}$$

Thus the correctness of the pipelined algorithm above has been shown. This proof is approached differently from those of the systolic arrays in [3], where the solution with space-time domain is obtained and then mapped to the structure of matrices. In this proof, the space-time coordinates are mapped to the coordinates indicating the stage of the partial product and the bit number for the equation. The solution is therefore obtained with these indices as domains. It is because these indices are more convenient for expressing the multiplication function as sequences of bits. We now let $A_{out} = \sum_{j=0}^{2n-1} 2^j A(n, j)$, $A_{in} = \sum_{j=0}^{2n-1} 2^j A_0(j)$, $B = \sum_{k=0}^{n-1} 2^k B(k)$, and $M = \sum_{k=0}^{n-1} 2^k M(k)$. The functionality of the pipeline can be described by $A_{out} = A_{in} + B \times M$. The data type is a bounded integer rather than a bit. The IPE element can thus be used at the next level without its implementation detail but only its functionality.

13.7 CONCLUDING REMARKS

We have presented a framework for general nonlinear systems with memory. An essential part of the semantics is a methodology for abstracting the behavior of such systems so they can be used as components at a higher level. The semantics of a particular system consists of:

1. An input mapping function from the abstract data type (e.g., bounded integers) to the domain of implementation (e.g., domain of indices i and j in the example above or the space-time domain).
2. A function which is the solution of the system of recursion equations that described the system.
3. An output mapping function from the domain of implementation to the abstract data type.

From an engineering point of view, the input and output mapping functions serve as precise interface specifications for the system.

The methodology can be applied to any system: linear, nonlinear, time-varying, history-dependent. We believe it provides, for the first time, a unified approach spanning the range from computer programs to linear transfer functions; from transistor circuits to high-level communicating processes.

ACKNOWLEDGMENTS

We wish to thank Lennart Johnsson for his help with the algorithm described in this paper.

This work is sponsored by System Development Foundation.

REFERENCES

[1] J. Backus, "Can Programming Be Liberated from the von Neumann Style? A Functional Style and Its Algebra of Programs," *Commun. ACM*, *21*(8):613–641 (Aug. 1978).

[2] R. E. Bryant, *A Switch-Level Simulation Model for Integrated Logic Circuits*, Massachusetts Institute of Technology, Cambridge Mass., Mar. 1981.

[3] M. C. Chen, "Space-Time Algorithm: Semantics and Methodology," Doctoral dissertation, Computer Science Department, California Institute of Technology, May 1983.

[4] A. B. Cremers and T. N. Hibbard, "Formal Modeling of Virtual Machines," *IEEE Trans. Software Eng.*, *4*:426–436 (1978).

[5] E. W. Dijkstra, "Cooperating Sequential Processes," in F. Genuys, ed., *Programming Languages*, Academic Press, New York, 1968, pp. 43–112.

[6] C. A. R. Hoare, "Communicating Sequential Processes," *Commun. ACM*, *21*(8):666–677 (Aug. 1978).

[7] J. J. Horning and B. Randell, "Process Structuring," *Comput. Surv.*, *5*(1):5–30 (Mar. 1973).

[8] H. T. Kung and C. E. Leiserson, "Algorithms for VLSI Processor Arrays," in C. Mead and L. Conway, *Introduction to VLSI Systems*, Addison-Wesley, Reading, Mass., 1980, Chap. 8.3.

[9] R. F. Lyon, "Two's Complement Pipeline Multipliers," *IEEE Trans. Commun.*, *COM-24*:418–425 (1976).

[10] Z. Manna, *Mathematical Theory of Computation*, McGraw-Hill, New York, 1974.

[11] C. Mead and L. Conway, *Introduction to VLSI Systems*, Addison-Wesley, Reading, Mass., 1980.

[12] R. Milner, *Models of LCF*, AIM-186/CS-332, Computer Science Department, Stanford University, 1973.

[13] J. L. Peterson, *Petri-net Theory and the Modeling of Systems*, Prentice-Hall, Englewood Cliffs, N.J., 1981.

[14] C. A. Petri, "Kommunikation mit Automaten," Doctoral dissertation, University of Bonn, Bonn, West Germany, 1962 (in German); also MIT Memorandum MAC-M-212, Project MAC, MIT, Cambridge, Mass.

[15] D. Scott, *Outline of a Mathematical Theory of Computation*, Oxford Monogr. PRG-2, Oxford University Press, Oxford, 1970.

[16] D. Scott and C. Strachey, in J. Fox, ed., *Toward a Mathematical Semantics for Computer Languages*, Polytechnic Institute of Brooklyn Press, New York, 1971.

[17] I. Sutherland and C. Mead, "Micro-electronics and Computer Science," *Sci. Am.*, *237*(3):210–229 (Sept. 1977).

[18] J. Wawrzynek and T. M. Lin, *A Bit Serial Architecture for Multiplication and Interpolation*, 5067:DF:83, Computer Science Department, California Institute of Technology, January, 1983.

14

An Overview of Signal Representation in Programs

GARY E. KOPEC

Fairchild Laboratory for Artificial Intelligence Research
Palo Alto, California

14.1 INTRODUCTION

Computer science research in programming methodology has led to the view that the fundamental activity in the development of well-structured programs is the recognition of *abstractions* [1]. In particular, two general kinds of abstractions have been identified: abstract *operations* and abstract *data types* [1–4]. An abstract operation corresponds to the notion of a procedure or subroutine which is supported in most contemporary programming languages. An abstract data type consists of a set of objects plus a set of primitive operations for creating and manipulating those objects. Examples of common data abstractions include fixed- and floating-point numbers, character strings, arrays, records, priority queues, and I/O streams [4].

Digital signal processing is based on a well-established body of mathematical theory which is explicitly used during program development. Perhaps the most basic feature of this theory is the central role of the concepts *signal* and *system*. This suggests that a well-structured signal processing program is one organized as a collection of "signal" and "system" abstractions. Similarly, a "signal processing programming language" is a language that provides explicit support for such abstractions.

Past attempts to define signal processing languages have been based on two main models of signal processing computation: *next-state simulation* [5] and *array processing*. Next-state simulation is the underlying computation model in stream-oriented block diagram programming languages [5–12] such as BLODI [6–8] and PATSI [5]. Array processing refers to the large and ill-defined collection of implementation techniques that exploit the separability of "kernel computation" and address arithmetic which characterizes many signal processing algorithms [13–15].

241

Signal processing algorithms frequently consist of a small number of *computation kernels* [13–15] which are repeatedly applied to many sets of arguments. The sequencing of the kernels and the data paths among them are independent of the values of the data. This algorithmic structure has been exploited in the design of highly parallel array processing hardware [16] and time-efficient software [13–15].

Next-state simulation and array processing have each been successful within some range of applications. Neither model, however, has supported the design of an effective general-purpose signal processing programming system. A detailed discussion of the limitations of these models may be found in [17]. The basic problem is that block diagram languages and array processing are attempts to model discrete-time systems as procedural abstractions that share a common algorithmic structure. Unfortunately, the class of "signal processing algorithms" is becoming increasingly heterogeneous. This observation, together with the growing recognition of the utility of the data abstraction concept, suggests exploring a different possibility—a signal processing language in which discrete-time signals and systems take the form of specialized data abstractions [17–21].

This chapter reviews three approaches to the representation of discrete-time signals as data abstractions. Two of these representations—the *streams* of block diagram languages and the *arrays* of array processing—are widely used in contemporary signal processing programming. The third representation was introduced in the recently proposed language SRL [21]. SRL is a signal representation language in which discrete-time signals take the form of abstract objects whose properties are explicitly designed to reflect those of the represented signals. Arrays, streams, and SRL signal objects are discussed in the context of a set of signal representation criteria which are motivated by elementary observations about the mathematics of signals. Of the three representations, only SRL signal objects appear to satisfy all the identified requirements.

The remainder of this chapter is organized as follows. Section 14.2 presents a set of signal representation criteria. These criteria are used in Sections 14.3 to 14.5 to assess the characteristics of streams, arrays, and SRL signal objects as signal representations. Section 14.3 reviews the use of streams in block diagram languages. Section 14.4 is a discussion of arrays. Section 14.5 discusses SRL.

14.2 SIGNAL REPRESENTATION REQUIREMENTS

A set of requirements for signal representation is motivated by three elementary observations about the mathematics of discrete-time signals and the notations commonly used to describe them. For simplicity and concreteness, these requirements will be stated for multidimensional discrete-time signals of the form

$$f : [0, N_1) \times \cdots \times [0, N_m) \mapsto \mathscr{R} \tag{14.1}$$

where $[0, N_i)$ is the set of integers $0, \ldots, N_i - 1$ and $\mathscr{R}$ is the set of real numbers.

The first observation about signals is that they are constant values whose mathematical properties are not subject to change. Mathematically, when a signal is "transformed" by a discrete-time system the result is a new signal and not a modification of the original one. For example, the sum of two signals x_1 and x_2 is a third signal whose samples are related to those of x_1 and x_2 by the relation

$$x_3(n) = x_1(n) + x_2(n).$$

The mathematical constancy of signals suggests that, in a program, signals should be represented by objects that are *immutable* [22]. The observable properties of a signal object should be established when the signal is created and remain fixed thereafter. A signal abstraction should include no operations that could be used to alter the observable properties of a signal.

The second observation about signals concerns the kinds of properties that distinguish one signal from another. Mathematically, two real-valued signals are equal if they have the same domain and their values are equal at each point of their domain. That is, if x_1 and x_2 are both signals with domain $\mathcal{D}$, then $x_1 = x_2$ if and only if

$$x_1(n) = x_2(n)$$

for each $n \in \mathcal{D}$. In effect, the identity of a signal is determined solely by the identity of its domain and the values of its samples.

The data abstraction embodiment of the notion of observable property is the concept of an *inquiry operation* [23]. The inquiry operations of a data type are the primitive operations used to extract information from objects of the type. They define the observable attributes by which individual members of the type are distinguished.

The mathematical definition of signal equality suggests that the inquiry operations of a signal abstraction should consist of a function that identifies the domain of a signal and a function that returns the value of an arbitrary sample. Specifically, there should be two operations, **dimensions** and **fetch**, which satisfy the following specifications. If x is a signal of the form (14.1) and $\mathcal{S}$ is an abstract object representing x, then

$$(\textbf{dimensions } \mathcal{S}) = (N_1, \ldots, N_m)$$

$$(\textbf{fetch } \mathcal{S} \; i_1, \ldots, i_m) = x(i_1, \ldots, i_m)$$

for each $(i_1, \ldots, i_m) \in \mathcal{D}$.

The third observation about discrete-time signals concerns the way in which signals are specified in ordinary mathematical notation. A class of signals is often defined by giving an expression for the value of the prototypical sample of the prototypical signal of the class. This expression contains a number of "free" parameters which define the space of represented signals. Each signal in the class corresponds to a specific set of values for the parameters. For example, the prototypical sine wave might be defined as the signal whose value at n is

$$x_{\omega, \phi}(n) = \sin(\omega \cdot n + \varphi) \tag{14.2}$$

where ω and φ are parameters representing the frequency and initial phase of the wave. A particular sine wave is characterized by its specific values for ω and φ.

The mathematical notation for signals suggests that the signal objects in a program should be grouped into signal classes, where each class is associated with a parameterized procedure for computing the value of an arbitrary sample of any signal in the class. A particular signal of the class should be created by binding these parameters to a set of specific values. The notion of creating a signal by fixing parameter values is similar to that of creating a *function closure* by pairing a function with an environment that contains bindings for some of its *free variables* [24,25].

The three requirements identified above—immutability, a specific set of inquiry operations, and signal classes—will be used in the following sections to assess the properties of streams, arrays, and SRL signal objects.

14.3 STREAMS AND BLOCK DIAGRAM LANGUAGES

The concept of a *stream* or *first-in-first-out queue* (FIFO) is familiar to most programmers, although perhaps not as a data type. Streams are commonly used to provide sequential access to the elements of a data set. For example, they are used in many contemporary programming languages as the basic file input/output facility [4,26,27]. Similarly, streams are often used for communication between concurrent processes [28–30]. For example, the *pipeline combinator* of the UNIX operating system [31] allows the quasi-parallel execution of separate programs which are connected, via streams, into a linear "assembly line."

In digital signal processing, streams are a natural signal representation for next-state simulation, the underlying computation model of block-diagram programming languages. This section reviews the concept of a stream and its use in next-state simulation. The discussion is carried out in the context of a simple stream abstraction and several examples of Lisp-based block diagram programs.

14.3.1 Simple Stream Abstraction

Figure 14.1 shows a partial set of operations for a simple stream abstraction. The operation **make-stream** creates and returns a new stream object. The operation **stream-put** appends an element to the end of a stream and **stream-get** removes an element from the front of a stream. The operation **stream-put-eos** appends a special *end of stream* token to the end of the stream. When returned by **stream-get**, the end-of-stream token signals that the stream will contain no additional elements. The

```
(make-stream)          ; create new (empty) stream
(stream-put val x)     ; append element to end
(stream-get x)         ; remove element from front
(stream-put-eos x)     ; append end-of-stream token
(stream-is-empty x)    ; test for elements present
```

Figure 14.1 Operations of a simple stream abstraction.

operation **stream-is-empty** is a predicate that returns **true** if the stream currently contains no elements.

14.3.2 Next-State Simulation

Next-state simulation is a generalization of the state-space representation of linear dynamical systems. A system is *n.s.s-realizable* if it can be represented by a set of equations of the form

$$s_k(n + 1) = \mathscr{F}_k(s_1(n), \ldots, s_N(n), x_1(n), \ldots, x_M(n)) \qquad k = 1, \ldots, N$$

$$y_p(n) = \mathscr{G}_p(s_1(n), \ldots, s_N(n), x_1(n), \ldots, x_M(n)) \qquad p = 1, \ldots, P \qquad (14.3)$$

$$s_k(0) = s_k^0 \qquad k = 1, \ldots, N$$

where $x_1, \ldots, x_M$, $y_1, \ldots, y_P$, and $s_1, \ldots, s_N$ are *input, output,* and *state* sequences, respectively. The functions $\mathscr{F}_1, \ldots, \mathscr{F}_N$ and $\mathscr{G}_1, \ldots, \mathscr{G}_P$ are *state-update* and *observation* functions and the values $s_1^0, \ldots, s_N^0$ are *initial conditions.*

The form of (14.3) suggests a simple algorithm for computing the output of an n.s.s-realizable system, using a stream representation for its input, output, and state sequences. The output samples are computed in order of increasing sample index (i.e., $n = 0, \ldots$) during successive passes through the body of a main loop. During each pass, samples $s_1(n), \ldots, s_N(n)$ and $x_1(n), \ldots, x_M(n)$ are removed from their respective state or input streams. Next, procedures that compute $\mathscr{F}_1, \ldots, \mathscr{F}_N$ and $\mathscr{G}_1, \ldots, \mathscr{G}_P$ are applied to these values to obtain $s_1(n + 1), \ldots, s_N(n + 1)$ and $y_1(n)$, $\ldots, y_P(n)$. Finally, the computed values are appended to the state and output streams. This cycle is repeated for each set of input samples. Before the iteration begins, the initial conditions are appended to the state streams.

14.3.3 Block Diagram Languages

Block diagram programming languages are based on the implementation of n.s.s-realizable systems by algorithms with the structure described above. In general, a block diagram language provides a set of primitive systems plus facilities for describing networks of these primitives and constructing their realizations.

Figure 14.2 shows the implementations of three simple block types (**adder**, **scalor**, and **unit-delay**) using **defprimtype**, the primitive block definition form of a hypothetical Lisp-based block diagram language. The facilities of this language are intended to be representative of those available in "real" block diagram languages. The first argument in each **defprimtype** form is the name of the block type being defined.

The definition of a two-input, single-output **adder** is given in Figure 14.2(a). The **:inputs** and **:outputs** clauses of the definition specify that the input and output "ports" of an **adder** are named **x1**, **x2**, and **y**, respectively. These ports are "connected" to streams when an instance of an **adder** is created. The **:fire** clause specifies the body of a 0-argument procedure (a *fire method*) which is invoked during each pass

```
(defprimtype adder
  :inputs (x1 x2)
  :outputs (y)
  :fire (stream-put   (+ (stream-get x1)
                         (stream-get x2))
                      y))
```

(a)

```
(defprimtype scalor
  :inputs (x)
  :outputs (y)
  :parameters (a)
  :fire (stream-put   (* a (stream-get x))
                      y))
```

(b)

```
(defprimtype unit-delay
  :inputs (x)
  :outputs (y)
  :state (s)
  :parameters (initial-value)
  :init (stream-put initial-value s)
  :fire (progn
          (stream-put (stream-get s) y)
          (stream-put (stream-get x) s)))
```

(c)

Figure 14.2 Definitions of primitive block types: (a) adder; (b) scalor; (c) unit delay.

through the n.s.s. computation cycle. The fire method of a block implements the state-update and observation functions and performs input-output operations on the streams connected to the block ports. The fire method of the **adder** retrieves a pair of elements from input streams **x1** and **x2** and appends the sum to output stream **y**.

Figure 14.2(b) shows an implementation of a single-input, single-output **scalor** (constant-gain amplifier). This example illustrates the concept of block *parameterization*. The **:parameters** option indicates that the class of **scalor** blocks is parameterized by **a**, the block gain factor. A value for this parameter is supplied whenever an instance of type **scalor** is created.

The **adder** and **scalor** are examples of *memoryless* systems implemented without the use of a state sequence. Figure 14.2(c) illustrates the use of a state stream to implement the memory of a **unit-delay**. The n.s.s realization of a delay is

$$s(n + 1) = x(n)$$

$$y(n) = s(n)$$

$$s(0) = s^0$$

where the state-update and observation functions are identity functions. The **unit-**

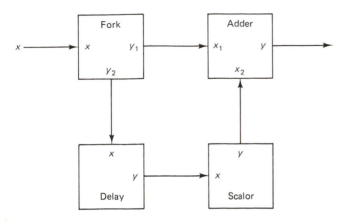

Figure 14.3 Block diagram of a first-order filter.

delay fire method removes $s(n)$ and $x(n)$ from the state and input streams and appends them to the output and state streams, respectively. The initial value s^0 (a parameter) is appended to the state stream by the **unit-delay** *init method*, a function that is called whenever an instance of a **unit-delay** is created.

The preceding examples illustrate the definition of primitive block types, defined by explicitly supplying a fire method written in a general-purpose programming language. The final example of this section illustrates the definition of a block diagram *network*, an interconnected collection of block operator instances. In a typical block diagram language, a fire method for a network is implicitly generated by a compiler that *sorts* the component blocks according to the data dependencies implied by the network topology.

Figure 14.3 shows a block diagram of a simple first-order transversal filter, defined by the equation

$$y(n) = x(n) + a \cdot x(n-1) \tag{14.4}$$

An implementation of this filter using **defnettype**, a hypothetical network definition form, is given in Figure 14.4. As indicated, a **first-order-filter** network contains four

```
(defnettype first-order-filter
    :inputs (x)
    :outputs (y)
    :parameters (a)
    :init (progn
            (part adder 'adder)
            (part fork 'fork)
            (part scalor 'scalor 'a a)
            (part delay 'unit-delay 'initial-value 0.0)
            (connect (≫ x) (≫ x fork))
            (connect (≫ y1 fork) (≫ x1 adder))
            (connect (≫ y adder) (≫ y))
            (connect (≫ y2 fork) (≫ x delay))
            (connect (≫ y delay) (≫ x scalor))
            (connect (≫ y scalor) (≫ x2 adder))))
```

Figure 14.4 Definition of first-order filter network type.

"parts," which are instances of types **adder**, **fork**, **scalor**, and **unit-delay**, respectively. (A **fork** block simply copies samples from its input port to both of its output ports.) The identity and properties of each network component are specified using a **part** form. The first argument to **part** is a name to be assigned to the block instance. The second argument is the type of the block. The remaining arguments specify the values of any block parameters. For example, the form

$$\text{(\textbf{part} delay 'unit-delay 'initial-value 0.0)}$$

indicates that each instance of a filter network contains a **unit-delay** block called **delay** whose **initial-value** parameter is 0.0. A **connect** forms asserts that a specified pair of block ports should be interconnected with a stream. The notation

$$(\gg \text{y1 fork})$$

is read "the (port) y1 of the (part) fork of the current block."

14.3.4 Appraisal of Streams and Block Diagram Languages

The properties of streams and block diagram operators with respect to each of the signal representation criteria are discussed below.

   Inquiry operations.   The stream inquiry operations (**stream-get** and **stream-is-empty**) do not support domain identification since the "length" of a stream is not a well-defined concept. Given a stream object, the only way to determine the length of the represented signal is by repeatedly invoking the operation **stream-get** until the end-of-stream token is encountered. This strategy will fail if **stream-put-eos** was not invoked previously.

   The strictly sequential access provided by **stream-get** does not satisfy the requirement of a method for obtaining the value of an arbitrary signal sample. The lack of "random access" makes streams inconvenient for algorithms, such as the fast Fourier transform (FFT), which are not n.s.s.-realizable. Furthermore, sequential streams are inappropriate as a general representation for multidimensional signals since there is no single access order which is suitable for all (or even most) multidimensional systems. The order in which samples of a multidimensional signal are computed greatly affects important performance attributes of a system, such as the dimension of its state space, its quantization noise and coefficient sensitivity properties, and the possible arithmetic parallelism [32].

   Immutability.   Each invocation of the inquiry operation **stream-get** removes and returns the "next" element from a stream. Since successive invocations of **stream-get** return different elements, streams are clearly mutable objects. However, since each stream element corresponds to a conceptually different sample of the represented signal, it is vacuously true that each signal sample has the same value every time it is observed. Nevertheless, the fundamental mutability of streams violates the signal immutability criterion.

Signal classes. The concept of a block type is similar to that of a signal class. The primary difference, of course, is that blocks are systems rather than signals. Nevertheless, system blocks illustrate the general kinds of facilities required by the signal class criterion. For example, the most important aspect of a primitive block type is its fire method. The fire method is a procedure that computes the output samples of any block in a class of blocks. Individual block instances are distinguished by the values of their parameters and the streams to which they are connected.

14.4 ARRAYS AND ARRAY PROCESSING

The term *array processing* is commonly used to refer to the large and ill-defined collection of implementation techniques that exploit the separability of "kernel computation" and address arithmetic that characterizes many signal processing algorithms. More generally, the term may be applied to any computation in which a discrete-time signal is represented by an array containing the values of its elements. With this extended definition, array processing languages include many which are not based on specialized processors, such as FANLZ [11] and ILS [33]. In addition, the broader definition encompasses many of the signal processing implementation techniques commonly used with general-purpose languages such as Fortran. This section considers array processing in this extended sense. As with streams, the properties of arrays with respect to the signal representation criteria will be established with reference to a specific array abstraction and example of an array operator.

14.4.1 Simple Array Abstraction

The concept of an array is supported as a data type in nearly all modern programming languages. Figure 14.5 shows a set of operations for a simple multidimensional array abstraction, intended to be typical of those commonly available. The operation **allocate-array** creates and returns a new array object whose dimensions are contained in the supplied list. The elements of an array with dimensions $(N_1, ..., N_m)$ have indices $i_1, ..., i_m$, where $i_k \in [0, N_k)$. The operation **array-dimensions** returns a list containing the dimensions of its array argument. The operation **array-fetch** returns the value of an array element, and **array-store** changes the value of an element.

A typical array processing operator is shown in Figure 14.6, in the form of a simple Lisp procedure. The function **array-add** sets each element of array **y** to the

```
(allocate-array dimensions)      ; create new array
(array-dimensions x)             ; return list of dimensions
(array-fetch x i-1...i-m)        ; fetch specified element
(array-store val x i-1...i-m)    ; store specified value
```

Figure 14.5   Operations of a simple array abstraction.

```
(defun array-add (x1 x2 y)
  (loop with n = (first (array-dimensions x1))
        for i from 0 below n
        do (array-store    (+ (array-fetch x1 i)
                               (array-fetch x2 i))
                    y    i)))
```
Figure 14.6 Simple array operator.

sum of the corresponding elements of arrays **x1** and **x2**. This procedure is typical of those used for array processing with general-purpose programming languages.

14.4.2 Appraisal of Arrays

Inquiry operations. The inquiry operations **array-fetch** and **array-dimensions** satisfy the requirements for signal inquiry operations.

Immutability. The operation **array-store** modifies its array argument and thus violates the signal immutability criterion. The mutability of arrays is often exploited in the technique of *in-place computation* [34], a method for decreasing the storage requirement of a program by using the same memory locations for both input and output. The signal immutability requirement implies that in-place algorithms may not be applied to arrays that represent discrete-time signals.

Signal classes. The array abstraction does not support the notion of parameterized classes of signals. The array representation of any particular signal is created by storing the numerical values of its samples in the array. There is no indication, in the resulting representation, of how those values are related to each other or how they were computed from the samples of other signals. Although different signals might be created using a common procedure (e.g., **array-add**), there is no mechanism for explicitly indicating this commonality. Array operators such as **array-add** fill no well-defined role analogous to the fire methods of block diagram operators.

14.5 THE SIGNAL REPRESENTATION LANGUAGE SRL

The basic concepts and facilities of SRL are explicitly motivated by the three observations about signals presented in Section 14.2. The fundamental activity in SRL programming is the implementation of signal types. A signal type is a representation for a class of signals that share a common procedure for computing the value of a sample. Instances of a signal type are created by fixing the values of the free variables of this procedure. The basic observable properties of a signal are the dimensions of its domain and the values of its samples. Finally, signal objects are immutable, so that these properties remain fixed after a signal is created.

14.5.1 Inquiry Operations on Signals

Figure 14.7 lists the basic inquiry operations on SRL signal objects. These operations are similar to those for arrays. The operation **signal-dimensions** returns a list containing the dimensions of its argument. The operation **signal-fetch** returns that sample of the (multidimensional) signal whose indices are **i-1**, ..., **i-n**.

In principle, the single-element access provided by **signal-fetch** is an adequate means of obtaining the samples of any signal. In practice, however, a problem arises because such access is not directly compatible with algorithms, such as the FFT, which are array oriented. Although a basic approach to eliminating this mismatch is easily identified (i.e., using buffers), formulating a detailed strategy requires addressing a number of difficult issues. For example, it may be impractical to permanently allocate a separate array to each signal that requires one for buffering. On the other hand, the desired permanence and immutability of signals may be compromised if the same array is used to buffer the samples of two or more signals. An important aspect of SRL is an array access discipline that allows storage to be reused while preserving signal immutability [21].

The **with-signal-values** form is an inquiry operation for obtaining an array containing the samples of a signal. The body of the form is a collection of expressions requiring access to the samples of signals **sig-1**, ..., **sig-n**. SRL executes the body after binding local variables **var-1**, ..., **var-n** to arrays containing the requested samples. The arrays are allocated and deallocated automatically by SRL. The dimensions of each array are the same as those of the corresponding signal. That is, the arrays contain ALL of the signal samples. User access to a signal value array within the body of the form is subject to the following two restrictions.

1. A signal-value array should be regarded as a temporary object that "disappears" when evaluation of the body terminates. Thus it should not be made part of any data structure that is accessible outside the (dynamic) scope of the **with-signal-values** form.

2. A signal value array should be regarded as "read-only" and must not be modified in any way. For example, operations that change the values of array elements or the dimensions of the array are prohibited.

The purpose of these restrictions is to support a variety of array management policies, effecting different trade-offs between time and space efficiency. For example,

```
(signal-dimensions signal)        ; return dimensions list
(signal-fetch signal i-1 ... i-n) ; return single sample
(with-signal-values ((var-1 sig-1)  ; array processing support
                  :
             (var-n sig-n))
     body)
```

Figure 14.7 SRL signal inquiry operations.

the first restriction permits a space-efficient stack-based allocation strategy. Similarly, the second restriction allows a signal value array to be retained by **with-signal-values** after it is first generated and reused every time the samples of a particular signal are requested.

14.5.2 Signal-Type Definitions

The primary activity in SRL programming is the implementation of signal types. Fundamentally, implementing a type involves supplying mechanisms for computing the dimensions and samples of any signal of the type. In concept, a signal type is similar to a Smalltalk *class* [35–37] or Zetalisp *flavor* [38].

Figure 14.8 illustrates the use of **defsigtype**, the SRL signal-type definition form, to define several simple classes of signals. The first argument to **defsigtype** is the name of the type being defined. The remaining arguments, consisting of alternating keywords and values, specify various optional properties shared by signals of the type.

Figure 14.8(a) shows an implementation of **sine-wave-signal**, the class of signals defined by (14.2). The **:parameters** option of **defsigtype** defines the names of the parameters by which individual signals of the type are distinguished. The parameters of a **sine-wave-signal** are its frequency (**omega**), initial phase (**phi**), and dimensions (**dimensions**).

The **:finder** option specifies the name to be given to an automatically generated function (the signal *finder*) which returns instances of the signal type. The arguments to a finder are the values of the signal parameters. For example, a representation for a 512-point sine wave with frequency .1 and initial phase 0.0 is returned by the finder invocation

(signal-sine-wave .1 0.0' (512)).

```
(defsigtype sine-wave-signal
  :parameters (omega phi dimensions)
  :finder signal-sine-wave
  :fetch ((n)
        (sin (+ (* omega n)
              phi))))

        (a)


(defsigtype sum-signal
  :parameters (x1 x2)
  :finder signal-sum
  :init (setq-my dimensions (min
                    (signal-dimensions x1)
                    (signal-dimensions x2)))
  :fetch ((n)
        (+ (signal-fetch x1 n)
          (signal-fetch x2 n))))

        (b)
```

Figure 14.8 Simple SRL signal-type definitions: (a) sine wave; (b) sum signal.

```
(defsigtype sum-signal
  :parameters (x1 x2)
  :finder signal-sum
  :init (setq-my dimensions (min
                              (signal-dimensions x1)
                              (signal-dimensions x2)))
  :values (with-signal-values   ((x1-vals x1)
                                 (x2-vals x2))
                           (let ((our-vals (allocate-array dimensions)))
                               (array-add x1-vals x2-vals our-vals)
                               our-vals)))
```

Figure 14.9   Array-based implementation of sum signal.

An important feature of the finder is that it maintains a database of signals which it has created and will return the same signal object each time it is invoked with a given set of arguments. This reflects the mathematical intuition that there is exactly one signal for each set of parameters.

The **:fetch** option specifies the argument list and name of a parameterized procedure (a *fetch method*) which returns a sample of any signal in the class being defined. The fetch method is invoked by the general inquiry operation **signal-fetch**. The parameters of the signal are accessed as free variables within the body of the fetch method. The fetch method of a **sine-wave-signal** is a straightforward implementation of (14.2).

The definition of **sine-wave-signal** is typical of the way 0-input signal generators are represented in SRL. Figure 14.8(b) shows an implementation of the class **sum-signal**, a simple binary combination of signals. The parameters of a **sum-signal** are **x1** and **x2**, the two signals whose sum is represented. The fetch method of **sum-signal** computes a requested sample by fetching the corresponding samples of **x1** and **x2** and adding them.

The signal-type definitions in Figure 14.8 are based on the use of fetch methods to compute individual signal samples. As discussed previously, SRL also supports array-oriented computation. Figure 14.9 shows an alternative implementation of **sum-signal** which is based on the use of arrays. The **:values** option specifies the body of a 0-argument procedure, the *values method*, which generates the signal value array returned by the inquiry operation **with-signal-values**. The **sum-signal** values method obtains the value arrays for **x1** and **x2** (using **with-signal-values**), allocates the output array (using **allocate-array**), and then computes and stores the output samples using the previously described array operator **array-add**.

14.5.3 Appraisal of SRL

SRL appears to satisfy all of the identified signal representation criteria.

Inquiry operations.   The SRL inquiry operations **signal-fetch** and **signal-dimensions** satisfy the requirements for signal inquiry operations.

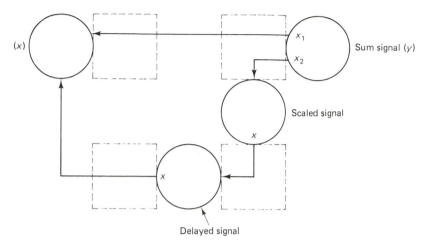

Figure 14.10   Signal graph for output of first-order filter.

Immutability.   SRL provides no operations that can be used to directly modify the observable properties of a signal. Although there are no mechanisms to enforce the restrictions on value array access described previously, the existence of a simple and well-defined set of conventions contributes to voluntary compliance.

Signal classes.   The fundamental activity in SRL programming is the implementation of signal types. A signal type is a representation for a class of signals that share a common parameterized procedure (fetch method) for computing the values of the samples.

The relationship between the signal-oriented focus of SRL and the system-oriented perspective of block diagram languages is illustrated by a simple example. Assuming the existence of an appropriate set of signal types and finders, an SRL representation for the output of the first-order filter defined in (14.4) could be created by evaluating the expression

(signal-sum x **(signal-scaled (signal-unit-delayed** x) a))

Figure 14.10 shows a signal graph that portrays the relationships among the various signals implied by this form. Comparing this graph with the filter block diagram in Figure 14.3 reveals that the roles of the nodes and arcs are reversed in the two graphs. This is similar to the situation in classical network theory, where signal flow graphs and block diagrams are duals of each other.

14.6 CONCLUDING REMARKS

Three signal representation criteria—immutability, a specific set of inquiry operations, and support for signal classes—were motivated by elementary observations about the mathematics of discrete-time signals. These criteria were used to assess the signal representations used in block diagram languages, array processing, and the signal representation language SRL. Streams are mutable and do not provide the required inquiry operations. The concept of an n.s.s block type, however, is similar to that of a signal class. Arrays provide appropriate inquiry operations but are mutable. Furthermore, current array processing languages do not explicitly support the concept of signal or system classes. The language SRL appears to satisfy all of the criteria identified.

REFERENCES

[1] B. Liskov, "An Introduction to CLU," CSG Memo 136, MIT Laboratory for Computer Science, Cambridge, Mass., Feb. 1976.

[2] J. Guttag, "The Specification and Application to Programming of Abstract Data Types," TR CSRG-59, Computer Systems Research Group, University of Toronto, Sept. 1975.

[3] W. Wulf, R. London, and M. Shaw, "Abstraction and Verification in Alphard: Introduction to Language and Methodology," Carnegie-Mellon University Tech. Rep., June 1976.

[4] B. Liskov, R. Atkinson, T. Bloom, E. Moss, C. Schaffert, R. Scheifler, and A. Snyder, *CLU Reference Manual*, TR-225, MIT Laboratory for Computer Science, Cambridge, Mass., Oct. 1979.

[5] B. Gold and C. Rader, *Digital Processing of Signals*, McGraw-Hill, New York, 1969.

[6] B. Karafin, "A Sampled-Data System Simulation Language," in F. Kuo and J. Kaiser, eds., *System Analysis by Digital Computer*, Wiley, New York, 1966.

[7] J. L. Kelley, C. Lochbaum, and V. A. Vyssotsky, "A Block Diagram Compiler," *Bell Syst. Tech. J.*, *40*(3): pp. 669–676 (May 1961).

[8] B. Karafin, "The New Block Diagram Compiler for Simulation of Sampled-Data Systems," *AFIPS Conf. Proc.*, Vol. 27, Spartan Books, New York, 1965, pp. 55–61.

[9] M. Dertouzous, M. Kaliske, and K. Polzen, "On-Line Simulation of Block-Diagram Systems," *IEEE Trans. Comput.*, C-18(4):333–342 (Apr. 1969).

[10] G. Korn, "High-Speed Block-Diagram Languages for Microprocessors and Minicomputers in Instrumentation, Control and Simulation," *Comput. Electr. Eng.*, *4*:143–159 (1977).

[11] W. Henke, "MITSYN—An Interactive Dialogue Language for Time Signal Processing," RLE TM-1, MIT Research Laboratory of Electronics, Cambridge, Mass., Feb. 1975.

[12] T. Crystal and L. Kulsrud, "Circus," CRD Working Paper 435, Institute for Defense Analysis, Princeton, N.J., Dec. 1974.

[13] L. Morris, "Automatic Generation of Time-Efficient Digital Signal Processing Software," *IEEE Trans. Acoust. Speech Signal Process.*, *ASSP-25*(1):74–78 (Feb. 1977).

[14] L. Morris and C. Mudge, "Speed Enhancement of Digital Signal Processing Software via Microprogramming a General Purpose Minicomputer," *IEEE Trans. Acoust. Speech Signal Process.*, *ASSP-26*(2):135–140 (Apr. 1978).

[15] L. Morris, "Time/Space Efficiency of Program Structures for Automatically-Generated Digital Signal Processing Software," *Record, 1977 IEEE Int. Conf. Acoust. Speech Signal Process.*, Hartford, Conn., May 1977.

[16] A. Salazar, ed., *Digital Signal Computers and Processors*, IEEE Press, New York, 1977.

[17] G. Kopec, "The Representation of Discrete-Time Signals and Systems in Programs," Ph.D. thesis, MIT, Cambridge, Mass., 1980.

[18] H. Gethoffer, A. Lacroix, and R. Reis, "A Unique Hardware and Software Approach for Digital Signal Processing," in *Record, 1977 IEEE Int. Conf. Acoust. Speech Signal Process.*, Hartford, Conn., May 1977.

[19] H. Gethoffer, K. Hoffmann, A. Lenzer, N. Roeth, and H. Waldschmidt, "A Design and Computing System for Signal Processing," in *Record, 1979 IEEE Int. Conf. Acoust. Speech Signal Process.*, Washington, D.C., Apr. 1979.

[20] H. Gethoffer, "SIPROL: A High-Level Language for Digital Signal Processing," *Proc. 1980 IEEE Int. Conf. Acoust. Speech Signal Process.*, Denver, Colo., Apr. 1980.

[21] G. Kopec, "The Signal Representation Language SRL," submitted for publication to the *IEEE Trans. Acoust. Speech Signal Process.*

[22] B. Liskov, A. Snyder, R. Atkinson, and C. Schaffert, "Abstraction Mechanisms in CLU," *Commun. ACM, 20*(8):564–576 (Aug. 1977).

[23] B. Liskov and V. Berzins, "An Appraisal of Program Specifications," in *Research Directions in Software Technology*, MIT Press, Cambridge, Mass., 1979.

[24] J. Allen, *The Anatomy of Lisp*, McGraw-Hill, New York, 1978.

[25] P. Winston and B. K. P. Horn, *Lisp*, Addison-Wesley, Reading, Mass., 1981.

[26] B. Kernighan and D. Ritchie, *The C Programming Language*, Prentice-Hall, Englewood Cliffs, N.J., 1978.

[27] K. Jensen and N. Wirth, *Pascal User Manual and Report*, Springer-Verlag, New York, 1974.

[28] C. A. R. Hoare, "Communicating Sequential Processes," *Commun. ACM, 12*:666–677 (Aug. 1978).

[29] G. Kahn and D. MacQueen, "Coroutines and Networks of Parallel Processes," in B. Gilchrist, ed., *Information Processing 77*, North-Holland, Amsterdam, 1977.

[30] K. Weng, "Stream-Oriented Computation in Recursive Data-Flow Schemas," Tech. Memo 68, MIT Laboratory for Computer Science, Cambridge, Mass., Jan. 1979.

[31] D. Ritchie and K. Thompson, "The UNIX Operating System," *Commun. ACM, 17*(7):365–375 (July 1974).

[32] D. Chan, "Theory and Implementation of Multidimensional Discrete Systems for Signal Processing," Ph.D. thesis, MIT, Cambridge, Mass., 1978.

[33] Signal Technology Inc., 15 West De La Guerra, Santa Barbara, CA 93101.

[34] A. Oppenheim and R. Schafer, *Digital Signal Processing*, Prentice-Hall, Englewood Cliffs, N.J., 1975.

[35] A. Kay and A. Goldberg, "Personal Dynamic Media," *IEEE Comput.*, Mar. 1977.

[36] D. Ingalls, "The Smalltalk-76 Programming System: Design and Implementation," *Conf. Rec. 5th Annu. ACM Symp. Principles of Programming Languages*, Tucson, Ariz., Jan. 1978.

[37] A. Kay, "Microelectronics and the Personal Computer," *Sci. Am.*, Sept. 1977.

[38] D. Weinreb and D. Moon, *Lisp Machine Manual*, MIT Artificial Intelligence Laboratory, Cambridge, Mass., July 1981.

15

Parallel and Pipelined VLSI Implementation of Signal Processing Algorithms

P. Dewilde, E. Deprettere, and R. Nouta

Delft University of Technology
Delft, The Netherlands

15.1 INTRODUCTION

Many important signal processing algorithms, especially filter algorithms, have very special structures or are even unstructured. The problem of realizing such structures with parallel and pipelined devices is nontrivial. Examples of popular algorithms are wave digital filters [1] and globally orthogonal cascade filters [2]. In many telecommunication applications it is essential to use such algorithms because they have very low sensitivity and no limit cycles, while achieving very steep spectral discrimination between stopbands and passbands. The algorithmic scheme for such filters is, compared with matrix operations, often unstructured, and the filter realization becomes difficult. Yet one wishes to obtain maximal throughput rates, and as a consequence one has to exploit available parallelism and pipelines to a maximum.

In Section 15.2 we present a classical scheduling method to discover which operations in a scheme may be performed in parallel and how much pipelining is possible between successive cycles. We illustrate the method with a fifth-degree wave digital filter.

Aside from difficulties with parallel/piped realizations, wave digital filters have other disadvantages which they share with many other filter structures (e.g., signals must be systematically scaled internally in order to keep the desired accuracy, and limit cycles may occur) [3]. These problems can be avoided altogether by using a new type of filter called an "orthogonal filter" [4–6]. Orthogonal filters realize an arbitrary stable transfer function with a global orthogonal transformation. By this we mean that the system map

$$\begin{bmatrix} \text{state} \\ \text{input} \end{bmatrix} \rightarrow \begin{bmatrix} \text{next-state} \\ \text{output} \end{bmatrix}$$

is an orthogonal (unitary) matrix.

Typically, the orthogonal map will be realized by a succession of Givens rotations. If rounding is done correctly, no limit cycles will occur nor will there be any need for scaling. It is necessary to realize the orthogonal filter as a cascade of elementary (degree 1 or 2) sections with controlled transmission zeros in order to achieve low sensitivity in the stopband (in the passband orthogonality itself takes care of sensitivity). In Section 15.3 we present the elementary orthogonal filter sections and study parallel/pipelined realizations. It will appear that a state-minimal realization does not allow for any pipelining. To achieve high throughput it is necessary to modify the basic scheme by introducing a limited amount of nonminimality. This will lead to cascaded filters which can be pipelined and achieve a throughput of one sample per clock cycle, provided that a piped realization for a CORDIC Givens rotation is available.

In Section 15.4 we discuss a method to realize a piped normalized CORDIC with minimal delay. One of the problems of the classical CORDIC realizations, discussed by Ahmed in Chapter 16, is the time wasted in the computation of angles and normalizations. We show that computation of angles may be avoided altogether by using an alternative angle representation and that normalization may be piped with the CORDIC operation itself. We present a CORDIC realization whereby its normalization cycle is internally piped with the main cycle, thereby reducing the total CORDIC processing time by a factor of at least 50% with respect to standard realizations.

15.2 PARALLELISM AND PIPELINING

A signal processing algorithm operates on samples that are inputted regularly each sample period. Given input samples and current state, the algorithm produces a new state and output samples. Each time the same algorithm is used to process the data. Operations occurring in one specific instance may sometimes be executed simultaneously. This we call "parallelism." If operations belonging to different instances of the algorithm are executed simultaneously, we have "pipelining." One may have parallelism and pipelining at different levels of operation: at the macro level or at the bit level. In this section we deal with parallelism at the macro level. Section 15.4 shows examples of parallelism at the bit level.

For ease of exposition we discuss a method to discover parallelism and pipelining using the wave digital filter algorithm depicted in Figure 15.1. The method is automatic and has been implemented in a set of computer programs [7]. However, a good understanding may be achieved pictorially and we shall follow that way here. The method is classical [8–10], but since it is effective and has not been used in the recent literature, it deserves at least a brief discussion.

The main tool in computing parallelism and pipelining for a given algorithmic situation is the precedence graph. In Figure 15.2 we construct a precedence graph for Figure 15.1 under the following (reasonable) assumptions:

1. Synchronous clocks and operations

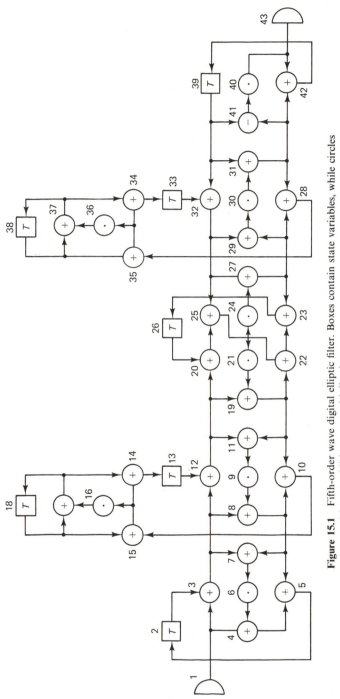

Figure 15.1 Fifth-order wave digital elliptic filter. Boxes contain state variables, while circles represent either an addition or a multiplication.

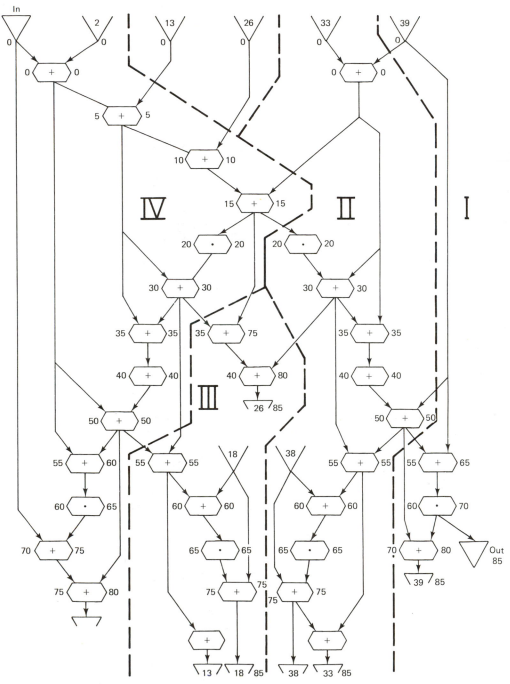

Figure 15.2 Precedence graph for Figure 15.1. The T nodes correspond to state variables (number are indicated). To each operation box there is assigned an E-level (at the left of the box) and an L-level (at the right). Initial E-levels of inputs and T nodes are zero. Final L-levels of output and next cycle T nodes are equal to the total cycle period. The processor assignment given by the assignment program is shown with Roman numerals.

260

2. A possibly unlimited number of devices which execute a multiply in 10 clock cycles and an add in 5 cycles (the method itself will deliver the correct number needed for time-optimal operations)

Operations always start with known variables represented by input variables and state variables. Nodes corresponding to a state are special and are called T nodes. Subsequent operations are placed recursively at their earliest possible position: following immediately after the operation(s) that produce their input data. In this scheme no time allowance is made for transport of data (it may also be included if necessary). The algorithm ends with the production of an output variable, and a new set of states, filling the T nodes for the next cycle of operation.

To each operation in the precedence graph one may assign two numbers: an E-level (for Earliest) and an L-level (for Latest). The E-level indicates at which point in time, relative to the start of the algorithm, a given operation may be executed at its earliest convenience (provided that one has unlimited computing power). More precisely: If $P(i)$ is the set of predecessors of operation i, one has

$$\text{E-level of } i = \max \left[\text{E-levels } P(i) + \text{ processing time of } P(i) \right] \qquad (15.1)$$

Similarly, the L-level of i indicates the latest time at which an operation i has to be started to assure minimal computation time for the cycle. More precisely: if $S(i)$ is the set of successors of operation i, we have

$$\text{L-level of } i = \min \left[\text{L-levels } S(i) - \text{ processing time of } i \right] \qquad (15.2)$$

Operations (15.1) and (15.2) are initialized—for the parallelism case—by

$$\text{E-level input} = 0 = \text{E-level T nodes}$$

$$\text{L-level output} = \text{L-level next-cycle T nodes}$$

$$= \text{total minimum processing time}$$

E-levels and L-levels for the case of Figure 15.1 are shown in Figure 15.2. The total cycle time is 85. E-levels and L-levels may be computed easily from the adjacency matrix of the precedence graph (see [3]).

From the precedence diagram in Figure 15.2 and the indicated E- and L-levels, a number of conclusions may be drawn:

1. If one only allows parallelism and no pipelining, the minimum processing time is 85 clock cycles.
2. There is always at least one path through the precedence graph which contains operations for which all E-levels and L-levels are equal. Such a path is called a "critical path." Its length in clock cycles is equal to the minimum total processing time.
3. From the precedence graph a concurrency factor may be deduced. This quantity is defined as the total number of operations divided by the number of operations on the longest path, which is equal to the number of levels on that

path plus one. For the algorithm shown, the concurrency factor is 2.3. It follows that one needs at least three processors to implement the algorithm. It is not hard to see that four will suffice for minimal time implementation (assignment to processing elements will be discussed).

4. The mere fact that a precedence diagram may be drawn is proof that the algorithm as given does not contain any deadlocks. During the assignment procedure, the precedence graph will be needed to avoid deadlocks there as well.

From Figure 15.2 one can infer that some T nodes (filter states) for the following cycle are already available for processing prior to the maximum length of the algorithm, while other T nodes in the current cycle are needed only at a very late stage. It follows that the next cycle may have started while the current one has not yet terminated. This is the essence of pipelining. The effect may be taken into account in the precedence graph by assigning nonzero E-levels to the terminal T nodes, in much the same way as before, and similarly, assigning L-levels to the initial T nodes. The terminated nodes for cycle n become the initial T nodes for cycle $n + 1$, and one may update their E-levels by substracting the length of cycle n. In this way negative E-levels may occur. They indicate how much time before the termination of cycle n the new states for cycles $n + 1$ are already available. This information is used to compress the new cycle in the previous. After a few repetitions of this procedure a stable situation is obtained which then produces the minimal pipelined cycle time as the length of a critical path in what has become a period of a piped precedence graph. Assignments may then be performed on this period in much the same way as is done on the original graph. In Figure 15.11 an example of this procedure is given (for the present case pipelining does not give much improvement: the cycle time is reduced from 85 to 80 with a throughput rate increase of 85/80). An estimate of the minimum period of a pipelined realization may often be obtained by finding a critical path between a state and its next occurrence (there need not be any such path, but in truly recursive situations there often is one). Such a path cannot be destroyed during compressions.

Once precedence is established and insight has been obtained in possible parallelism/pipelining, one must assign operations to the available hardware. These assignment and scheduling procedures are not as straightforward as the construction of the precedence diagram. A vast literature is available on the subject (see, e.g., [8–10]). Moreover, the problem seems, in this generality, NP complete. In practical cases, however, fast algorithms may be constructed and fast, intelligent (but possibly not optimal) choices based on the precedence diagram may be made. The assignment question is then divided into two main parts which are treated in an independent manner:

1. In what sequence should operations be assigned to a processor; that is: What is a good time for assignment? If several operations are ready for assignment, which one is to be assigned first?

2. Where, in which processor, must an operation be placed so as to minimize exchange of data between processing units?

A standard set of assumptions ease the discussion.

1. Exchange of data does not cost any time.
2. One is interested in minimizing communication between processors.
3. One desires minimal time operations.

The question of "when" to schedule a given node has been examined in [8]. To find the optimal solution, exhaustive and time-consuming search is needed. A simple suboptimal algorithm is proposed by Hu [9].

1. A node must be scheduled as soon as possible.
2. A node with lowest L-level has precedence.

Hu's algorithm may lead to scheduling errors, because it may sometimes be useful to delay an operation in order to allow some later but more urgent operations to be performed first. Hu's algorithm may be improved by adding the following rule:

> If a given operation i which is ready for scheduling actually blocks a later operation with lower L-level (hence a more urgent operation), operation i must be postponed.

The next question is "where" (in which processor) to schedule an operation. If we wish to minimize time wasted by exchanging data, minimization of the number of exchanges between processing elements becomes important. Here again, an intelligent heuristic algorithm is to be preferred to an exhaustive search. Such an algorithm will assign operations on the basis of a measure of affinity between an operation and a processor. The problem is the determination of a correct affinity measure. One must take into account:

1. The number of paths connecting the processor to other processors, ingoing as well as outgoing
2. The weight already assigned to one processor

The determination of the number of paths from one operation to another (even those connecting one operation to another in the next cycle) may be done efficiently on the incidence graphs (see [10]). The method computes numbers C_{ij} and T_i, where C_{ij} is the number of paths between nodes i and j, while T_i is the total number of paths entering node i. A good measure for affinity proved to be

$$\text{affinity} = \frac{C_{ij}}{T_i} + \frac{C_{ji}}{T_j} \tag{15.3}$$

On the basis of that measure, the assignment shown in Figure 15.2 was made automatically for the algorithm of Figure 15.1. Our experience is that machine-produced assignments are superior to the hand-produced.

15.3 PARALLEL/PIPELINED ORTHOGONAL FILTERS

Globally orthogonal filters when realized in cascade have very desirable properties: low passband and stopband sensitivity, resulting in very high attenuation values. They are much better in that sense than are directly cascaded filters, even though the cascaded pieces may be realized in an orthogonal fashion [11]. However, one has to pay a heavy price for their quality: there is hardly any parallelism or pipelining possible in the minimal version. Luckily, by using nonminimal versions of the filters we shall show that it is possible to obtain high throughput rates by pipelining while keeping the desirable global orthogonality.

A globally orthogonal filter may use only orthogonal transformations, the most elementary of which are the Givens rotors, conveniently depicted in Figure 15.3. Neither summators nor branching is allowed, because such operations are not orthogonal (only a branch at the very end is acceptable). The archetype of a globally orthogonal filter is a Levinson shaping filter, depicted in Figure 15.4. This filter realizes an autoregressive transferfunction and may easily be pipelined, as is shown in the precedence diagram of Figure 15.5 where two piped cycles are shown. If each orthogonal transformation takes one cycle, then a throughput rate of one per two cycles is achieved. This structure is nicely realized by means of a CORDIC rotor. In case the piped CORDIC algorithm described in Section 15.4 is used, a throughput rate of 1 per clock cycle may be obtained on a multiplexed signal.

For the high-quality selective filters that are used in telecommunication it is necessary to realize ARMA orthogonal filter structures. It has been shown in [4,5] that any transfer function may be realized by piecing together the three types of building blocks shown in Figure 15.6. An example of the overall architecture obtained by cascading sections is shown in Figure 15.7 (the example is an orthogonal realization of a ninth-order elliptic filter).

The precedence diagram corresponding to Figure 15.7 is shown in Figure 15.8. From it we learn that it is not possible to produce a pipeline for this type of filter, because there is a critical path from one state to the same state in the next cycle (in the example the critical path goes from state 1 to state 1). The remedy to the

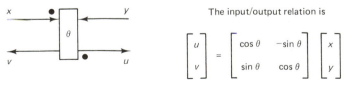

The input/output relation is

$$\begin{bmatrix} u \\ v \end{bmatrix} = \begin{bmatrix} \cos\theta & -\sin\theta \\ \sin\theta & \cos\theta \end{bmatrix} \begin{bmatrix} x \\ y \end{bmatrix}$$

Figure 15.3 Givens rotor.

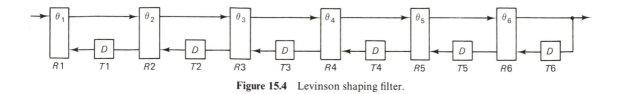

Figure 15.4 Levinson shaping filter.

situation is the replacement of the basic cells shown in Figure 15.6 by a collection of possibly nonminimal cells obtained by cascading a degree-reducing type II or type III cell followed by a dummy type I cell (Figure 15.9). In this way the pipelining advantages of the Levinson cell are "injected" into the other cells (Figure 15.10). It may seem at first strange that a trivial modification of the basic cell leads to a pipeline filter. However, the precedence diagram in Figure 15.11 shows that the longest critical path between a state and its next value is 6, instead of 18, making an increase in throughput rate for this type of filter 18/6.

For larger filters the gain becomes spectacular: the minimal version will have a critical path of length $2n + 2$ (with n the degree of the filter) while the nonminimal

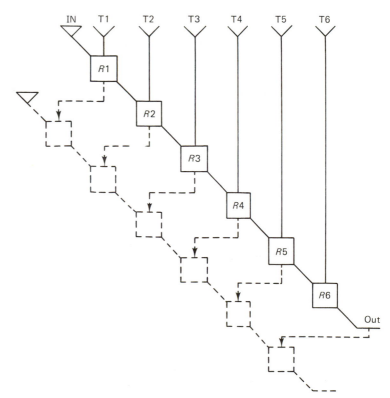

Figure 15.5 Precedence diagram for the Levinson shaping filter showing the possibility for pipelining. If the Givens rotor is executed as a piped CORDIC, a throughput rate of one sample per clock cycle can be achieved in a multiplexed environment.

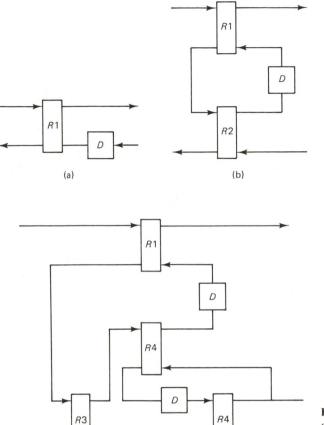

Figure 15.6 Building blocks for cascade global orthogonal filter realizations: (a) type I: AR section; (b) type II: real degree 1 section; (c) type III: real degree 2 section.

version's length remains at 6, resulting in a throughput increase of $(2n + 2)/6$. If a throughput of one sample per clock cycle is desired, use must be made of a piped CORDIC module whereby pipelining between modules is done at the bit level. With this type of doubly piped CORDIC modules, highly modular realizations for any transfer function are obtained.

15.4 THE DOUBLY PIPED CORDIC MODULE

To enhance the throughput both of an orthogonal filter and of an array of CORDIC processors, it is necessary to use a doubly piped CORDIC module. By doubly piped we mean that not only are the subsequent CORDIC rotations piped at the algorithmic level, but the internal operations inside each CORDIC layer are

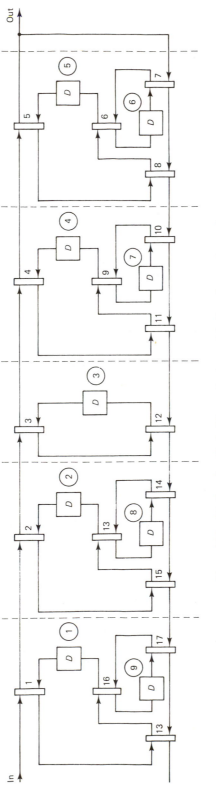

Figure 15.7 Ninth-order elliptic filter realized as a globally orthogonal cascade.

267

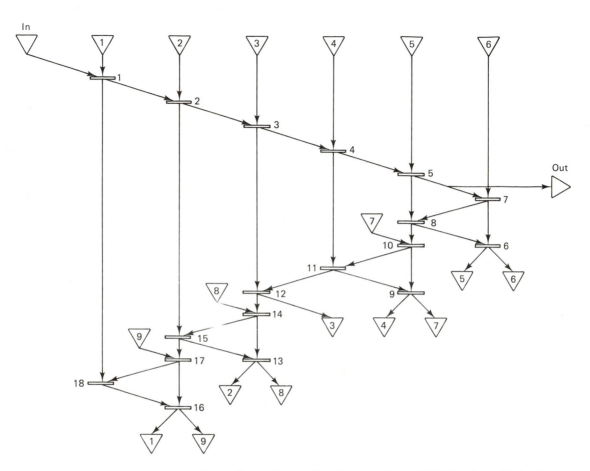

Figure 15.8 Precedence diagram for the filter shown in Figure 15.7. There is a critical path from state 1 to next state 1 showing that pipelining is not possible.

piped at the bit level. This improvement, together with the use of a representation for the angle consisting of an encoding of the sequence of elementary CORDIC rotations, will make the CORDIC fast enough to allow clockrate throughputs.

To illustrate the basic ideas we concentrate on the case where the CORDIC is used exclusively for normal (not hyperbolic) rotations. The more general case is a less than straightforward but working extension, and is described by Deprettere and Udo [12].

The typical ith-level CORDIC operation is an elementary rotation over an angle θ_i with $\tan \theta_i = 2^{-i}$:

$$x' = x - \sigma_i 2^{-i}$$
$$y' = \sigma_i 2^{-i}x + y$$

$$(15.4)$$

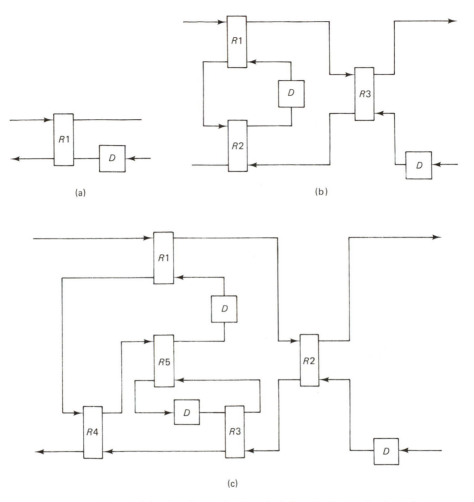

(a)

(b)

(c)

Figure 15.9 Nonminimal orthogonal cells suited for pipeline realization of an arbitrary ARMA transfer function: (a) type I: AR section; (b) type II: real degree 1 pipeline section; (c) type III: real degree 2 pipeline section.

where $\sigma_i = \pm 1$. The initial rotation is always taken over an angle of $\pm 90°$ represented by $\sigma_0 = \pm 1$. The sequence σ_i $(i = 0, \ldots, n)$ is representative for the total angle and may be used as an encoding of its value. Two types of operation are typical: one in which a two-dimensional vector is given and the angle sequence and norm are to be computed, and the other where the sequence σ_i is given and used to rotate a given vector over the angle represented by the sequence. In almost no signal processing algorithm is one interested in the angle itself—the encoding σ_i will never leave the machine.

The operations (15.4) are "unnormalized." The end result (after going through

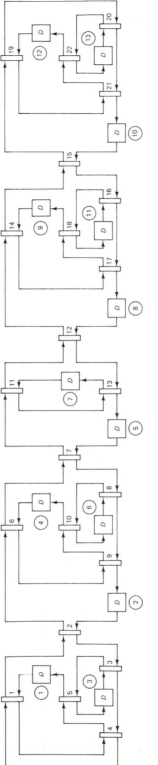

Figure 15.10 Nonminimal realization of a ninth-order elliptic filter in a globally orthogonal structure.

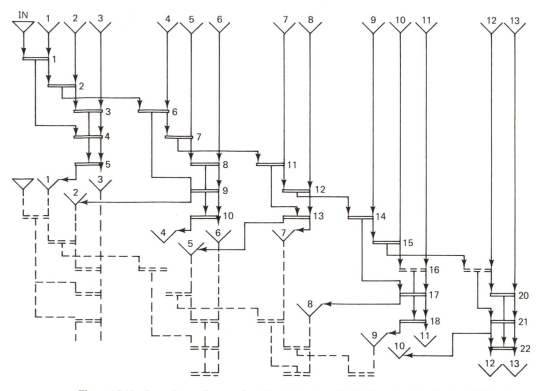

Figure 15.11 Precedence diagram for the ninth-order elliptic filter showing the possibility of obtaining piped structures. The longest critical path is 6 units as compared to 18 in Figure 15.8.

n piped sections) is too big by a factor $1/K_n$, where K_n is only dependent on the word length n. It may be expressed by the product

$$K_n = \prod_{i=1}^{n} (1 - \varepsilon_i 2^{-i})$$

where $\varepsilon_i = 0$ or 1. Product of a vector $[x \quad y]^T$ with $(1 - \varepsilon_i 2^{-i})$ has the same complexity and necessitates the same hardware as (15.4):

$$x' = x - \varepsilon_i 2^{-i} x$$
$$y' = y - \varepsilon_i 2^{-i} y \qquad (15.5)$$

The normalization may be executed after all (15.4) operations are terminated, or interleaved with them. Interleaving the normalization (15.5) with the operation (15.4) will allow us to reduce the total duration of the CORDIC sequence, as we shall soon demonstrate. The basic configuration of the ith CORDIC layer is shown in Figure 15.12. A CORDIC pipe will be a cascade of these layers. This is the macro-level pipe mentioned earlier.

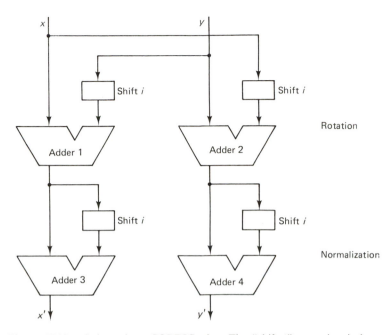

Figure 15.12 *i*th layer in a CORDIC pipe. The "shift *i*" operation is but an interconnect.

There are many ways to realize the adders of Figure 15.12. If one follows the philosophy of an array multiplier, each adder is a linear array of full-adders (FAs) with a carry rippling from one FA to the next. In an array multiplier this layer structure is advantageous because the carry in one layer is piped with the carry in the next layer—or may even be passed on to the next layer. Here, however, due to the presence of shifts, one cannot have the carry of the next layer rippling in parallel with the previous one. For instance and with reference to Figure 15.12, the shift *i* in adder 1 causes the *i*th bit of *y* to be an input of the first FA and the carry can start its ripple only after the carry that produces *y* in the previous layer has reached position *i*. It follows that a direct realization of Figure 15.12 with full-adder layers would result in an overall logical depth of $n + i$ full-adders per layer. The last layer (with $i = n - 1$) determines the overall delay, because the subsequent layers are piped with each other.

The overall logical depth per normalized section may be reduced to *n* by modifying the architecture of the shift + adder combination. In Figure 15.13 we show the idea for the two's-complement arithmetic case. Let the *i* top bits of *x* be denoted $\lceil x \rceil$; then depending on the value of the sign bit of *y* and the carry *c* from the remaining full-adders, we shall have that the top *i* bits of the result must be either $\lceil x \rceil$, $\lceil x \rceil + 1$, or $\lceil x \rceil - 1$. The circuits indicated by the blocks SC compute these three quantities directly. They are roughly equal to a FA because they consist of two independent half-adders. They ripple in pipe with the production of *x*. When

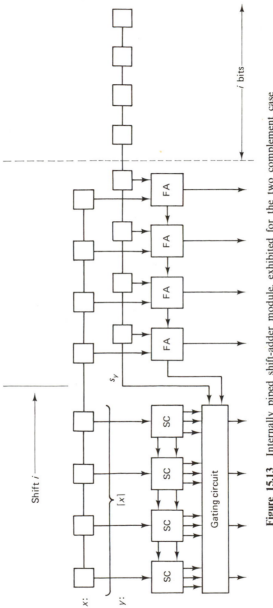

Figure 15.13 Internally piped shift-adder module, exhibited for the two complement case with $n = 8$ and $i = 4$.

273

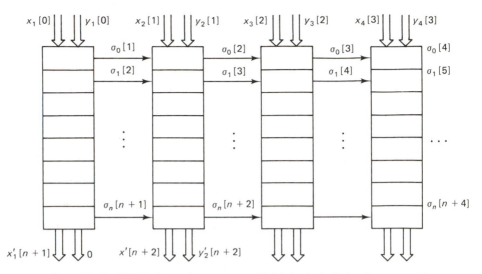

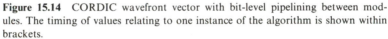

Figure 15.14 CORDIC wavefront vector with bit-level pipelining between modules. The timing of values relating to one instance of the algorithm is shown within brackets.

s_y and c become available (they are produced by circuits that pipe directly with y), a decision is made between the three possibilities and the result gated to the output. The result becomes available one FA delay after x and y are available themselves. The circuit shown in Figure 15.13 typically replaces the normalization adders of Figure 15.12, but may also be used elsewhere in the pipe. Many variations on the basic solution are possible: for instance, one in which the carries are shifted to the following layer. With this type of architecture, the internal normalization has an infinitesimal effect on the timing but does cost additional hardware (an extra layer of FAs and SCs per stage). For the hyperbolic case, the same overall strategy may be followed [12].

The doubly piped CORDIC will normally be latched at each level. This allows for a bit-level pipeline between CORDIC modules, as is exhibited in Figure 15.14 for the case of a linear wavefront processor, where the first CORDIC computes an angle encoded by the $\{\sigma_i\}$ sequence, which is then propagated to subsequent modules, a basic operation in a QU factorization or in an eigenvalue solution based on Givens rotation.

15.5 CONCLUDING REMARKS

The design of high-quality filters and signal processing algorithms requires task-scheduling methods in order to achieve high throughputs without sacrificing quality. A central role is played by calculus on precedence graphs and by task assignment procedures. A careful application of the method shows that one may achieve—in multiplexed circumstances—a throughput of one sample per clock period if use is made of bit-level pipelined modules such as the piped CORDIC.

In a structured design methodology the decomposition of an algorithm into functional units operating in parallel or in a pipe seems to be an essential step which can be repeated at subsequent levels of abstraction. We obtain a map from the topmost algorithmic description to a structure on silicon by executing consecutive applications to the parallelism/pipeline assignment. For example, in Figure 15.11, the functional units are CORDIC units which are piped among themselves, while internally the CORDIC algorithm is itself a piped structure. The descent of one level of abstraction to a lower one is done by ignoring the (distributed) control structure that links the decomposed elements together. At each level one may distinguish two layers: a functional layer which describes what the intended purposes of the device under design are, and an implementation layer which gives an implementation diagram appropriate for that level. For example, for the algorithmic description level, Figure 15.10 is functional while Figure 15.11 is the implementation. The transition from functional to implementation is executed by means of the parallel/-pipe assignment. Descending one level, the diagram of Figure 15.11 becomes functional, and implementation will tell how the devices are actually implemented (CORDIC) and the transport between module executed (bus transport) (Figure 15.14). Continuing this way, one arrives at the logic level, the circuit level, and finally the implementation level.

With respect to the realization of an overall pipelined CORDIC, the following remark is in order. We have shown in Section 15.4 that it is possible to pipe rotation and normalization in a single layer resulting in almost no loss of time for the normalization. One may think that the same technique can be used to pipe subsequent CORDIC layers as in an array multiplier. This, however, is not quite possible. The ripple mechanism of the CORDIC does not allow for overall piping, although individual sections may be piped with each other. Given the constraints on the CORDIC algorithm, it is an interesting topic of research to compare the merits of CORDIC to those of multiplier implementation as far as maximal throughput is concerned. Application of the CORDIC module to wavefront arrays [13] is also being investigated jointly by the Delft University of Technology and the University of Southern California.

REFERENCES

[1] A. Fettweis, "Digital Filter Structures Related to Classical Filter Networks," *Arch. Elektr. Uebertrag., 25*:79–89 (1971).

[2] E. Deprettere and P. Dewilde, "Orthogonal Cascade Realization of Real Multiport Digital Filters," *Circuit Theory Appl., 8*:245–272 (1980).

[3] R. Nouta, "Studies in Wave Digital Filter Theory and Design," Ph.D. thesis, Delft University of Technology, 1980.

[4] E. Deprettere and P. Dewilde, "Orthogonal Cascade Realization of Real Multiport Digital Filters," *Circuit Theory Appl., 8*:245–272 (1980).

[5] P. Dewilde, "Stochastic Modeling with Orthogonal Filters," *in Outils et modèles mathématiques pour l'automatique, l'analyse de systèmes et le traitement du signal*, Vol. 2, CNRS, Paris, 1982, pp. 331–398.

[6] P. Dewilde and H. Dym, "Schur Recursions, Error Formulas and Convergence of Rational Estimators for Stationary Stochastic Sequences," *IEEE Trans. Inf. Theory, IT-27*(4):446–461 (July 1981).

[7] R. S. Martens, "Een hulp voor het Optimaal Implementeren van Digitale Filters in Parallelle Hardware," Delft University of Technology, Network Theory Tech. Rep. 30, 1980.

[8] C. V. Ramamoorty, K. M. Chandy, and M. J. Gonzales, "Optimal Scheduling Strategies in a Multiprocessor System," *IEEE Trans. Comput., C-21*(2):137–146 (Feb. 1972).

[9] T. C. Hu, "Parallel Sequencing and Assembly Line Problems," *Oper. Res., 9*:841–848 (Nov. 1961).

[10] D. M. Schuler and E. G. Ulrich, "Clustering and Linear Placement," *Proc. ACM IEEE Design Autom. Workshop*, June 26–28, 1972, Dallas, pp. 250–256.

[11] R. A. Roberts and C. T. Mullis, "Digital Signal Processing Structures for VLSI," *Proceedings*, USC Workshop on "VLSI and Modern Signal Processing," pp. 83–88.

[12] E. Deprettere and R. Udo, "The Pipeline CORDIC," Internal Rep. Network Theory Section, Delft University of Technology, 1983.

[13] S. Y. Kung, K. S. Arun, R. J. Gal-Ezer, and D. V. Bhaskar Rao, "Wavefront Array Processor: Language, Architecture, and Applications," *IEEE Trans. Comput., C-31*(11):1054–1066 (Nov. 1982).

16

Alternative Arithmetic Unit Architectures for VLSI Digital Signal Processors

Hassan M. Ahmed

Codex Corporation
Mansfield, Massachusetts

16.1 INTRODUCTION

Digital signal processing is a term encompassing many different techniques for transforming numbers, usually digital samples of analog signals, into digital samples of a more desirable signal [the signals may be one dimensional (e.g., human speech); two-dimensional, as in the case of images; or of still higher dimensionality]. A digital signal processor may be generically viewed as shown in Figure 16.1. Input circuits accept the digital samples while output circuits deliver the transformed samples. The flow of data transits an arithmetic unit where the desired transformations take place. The specific nature of these transformations (e.g., low-pass filtering, adaptive filtering), as well as the data flow, occur under the auspices of a controller. This chapter is concerned primarily with study of the arithmetic unit. Although such study may be done in isolation, we will see that the performance of the digital signal processor is strongly dependent on the combined operation of all four blocks and in particular, the interactions of the AU and controller.

Not all digital signal processing involves the use of arithmetic in the usual sense. For example, many digital codes can be implemented solely through the use of logical operations [1] (e.g., exclusive-or) and never require the traditional arithmetic operations, such as addition and multiplication. (We ignore the obvious connection between arithmetic and logical operations.) For the purposes of this chapter we limit ourselves to signal processing tasks that actually require arithmetic in our traditional sense. These might include speech synthesis and digital filtering.

Signal processing tasks vary considerably in nature and purpose; however, one common element is the need for performing a very large number of operations very quickly. Whereas fast, high-throughput machines have been synonymous with large

277

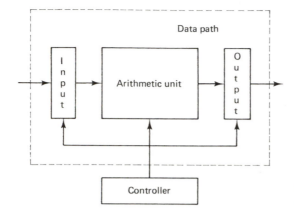

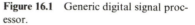

Figure 16.1   Generic digital signal processor.

physical volume and high power dissipation, very large scale integration (VLSI) offers the ability to provide more computing power in a given volume than ever before. However, typical throughputs required of machines to implement even audio band algorithms (e.g., speech analysis or digital communications over telephone lines) challenge the capabilities of the most modern VLSI microprocessors for digital signal processing.

Almost all present-day VLSI signal processing chips are custom designed and laid out to achieve the tightest packing and highest throughput, both of which are pushing the limits of technology. Although throughput is influenced by circuit design, it can also be dramatically affected with clever architectures. In this chapter we show various arithmetic unit architectures which are suited to different signal processing algorithms, with the idea that a good match of the AU architecture to the problem will improve throughput without exploiting the "last nanosecond" the technology affords. In addition to the traditional multiplier, we examine the CORDIC [3] and convergence computation [4] algorithms for performing elementary arithmetic operations. We present some alternative architectures for each type of arithmetic unit which should be included in a designer's repertoire of DSP building blocks. The appropriate one is chosen based on throughput and area considerations as well as the type of algorithms to be implemented.

16.2 DEPENDENCE OF ARITHMETIC AND CONTROL

One often hears a statement similar to the following: "One million multiply operations per second are required to implement a particular algorithm in real time." The impact of such a statement on the arithmetic unit is strongly dependent on the control features of the signal processor. Although controller design is beyond the scope of this chapter, we will give a simple example of how it affects processor throughput. Further exploration is left to the reader.

Consider the programmable digital signal processor of Figure 16.2. We wish to

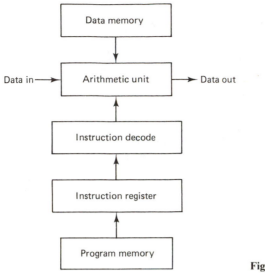

Figure 16.2   Programmable DSP.

determine the maximum allowable latency in the multiplier to achieve 1 million multiplications per second in a long sequence of multiply operations. First examine the simple processor strategy of Figure 16.3(a), in which each instruction is decoded and then the operation takes place. For simplicity, assume that the decoder and multiplier times are equal. Then, the multiplier must operate at 500 ns/multiply to achieve the desired throughput.

Next consider the pipelined strategy of Figure 16.3(b), in which an instruction is decoded during the execution cycle of the previous instruction. The desired

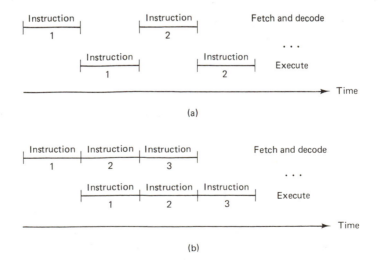

Figure 16.3   Controller/arithmetic unit interaction.

throughput is now achieved with a multiplier capable of 1000 ns/multiply. A remarkable impact on the design and layout of the DSP chip is realized with a simple change in control strategy. The latter technique is not without its problems; for example, the reader should consider how a conditional branch instruction might be implemented in such a pipelined structure (fortuitously, program control is usually limited during computation sequences in signal processing, thus facilitating the pipelining).

This simple example shows the interrelation of arithmetic unit and controller design. There are many more subtle points, such as address manipulation, that limit the performance of a signal processor and establish a bound on throughput regardless of AU speed.

The remainder of this chapter will be concerned with the design of arithmetic units. Little regard will be paid to control aspects; however, the foregoing example should illustrate that remarkable throughput enhancements can often be obtained through clever architectural changes rather than brute-force circuit speed. In the same spirit, we will examine how clever arithmetic unit architectures can provide large throughputs.

16.3 FUNDAMENTAL OPERATIONS IN SIGNAL PROCESSING

There are at least four major commercial offerings of VLSI signal processors: the Texas Instruments TMS320 [5], Bell Laboratories DSP [6], NEC 7720 [7], and the AMI 2811 [8]. A number of nonintegrated processors have also been reported in the literature (see, e.g., [9–11]). All of these machines have a fast multiply and accumulate facility that forms a substantial portion of the arithmetic unit. Such similarity arises from the realization that the multiply and add operation is fundamental to all signal processing algorithms involving digital filtering. For example, the filter structure of Figure 16.4 is defined by

$$Y_T = \sum_{K=0}^{n-1} C_K X_{T-K}$$

where Y_T is the filter output at time T, X_T the filter input at time T, and C_K the

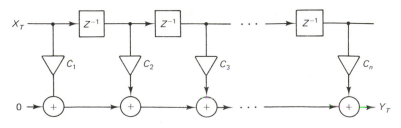

Figure 16.4 Tapped-delay-line filter.

filter coefficient. It is clear that multiply and accumulate is the fundamental operation. In fact, the assembly language code of a signal processor to implement this filter might be

```
CLEAR      ACC                    ; Clear accumulator
MAC        C₀, Xᴛ                 ; ACC ← ACC + C₀ * Xᴛ
MAC        C₁, Xᴛ₋₁               ; ACC ← ACC + C₁ * Xᴛ₋₁
  ⋮
MAC        Cₙ₋₁, Xᴛ₋ₙ₊₁          ; NOW ACC = Yᴛ
```

However, multiplication is merely one of many fundamental operations describing digital signal processing algorithms. Trigonometric quantities, particularly vector rotations, are basic to some filtering structures involving exact least-squares modeling and matrix algebra, as well as other algorithms, such as the discrete Fourier transform. Consider, for example, the square-root-normalized ladder filter [2] shown in Figure 16.5. Each filter stage is described by

$$\rho_{n+1,\,T} = \sqrt{1 - v_{n,\,T}^2}\ \sqrt{1 - n_{n,\,T-1}^2}\ \rho_{n,\,T} + v_{n,\,T}\, n_{n,\,T-1}$$

$$v_{n+1,\,T} = \frac{v_{n,\,T} - \rho_{n+1,\,T}\, n_{n,\,T-1}}{\sqrt{1 - \rho_{n+1,\,T}^2}\ \sqrt{1 - n_{n,\,T-1}^2}}$$

$$\eta_{n+1,\,T} = \frac{n_{n,\,T-1} - \rho_{n+1,\,T}\, v_{n,\,T}}{\sqrt{1 - \rho_{n+1,\,T}^2}\ \sqrt{1 - v_{n,\,T}^2}}$$

Indeed, multiplications are necessary to describe these equations, however, by employing some manipulations discussed in [12], these equations may be reexpressed as:

$$\begin{bmatrix} \rho_{n+1,\,T} & v_{n,\,T}^* \\ \eta_{n,\,T}^* & \text{don't care} \end{bmatrix} = \begin{bmatrix} \sin\theta_v & -\cos\theta_v \\ \cos\theta_v & \sin\theta_v \end{bmatrix}^T \begin{bmatrix} \rho_{n,\,T} & 0 \\ 0 & 1 \end{bmatrix} \begin{bmatrix} \sin\theta_\eta & -\cos\theta_\eta \\ \cos\theta_\eta & \sin\theta_\eta \end{bmatrix}$$

$$\begin{bmatrix} v_{n+1,\,T} & \eta_{n+1,\,T} \end{bmatrix} = \begin{bmatrix} 1 & 0 \end{bmatrix} \begin{bmatrix} \sinh\theta\rho & \cosh\theta_\rho \\ \cosh\theta_\rho & \sinh\theta_\rho \end{bmatrix} \begin{bmatrix} v_{n,\,T}^* & \eta_{n,\,T}^* \\ 0 & 0 \end{bmatrix}$$

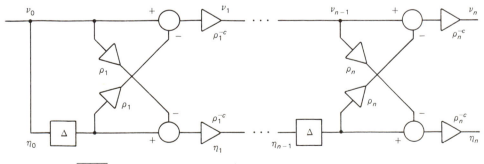

Notes: $\rho^c = \sqrt{1 - \rho^2}$, $\rho^{-c} = 1/\rho^c$

Figure 16.5 Ladder filter.

where

$$\theta_v = \cos^{-1} v_{n,\,T} \qquad \theta_\eta = \cos^{-1} \eta_{n,\,T-1} \qquad \theta_\rho = \tanh^{-1} 1/\rho_{n+1,\,T}$$

Now the fundamental equation structure is clearly seen to consist of two dimensional vector rotations. The CORDIC algorithms [3] provide a convenient method for realizing the rotations, while a simple multiplier does not.

Similarly, it is relatively simple to show that other popular algorithms such as the DFT, FFT, complex filter and LMS adaptive equalizer are readily described with rotations. The detailed arguments are provided in [12,13].

Still another common algorithm is the u-law compression [14] used in telephony applications. While multipliers or CORDIC blocks can compute this conversion, it is most simply done with the convergence computation method (CCM) of Chen [4].

The message we wish to impart from this brief discussion is that although fast multipliers are prevalent in DSP architectures, they are not necessarily suited to all problems. It is desirable to have a host of arithmetic capabilities, of which multipliers, CORDIC operations, and CCM form a rich set. In fact, the CORDIC and CCM are intimately linked [15] and multiplication is a special case of both of these, so there is hope that a single hardware structure can be used to implement all three techniques.

We will now present architectures for multipliers, CORDIC, and CCM arithmetic units and then compare their relative performances. Only fixed-point operations will be considered since it is relatively straightforward to demonstrate that floating-point calculations can be realized as fixed-point calculations on the mantissas of the operands plus some manipulation of the exponents.

16.4 ARCHITECTURES FOR FAST MULTIPLICATION

Much attention has been paid in the past to the realizaton of fast multiplication circuitry. We will present some strategies for bit parallel multiplication and illustrate how throughput is dramatically affected by the architecture. The reader is referred to [15,16] for bit serial multiplication methods which are not addressed in this chapter. Bit parallel operations often provide a better throughput in a given area than do their bit serial counterparts [15].

16.4.1 Indirect Multiplication Algorithms

Indirect multiplication is reminiscent of the usual shift and add techniques learned in grade school. Consider the structure of Figure 16.6, in which X and Y are the two n-bit operand registers containing the magnitudes of the multiplier and multiplicand, while the n-bit accumulator, A, holds the partial sums of the multiplication. Two flip-flops hold the signs of the operands. Define Y_i as the ith bit of Y and let

$$X \wedge Y_i = \begin{cases} X & Y_i = 1 \\ 0 & Y_i = 0 \end{cases}$$

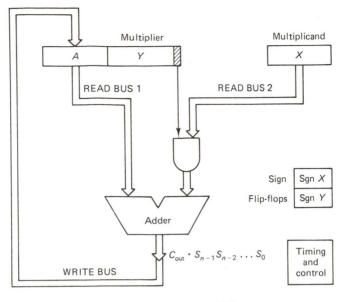

Figure 16.6   Indirect multiplier.

be an n-bit result from the $n \times 1$ logical AND operation. The product, XY, is computed as follows:

$$\text{For} \quad i = 0 \text{ to } n - 1 \text{ begin};$$
$$C_{out} \cdot S \leftarrow A + (X \wedge Y_0) \qquad \text{/*addition*/}$$
$$A.Y \leftarrow C_{out} \cdot S \cdot Y_{n-1} Y_{n-2} \cdots Y_{i+1} \qquad \text{/*right shift*/}$$
$$\text{end};$$

where S and C_{out} are the sum and carry out from the adder and A.B denotes the concatenation of A with B. The 2n-bit product appears in A.Y at the end of the multiplication. Notice that the right shift in the algorithm above is completely analogous to the left shift in the grade school method.

A total of n cycles are required per multiplication in this indirect scheme. For example, a 500-ns multiply operation on 16-bit quantities would require a cycle time of 31 ns. A cycle includes the time to access the registers, perform an addition and shift, and finally store the result back in the A and X registers. The circuit speeds necessary to achieve a particular cycle time are largely dependent on the number of buses constructed and this is smallest for the three bus case shown in Figure 16.6 where the two adder operands are fetched simultaneously on separate buses and the result is written back on still another bus.

We can speed up this shift and add technique by operating on many bits per cycle rather than just one. This is known as multiple bit scanning. Consider the n-bit quantities to be partitioned into nonoverlapping k-bit blocks. Without loss of generality, we take n to be an integral multiple of k. Let m_i be the value of the ith block. Then the multiplication algorithm may be written:

```
for     i = 0 to (n/k) − 1 begin;
        Cout . S ← A + mi X
        A.Y ← Cout . S . Yn−1 Yn−2 ... Yi+k
end;
```

where the right shift is now k bits instead of one. Notice that

$$m_J X = \sum_{i=0}^{k-1} b_{i, j} 2^i X$$

where $b_{k-1, j} b_{k-2, j} \cdots b_{0, j}$ is the jth block of k bits. Therefore the product, $m_i X$, is readily implemented with a $k + 1$ operand adder. Only multiples of the multiplicand that can be produced by shifting are required. By operating on k bits at once, this scheme requires only $[n/k] + 1$ cycles to complete a multiplication rather than n, however, at the expense of more sophisticated hardware. It is noteworthy that the multiple scanning method is equivalent to a radix $r = 2^k$ multiplication.

We have seen that each bit in a non overlapped multiple bit scanning multiplier contributes a term to the partial product. We can reduce the number of additions required by scanning overlapped blocks of multiple bits and exploiting some properties of bit strings. We reduce execution time by shifting across long zero strings in the multiplier. For example, consider a bit string of k consecutive 1's.

$$\ldots 0, \underbrace{1, 1, 1, \ldots, 1}_{k}, 0 \ldots$$

Using the fact:

$$2i + k − 2i = 2i + k − 1 + 2i + k − 2 + \cdots + 2i + 1 + 2i$$

the string may be recoded as:

$$\ldots 1, \underbrace{0, 0, \ldots, 0}_{k-1}, \bar{1}, 0$$

where $\bar{1}$ is used to denote -1. Now, k consecutive additions have been replaced with a single addition and a single subtraction through multiplier recoding. It is precisely this property that is useful in recoded multipliers (see, e.g., [17–19]).

16.4.2 Recoded Multipliers

We have just shown a simple recoding technique for a string of bits, which reduces the number of additions required in a multiplication. This property, known as string recoding, is the basis of many multiplication schemes, including the celebrated method of Booth [20]. Let $\mathbf{B}'$ be a string of n binary digits and let $\mathbf{D}$ be the recoded string consisting of n digits chosen from the alphabet $\{1, 0, \bar{1}\}$. Then if $\mathbf{B} = 0 \cdot B' \cdot 0$, we have

$$D_i = \begin{cases} 0 & B_i = B_{i-1} \\ 1 & B_i < B_{i-1} \\ \bar{1} & B_i > B_{i-1} \end{cases}$$

is the ith digit of D. Clearly, the recoding of an entire string may be performed in a bit parallel fashion very rapidly.

The purpose of recoding is to generate a string $\mathbf{D}$ which has more zero digits than $\mathbf{B}$ (recall that these do not contribute to the multiplication), thus enhancing the multiplier speed. In general, the recoding is effective on strings, $\mathbf{B}$, which do not have many isolated 1's.

The Booth multiplication algorithm may be stated as follows:

```
for    i = 0 to n − 1 begin;
       if Y₀ = Y₁ then PP ← right shift (PP)
       if Y₁ < Y₀ then PP ← right shift (PP + X)
       if Y₁ > Y₀ then PP ← right shift (PP − X)
end;
```

where PP is the partial product stored in A.Y. This algorithm operates directly on two's complement numbers without the need for conversion to sign-magnitude form.

16.4.3 Array Multipliers—The Braun Multiplier

Consider two unsigned integers, $\mathbf{A} = a_{n-1}a_{n-2} \cdots a_0$ and $\mathbf{B} = b_{m-1}b_{m-2} \cdots b_0$ having magnitudes

$$|\mathbf{A}| = \sum_{i=0}^{n-1} a_i 2^i$$

$$|\mathbf{B}| = \sum_{i=0}^{m-1} b_i 2^i$$

We desire the product, $\mathbf{P}$, such that

$$|\mathbf{P}| = |\mathbf{A}||\mathbf{B}| = \sum_{i=0}^{m-1} \sum_{j=0}^{n-1} (a_i b_j) 2^{i+j}$$

$$= \sum_{k=0}^{m+n-1} P_k 2^k$$

The latter equation has the interesting property that each P_k consists of a summation of single-bit quantities, $a_i b_j$, all of which can be formed through simple

AND gates. An array implementation of the multiplication is shown in Figure 16.7 for $m = 5$ and $n = 4$. This arrangement, known as the Braun multiplier [21], is constructed from full-adder cells whose inputs are the single-bit products $a_i b_j$. The structure is very regular and suitable for integration to the extent that it consists of $m(n-1)$ identical full-adder cells as well as mn AND gates.

We can estimate the performance of the array by analyzing the longest propagation path. The multiplication time, T, is given by

$$T = T_a + (n + m - 2)T_f$$

where T_a and T_f are the delay times through the AND gate and full-adder, respectively. Choosing an n-channel MOS technology and utilizing gates of identical drive, we have $T_f = 2T_a$, yielding

$$T = (2n + 2m - 3)T_a$$

Thus a 500-ns 16×16 multiplier requires a gate delay of 8 ns. When we consider that each full-adder fans out to only two identical units in close physical

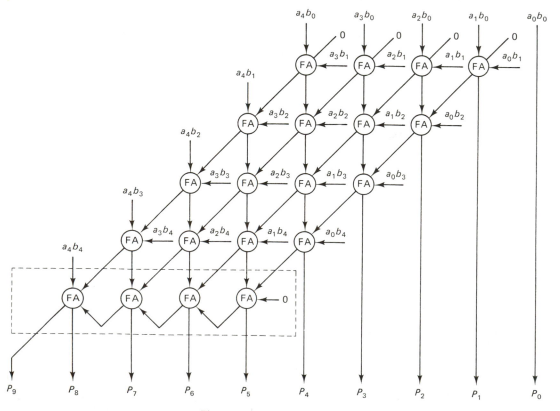

Figure 16.7   Braun multiplier. FA, full-adder.

proximity on a chip, this is a much more realistic performance requirement than the 31-ns cycle time of the indirect multiplier discussed earlier.

In scrutinizing the performance of this multiplier, it is readily seen that a delay of $(m - 1)T_f$ is incurred in the bottom row of the adders due to carry propagation. This delay can be reduced to a constant (e.g., $10T_a$) by adding full carry lookahead to the lower adder row. The multiplication time is now

$$T = (2n + 10)T_a$$

The 500-ns multiplier can now be realized with 12-ns gates—however, at the expense of carry lookahead hardware.

The Braun multiplier operates on unsigned quantities, thus necessitating pre-complementation hardware to handle radix complement number representations (e.g., one's complement and two's complement). We would much prefer a structure that operated on the numbers directly, and this will be the subject of the next section.

16.4.4 The Baugh–Wooley Two's-Complement Multiplier

The subject of array multipliers that operate directly on two's-complement numbers has been discussed by many authors, notably Pezaris [22] and Baugh and Wooley [23]. We illustrate the latter method, depicted in Figure 16.8 for the case of $m = 6$, $n = 4$. Again, the multiplication is computed as the accumulation of a collection of summands. The innovation in the method of Baugh and Wooley is that all the summands are of positive sign, thus requiring only simple addition hardware, in spite of the signed operands.

Again begin with two binary vectors $\mathbf{A} = a_{m-1}a_{m-2} \cdots a_1 a_0$ and $\mathbf{B} = b_{n-1}b_{n-2} \cdots b_1 b_0$ in two's-complement form. Their values, denoted A and B, are

$$A = -a_{m-1}2^{m-1} + \sum_{i=0}^{m-2} a_i 2^i$$

$$B = -b_{n-1}2^{n-1} + \sum_{i=0}^{n-2} b_i 2^i$$

and their product, $\mathbf{P} = P_{m+n-1}P_{m+n-2} \cdots P_1 P_0$ has value

$$P = -P_{m+n-1}2^{m+n-1} + \sum_{i=0}^{m+n-2} P_i 2^i$$

$$= a_{m-1}b_{n-1}2^{m+n-2} + \sum_{i=0}^{m-2}\sum_{j=0}^{n-2} a_i b_j 2^{i+j} - \sum_{i=0}^{m-2} a_i b_{n-1}2^{n-1+i} - \sum_{i=0}^{n-2} a_{m-1}b_i 2^{m-1+i}$$

Baugh and Wooley reexpressed the two subtractions as additions of two's-complement numbers that are readily formed. The crucial observation is that a

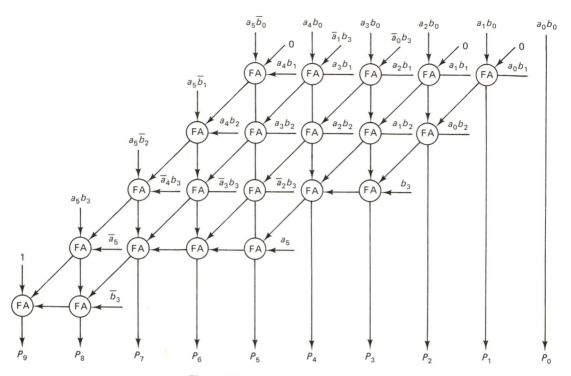

Figure 16.8 Baugh–Wooley multiplier. FA, full-adder.

number $\mathbf{X} = x_{k-1}x_{k-2} \cdots x_1, x_0$ having value

$$X = -x_{k-1}2^{k-1} + \sum_{i=0}^{k-2} x_i 2^i$$

can be negated as follows:

$$-X = -(1 - x_{k-1})2^{k-1} + \sum_{i=0}^{k-2} (1 - x_i)2^i + 1$$

Applying this fact to the equations above, we can rewrite the product simply in terms of additions (i.e., no subtractions):

$$P = a_{m-1}b_{n-1}2^{m+n-2} + \sum_{i=0}^{m-2} \sum_{j=0}^{n-2} a_i b_j 2^{i+j}$$

$$+ \left[-2^{m+n} + 2^{m+n-1} + (\bar{a}_{m-1} + \bar{b}_{n-1})2^{m+n-2} + b_{n-1} + \sum_{i=0}^{m-2} \bar{a}_i b_{n-1} 2^i \right.$$

$$\left. + a_{m-1} + \sum_{i=0}^{n-2} a_{m-1} \bar{b}_i 2^i \right]$$

The latter expression provides the means for obtaining P solely with additions and is the basis of the Baugh–Wooley multiplier.

Figure 16.8 shows an array implementation of the multiplier with full-adder cells for $m = 6$, $n = 4$. In general, the array requires $m(n - 2) + 3$ full-adders and the multiply time is

$$T = 2T_a + (m + n)T_f$$
$$= 2(m + n + 1)T_a$$

When $m = n$, the bottom row of the array becomes simpler and full-carry lookahead then provides a substantial speed improvement. If the carry lookahead adder has addition time $10T_a$, then

$$T = 2(n + 6)T_a$$

The gate speeds and hardware required to implement this multiplier are comparable to the Braun array, which operates only on unsigned integers. When a two's complementer, typically having a delay of $2nT_a$, is added to the Braun array we see that the Baugh–Wooley arrangement is superior both in speed and chip real estate.

16.4.5 Other Multiplication Strategies

Two additional methods of multiplication that have been successfully employed in signal processing applications are based on logarithmic processing and on residue arithmetic. In the former case we observe that

$$A \cdot B = \log^{-1}(\log A + \log B)$$

Therefore, by storing values of $\log x$ and $\log^{-1} x$ in tables, this scheme requires only one addition and three table lookups. Limitations to logarithmic processing arise due to very large lookup tables being required when high precision must be achieved, thus limiting both speed and area. The reader is referred to [24,25] for further information.

Residue arithmetic is of great interest in signal processing systems because all bits of a residue representation are of equal weight. No carry information is necessary as there is no least or most significant bit, thus realizing very fast arithmetic operations. For example, let $P = P_1, P_2, \ldots, P_k$ be a set of moduli whose elements are relatively prime. The unique residue representation of $X \in [-N/2, N/2)$ is the K-tuple:

$$\mathbf{X} = X_1, X_2, \ldots, X_k$$

where

$$X_i = \begin{cases} X_i \bmod P_i & X_i \geq 0 \\ P_i - (|X_i| \bmod P_i) & X_i < 0 \end{cases}$$

and

$$N = \sum_{i=1}^{k} P_i$$

A composition of two numbers X and Y with representations

$$\mathbf{X} = x_1 x_2 \cdots x_k$$

$$\mathbf{Y} = y_1 y_2 \cdots y_k$$

denoted $X \cdot Y$ (where $\cdot$ denotes addition, multiplication, or subtraction) is simply

$$\mathbf{X} \cdot \mathbf{Y} = (x_1 \cdot y_1 \quad x_2 \cdot y_2 \quad \cdots \quad x_k \cdot y_k)$$

Since each residue digit is computed independently, this method of arithmetic is very fast and conducive to high parallelism. Residue arithmetic is, however, not free of problems. For example, numbers are restricted in dynamic range to $[-N/2, N/2)$, which is often less than the range afforded by the binary representation. The conversion of binary quantities to a residue representation is often done with table lookup, which can become the limiting speed factor. Perhaps the greatest drawback of this approach is that the lack of a most or least significant bit makes operations such as division, scaling, rounding, and magnitude comparison very cumbersome. Situations where such operations occur with high frequency involve multiple conversions of the number representation and the advantage of residue arithmetic is lost. Further information is available in [26–28].

Both the logarithmic processing and residue arithmetic schemes provide unconventional alternatives to the construction of arithmetic units for signal processors. However, owing to the associated difficulties mentioned, they are uncommon in general-purpose signal processors and thus we have not concentrated on them. The reader with a specific algorithm in mind may find the techniques very useful, however, and is directed to the references provided. We proceed next to some alternatives to the traditional AU, which are well suited to general-purpose signal processing.

16.5 THE CORDIC ALGORITHMS

We saw in an earlier section that some signal processing algorithms are best described by generalized vector rotations rather than simply multiplications. The CORDIC (for coordinate rotation digital computer) algorithms provide an efficient, essentially bit-recursive method for computing two-dimensional vector rotations with simple shift and add operations.

For coordinate systems, parametrized by a quantity m, in which the norm, R, and angle, Φ, of a vector $\mathbf{X} = (x_0, y_0)$ are defined by

$$R = \sqrt{x_0^2 + m y_0^2}$$

$$\Phi = m \tan^{-1} \frac{y_0 \sqrt{m}}{x_0}$$

the iterative CORDIC algorithm is given by [29]

$$\mathbf{X}_{i+1} = \begin{bmatrix} 1 & -\mu_i \delta_i \\ \mu_i \delta_i & 1 \end{bmatrix} \mathbf{X}_i \tag{16.1}$$

$$z_{i+1} = z_i - \alpha_i$$

where

$$i = \frac{1}{m} \tan^{-1}(\delta_i \sqrt{m})$$

and $\{\delta_i\}$ is a set of arbitrary constants. These equations specify a rotation of $\mathbf{X}_i$ through an angle $\mu_i \alpha_i$ to $\mathbf{X}_{i+1}$, where $\mu_i = \pm 1$ is the direction of rotation. The total rotation at step i which has been applied to the initial vector, $\mathbf{X}_0$, is accumulated in the auxiliary variable z_i. By initializing $z_0 = \theta$ and choosing the rotation sequence $\{\mu_i \alpha_i\}$ such that $z_n \to 0$, we obtain a rotation of X_0 through θ. Similarly, by setting $z_0 = 0$ and choosing $\{\mu_i \alpha_i\}$ to rotate $\mathbf{X}_0$ to a known destination $\mathbf{X}_n$, we accumulate the angle between $\mathbf{X}_0$ and $\mathbf{X}_n$ in z_n. For example, when $\mathbf{X}_n$ lies along the abscissa, $z_n = \tan^{-1}(y_0/x_0)$, thus obtaining an inverse tangent function in the particular coordinate system selected by the value of m. The circular, linear, and hyperbolic coordinate systems, corresponding to $m = 1, 0,$ and -1, respectively, are of special interest. Their associated functions are shown in Figure 16.9.

The key to a simple implementation of the CORDIC recursions lies in the choice of $\{\delta_i\}$. When these are integral powers of the machine radix (e.g., $\delta_i = 2^{-i}$),

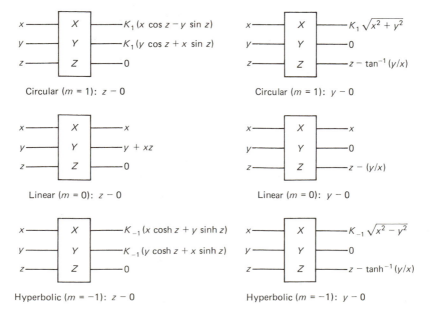

Figure 16.9 CORDIC functions.

the scaling by δ_i reduces to a mere shift and add operation. Guidelines for the choice of $\{\delta_i\}$ and $\{\alpha_i\}$ are given in [14,29]. Clearly, we must choose $\{\alpha_i\}$ to be a nonincreasing sequence for reasonable convergence of the algorithm. Details on the convergence behavior of the algorithm may be found in [14].

When $\{\delta_i\}$ is chosen as suggested, the CORDIC arithmetic unit assumes the simple form of Figure 16.10. The x and y channels provide the updates of $\mathbf{X}_i$ to $\mathbf{X}_{i+1}$ according to equation (16.1), while the z channel accumulates the rotation using values of $\{\alpha_i\}$ obtained from a table. Typically, n values of δ_i and hence α_i suffice for n-bit quantities. These are chosen a priori and stored. All rotation angles θ are then approximated by choosing $\{\mu_i\}$ (recall that $\mu_i = \pm 1$) such that

$$\theta = \sum_{i=0}^{n-1} \mu_i \alpha_i$$

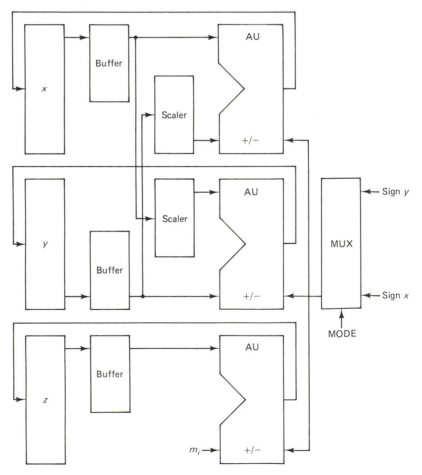

Figure 16.10   CORDIC arithmetic unit.

that is, the CORDIC equations define a bang-bang control system for rotating X_0 to $\mathbf{X}_n$ through a sequence of rotations having predetermined magnitudes but dynamically chosen directions.

16.5.1 Idiosyncracies of the CORDIC Method

We have ignored an important property of (16.1) until now. In reality, it not only represents a rotation of $\mathbf{X}_i$ to $\mathbf{X}_{i+1}$ but also a scaling of the vector, that is,

$$|\mathbf{X}_{i+1}| = K_i|\mathbf{X}_i|$$

where

$$K_i = \sqrt{1 + m\delta_i^2}$$

Therefore, in n iterations we have

$$|\mathbf{X}_n| = K|\mathbf{X}_0|$$

where

$$K = \prod_{i=0}^{n-1} K_i$$

Consequently, the functions obtained by vector rotation are scaled by a known, spurious factor K as shown in Figure 16.9. We must search for a technique to normalize the effect of this constant.

Still another problem with the CORDIC algorithm is its limited region of convergence. Not all possible rotation angles in the coordinate system of interest converge to the desired result. The choice of $\{\delta_i\}$ markedly affects the region of convergence.

A number of schemes have been proposed in the literature [29,30] to expand convergence region and eliminate spurious scaling. They all must test $\mathbf{X}_i$ to determine if a problem exists, and then take corrective action through additional calculations. These schemes generally impose a substantial speed overhead on the calculation [15]. An efficient method for simultaneously enlarging convergence region and normalizing the effect of scale factors was presented in [14]. It incurs very little speed overhead and operates within the framework of the basic CORDIC equations, thus necessitating no special hardware. The reader is referred to [14] for the details.

16.5.2 Applications of CORDIC Arithmetic Units
in Signal Processing

Many signal processing algorithms can be cast into a framework involving the rotations computed by the CORDIC algorithms. Some obvious applications include methods with complex arithmetic, for example, discrete Fourier transform or the complex filtering and adaptive equalization tasks of a modem. Their implementations, detailed in [12,13], rely on the expression of complex multiplications as circular rotations.

Rewriting an algorithm with the CORDIC rotation operations is not always obvious, as was seen with the ladder filter equations in an earlier section. Through certain interpretations, we were able to express the recursions of a filter stage as a sequence of circular and hyperbolic rotations. Figure 16.11 shows the implementation with CORDIC AUs (which normalize the scale constant). Only two AUs and five AU operations are necessary to achieve the ladder filter computations, which would have been very cumbersome in a traditional multiplier structure.

16.5.3 Enhancement of CORDIC Speed

We showed that the indirect multiplication architectures could be significantly enhanced in throughput using, for example, array multipliers. Our modifications were based on the fact that the individual summands required in a multiplication may all be generated in parallel. This is unfortunately not the case with division and CORDIC operations, in which each successive iteration relies on the sign of the previous result. Hence our scope for speeding up these operations is much more limited.

We have already noted that signal processing algorithms consist of repetitive operations that may be pipelined. We exploit that fact to construct the pipelined CORDIC AU architecture of Figure 16.12. The area penalty incurred in this structure is not high when the dynamic storage property of an MOS technology is employed. Local interconnections also reduce stage loading. The parallel shifter of Figure 16.10 undoubtedly consumes the most chip real estate in a conventional CORDIC AU. It is distributed among the stages in the pipelined AU, each shifter being capable of one or two different shift values determined by the preselected $\{\delta_i\}$. Hence we do not incur a very significant area penalty in the realization of this large block of circuitry; however, we gain speed by reducing shifter delay per stage.

If a conventional CORDIC block requires time $T = nT_c$ to complete an operation (where T_c = time for one CORDIC iteration), we have that the latency, L, and period, T, of the pipelined AU are

$$L = nT_c'$$

$$T = T_c'$$

where $T_c' < T_c$ due to a reduction in shifter delay. In a long sequence of CORDIC operations, we are operating at an effective rate of T_c' per operation rather than nT_c.

The CORDIC arithmetic unit is also very useful in array processor architectures (e.g., systolic arrays) for performing matrix algebra operations that are common in signal processing, for example, matrix factorization, as detailed in [31,32].

It is possible to speed up the unpipelined CORDIC structure of Figure 16.9 within the framework of (16.1), owing to the convergence properties of the algorithm. We cannot present the details appearing in [14] due to lack of space, so we state the important results here. We will actually derive a hybrid scheme that utilizes normal CORDIC hardware in conjunction with a multiplier.

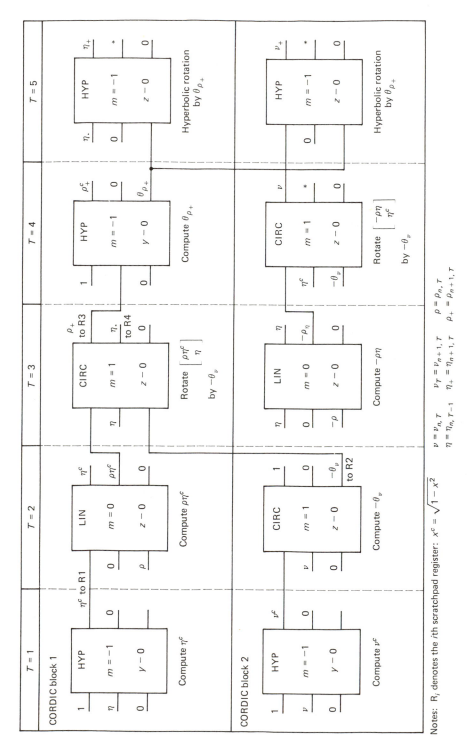

Figure 16.11 Ladder filter implementations.

Notes: R_i denotes the ith scratchpad register: $x^c = \sqrt{1 - x^2}$ $\nu = \nu_{n,T}$ $\nu_T = \nu_{n+1,T}$ $\rho = \rho_{n,T}$
$\eta = \eta_{n,T-1}$ $\eta_+ = \eta_{n+1,T}$ $\rho_+ = \rho_{n+1,T}$

295

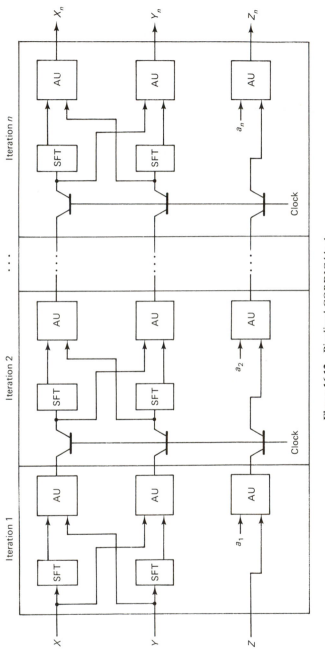

Figure 16.12 Pipelined CORDIC block.

First, we note that the vector $\mathbf{X}_m$ is within z_m, of $\mathbf{X}_n$ in m steps [29]. Choosing $\{\alpha_i\}$ to be a decreasing sequence of n angles, we stop the CORDIC recursions after $m < N$ steps, which leaves a small residual rotation, z_m. Truncating the Taylor series of $\sin z_m$ and $\cos z_m$ to a single term is justified for sufficiently small z_m, thus leading to the approximation

$$\cos z_m \simeq 1$$

$$\sin z_m \simeq z_m$$

We now obtain $\mathbf{X}_n$ from $\mathbf{X}_m$ with a single rotation of z_m as follows:

$$\mathbf{X}_n = \begin{bmatrix} \cos z_m & -\sin z_m \\ \sin z_n & \cos z_m \end{bmatrix} \mathbf{X}_n$$

$$\simeq \begin{bmatrix} 1 & -z_m \\ z_m & 1 \end{bmatrix} \mathbf{X}_m$$

Therefore, the final rotation is obtained with two multiplications by small multipliers (which could be approximated with a shift operation, i.e., $z_m \simeq 2^{-k}$), yielding an execution time of

$$T = mT_c + 2T_m$$

where T_m is the multiply time. The major result which lends credibility to this hybrid CORDIC/multiplier strategy is that for n-bit numeric precision, the Taylor series truncation is justified for [14]:

$$m \to \frac{n+1}{2}$$

that is, roughly one-half the CORDIC iterations may be eliminated. Other such hybrid methods are also given in [14].

Remark. The data flow in this hybrid scheme is from the CORDIC block to the multiplier. These two units can be pipelined for greater efficiency, such that the CORDIC hardware starts another operation after passing results from the previous operation to the multiplier.

16.6 THE CONVERGENCE COMPUTATION METHOD

The convergence computation method (CCM) is an iterative scheme for computing certain elementary functions [4]. In order to evaluate $z_0 = f(x)\big|_{x=x_0}$, introduce a variable y to form the convergence function $F(x, y)$ satisfying:

1. $\exists$ a known $y = y_0$ such that $F(x_0, y_0) = z_0$

2. $\exists$ a convenient transformation $(x_k, y_k) \xrightarrow{G} (x_{k+1}, y_{k+1})$ such that $F(x_k, y_k)$ is invariant $\forall\, k \geq 0$

3. $\exists$ a known destination x_w reachable through G from x_0. The resulting y_w reached through G satisfies

$$y_w = F(x_w, y_w) = z_0$$

and is the desired result.

For example, if $f(x) = we^x$, let $F(x, y) = ye^x$ with $y_0 = w$, thus satisfying condition 1. Next choose the transformations

$$x_{k+1} = x_k - \ln a_k$$

$$y_k + 1 = y_k a_k$$

Clearly, $F(x_k, y_k) = F(x_{k+1}, y_{k+1}) \ \forall \ k \geq 0$, satisfying condition 2.
Finally, let $x_w = 0$. Then

$$x_w = x_0 - \sum_1^w \ln a_k \to \ln \prod_1^w a_k = x_0$$

and

$$F(x_w, y_w) = y_w = y_0 \prod_1^w a_k$$

$$= y_0 e^{x_0}$$

satisfying condition 3 and yielding the desired result.

To simplify implementation, the a_k are chosen to have the form (for a radix-2 machine)

$$a_k = 1 + 2^{-m_k}$$

resulting in shift and add operations only in the y transformations. The choice of m is detailed in [4]. It affects the rate of convergence of the algorithm, which requires as few as $n/4$ iterations for n-bit precision. A machine architecture is shown in Figure 16.13, which consists of the shift and add circuitry together with a small lookup table holding values of $\ln a_k$. Notice the similarity with the CORDIC method, in which the $\{\delta_i\}$ were chosen to be powers of the machine radix and a table was required for $\{\alpha_i\}$.

16.7 COMPARISON OF ARITHMETIC UNIT ARCHITECTURES

We have presented several examples of arithmetic units and some algorithms to which they are well matched. A direct comparison of implementation of these structures as well as their performance on benchmark programs will now be presented. Operands are assumed to be n bits wide. A word of caution, however, that the

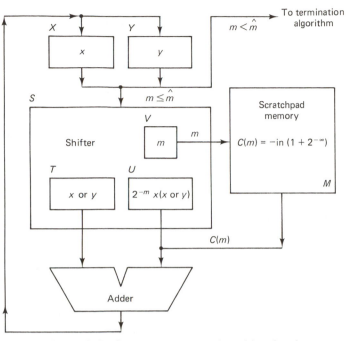

Figure 16.13   Convergence computation arithmetic unit.

outcome of comparisons can often vary considerably with the assumptions that are made in constructing them.

First consider the chip real estate required for a VLSI implementation of each of the units. Relative areas are given in Figure 16.14 with the indirect, single-bit scanning multiplier taken as a baseline. Our baseline architecture, shown in Figure 16.6, consists of an n-bit adder, three registers, and two flip-flops. In order to handle a two's-complement representation, we must include a precomplementer, which is most readily realized as a one's complementer plus a carry-in to the least significant bit of the adder.

We have assumed that a 1-bit slice of the adder together with its buses is equivalent to three times the area of a storage cell with its buses. This assumption is valid for an nMOS technology. The size of the parallel shifter required in the CORDIC arithmetic unit is obtained from [14,30], appropriately adjusted for comparison in an identical technology to the remaining AUs. Routing area adjustments are made for the array multipliers, which have many short interconnections. All implementations are based on bit parallel arithmetic; however, the parallel CORDIC AU also executes all three of the CORDIC equations in parallel, whereas the sequential CORDIC AU computes these equations in succession. Notice that the relative areas of the pipelined CORDIC AU and the parallel CORDIC AU both grow linearly with n. One might expect the pipelined unit's relative area to grow as n^2; however, this is not the case because the distributed shifter area does not grow

AU type	Relative area	Example: $n = 16$
Indirect multiplier	1.0	1.0
Braun multiplier	$\dfrac{9n - 7}{14}$	9.7
Baugh–Wooley multiplier	$\dfrac{9}{14}(n - 1)$	9.6
Parallel CORDIC	$\dfrac{n + 20}{10}$	3.6
Sequential CORDIC	$\dfrac{n + 14}{14}$	2.1
Pipelined CORDIC	$\dfrac{10n}{7}$	23
CCM	$\dfrac{n + 30}{35}$	1.3

Figure 16.14 Relative area requirements for various arithmetic units. All operands are n bits.

with n. (Note that the absolute rather than the relative areas of both these schemes, grow as n^2.) Finally, we have not included the logarithmic processing or residue arithmetic multipliers since their area is largely memory limited.

The relative areas listed in Figure 16.14 do not include the control circuits to sequence the AUs (e.g., the iterations of the indirect multiplier or CORDIC); however, this area is comparable for the different strategies. We see that the CORDIC and CCM arithmetic units are smaller than the array multipliers. They also provide a richer complement of functions—however, at a much reduced execution speed. The relative speeds, T1, of the various AUs for a variety of operations is given in Figure 16.15. This comparison is based on a register access being possible in one-half the amount of time required to propagate through a single bit slice of an adder. Functions not directly computable by the AU are calculated with a series expansion. For example, we have chosen a seven-term truncated Taylor series for computing trigonometric functions with a multiplier AU. Notice that the multipliers' performances are constant across the table because these are relative speeds compared to the baseline multiplier. The speed of each multiplier relative to the baseline case is fixed. Finally, since signal processing algorithms often execute a large number of similar operations in succession, we have also provided the relative execution time per operation for a sequence of identical operations, T2. The pipelined structure benefits here.

Our conclusion from such a comparison points to the fact that the CORDIC and CCM methods provide the most functional flexibility for the real estate consumed. They are also no slower than indirect multiplication. However, if an application simply calls for tapped-delay line type filtering at a high rate, the array multiplier provides the best throughput per area.

Operation	Multiplication		Trigonometric function		Vector rotation		$\sqrt{X^2 + Y^2}$	
Arithmetic unit	T1	T2	T1	T2	T1	T2	T1	T2
Indirect multiplier	1.0	1.0	1.0	1.0	1.0	1.0	1.0	1.0
Braun multiplier	$\dfrac{5}{2n}$	$\dfrac{5}{2n}$	$\dfrac{5}{2n}$	$\dfrac{5}{2n}$	$\dfrac{5}{2n}$	$\dfrac{5}{2n}$	$\dfrac{5}{2n}$	$\dfrac{5}{2n}$
Baugh–Wooley multiplier	$\dfrac{2}{n}$	$\dfrac{2}{n}$	$\dfrac{2}{n}$	$\dfrac{2}{n}$	$\dfrac{2}{n}$	$\dfrac{2}{n}$	$\dfrac{2}{n}$	$\dfrac{2}{n}$
Parallel CORDIC	1.3	1.3	0.2	0.2	0.08	0.08	0.2	0.2
Sequential CORDIC	3.9	3.9	0.6	0.6	0.24	0.24	0.6	0.6
Pipelined CORDIC	1.3	1.3	0.2	$\dfrac{0.2}{n}$	0.08	$\dfrac{0.08}{n}$	0.2	$\dfrac{0.2}{n}$
CCM	1	1	0.6*	0.6*	0.24*	0.24*	0.5	0.5

*Assumes complex CCM [14].

Figure 16.15   Relative speeds of the arithmetic units. All operands are n bits.

Remark. Note that Figure 16.15 indicates that the CORDIC unit computes a vector rotation more efficiently than a trigonometric function. Actually, both of these operations require the same amount of absolute time with the CORDIC unit. Remember, however, that Figure 16.15 shows performance relative to the baseline multiplier, which requires more time to perform a vector rotation than to compute a trigonometric quantity. Hence the relative performance of the CORDIC is better for the former operation.

16.8 CONCLUDING REMARKS

In our isolated study of arithmetic units for digital signal processors, we found that high throughput rates on operations including multiplication, square roots, and vector rotations were desirable. Alternative architectures for fast multiplication were presented, followed by implementations of the CORDIC and CCM algorithms for computing elementary functions. The study culminated in the relative performances of Figures 16.14 and 16.15, which provide guidelines for selection of the AU best suited to a particular class of tasks.

Our results clearly show that the CORDIC and CCM methods are not useful when fast multiplication is the only operation of interest because they consume a significant amount of area for rather mediocre throughput compared with the baseline multiplier. The Baugh–Wooley array multiplier provides the most rapid operation. It also has a better throughput per unit area than our baseline case. The array multiplier provides the most rapid multiplication of all the schemes presented. In fact, in a long sequence of operations, a pipelined array multiplier would be the best choice.

When the operations of interest include trigonometric functions or logarithms as well as multiplication, the CORDIC and CCM techniques become very attractive since they will outperform a multiplier that is computing a truncated series expansion. The throughput/area trade-off is tremendous when the algorithms of interest contain the CORDIC or CCM functions (i.e., when the latter are well matched to the problem).

It is important to note that the utility of the CORDIC or CCM lies in their generality. It is usually possible to find a more efficient scheme to compute a particular function (e.g., square root) if this is all that is required. However, a digital signal processer intended for a variety of applications can benefit from the flexibility offered by the CORDIC AU.

We conclude this chapter by once again cautioning the reader about the interactions between AU and controller, which have been ignored here.

REFERENCES

[1] W. W. Peterson and E. J. Weldon, Jr., *Error-Correcting Codes*, MIT Press, Cambridge, Mass., 1972.

[2] D. T. Lee and M. Morf, "Recursive Square Root Ladder Estimation Algorithms," *Proc. 1980 ICASSP*, Apr. 1980, pp. 1005–1017.

[3] J. E. Volder, "The CORDIC Trigonometric Computing Technique," *IRE Trans. Electron. Comput.*, EC-8(3):330–334 (Sept. 1959).

[4] T. C. Chen, "Automatic Computation of Exponentials, Logarithms, Ratios and Square Roots," *IBM J. Res. Dev.*, July 1972, pp. 380–388.

[5] S. J. Magar et al., "A Microcomputer with Digital Signal Processing Capability," *Proc. 1982 Int. Solid State Circuits Conf.*, 1982, pp. 32–33.

[6] J. R. Boddie et al., "DSP: Architecture and Performance," *Bell Syst. Tech. J.*, 60(7), Part 2:1449–1462 (1981).

[7] Y. Kawakami et al., "A Single Chip Digital Signal Processor for Voiceband Applications," *Proc. 1980 Int. Solid State Circuits Conf.*, 1980, pp. 40–41.

[8] *Signal Processing Peripheral Reference Manual*, available from American Microsystems, Inc., Santa Clara, Calif.

[9] J. V. Harshman, "Architecture of a Programmable Digital Signal Processor," *Proc. Natl. Telecommun. Conf.*, Dec. 1974, pp. 496–500.

[10] H. Aiso et al., "A Very High Speed Microprogrammable, Pipelined Signal Processor," *Proc. IFIP Congr.*, Aug. 1974, pp. 60–64.

[11] C. V. W. Armstrong et al., "A Multimicroprocessor Array Processor for Radar Signal Processing," *Proc. 6th ACM Sigarch Symp.*, Philadelphia, May 1979.

[12] H. M. Ahmed, D. T. Lee, M. Morf, and P. H. Ang, "A VLSI Speech Analysis Chip Set Based on Square Root Normalized Ladder Forms," *Proc. 1981 ICASSP*, Mar. 1981, pp. 648–653.

[13] H. M. Ahmed and M. Morf, "Synthesis and Control of Signal Processing Architectures," *Proc. 1981 VLSI Int. Conf.*, Edinburgh, Aug. 1981.

[14] H. M. Ahmed, *Signal Processing Algorithms and Architectures*, Ph.D. thesis, Dept. of Electrical Engineering, Stanford University, June 1982.

[15] R. F. Lyon, "Two's Complement Pipeline Multipliers," *IEEE Trans. Commun.*, COM-24:418–425 (Apr. 1976).

[16] J. Kane, "A Low Power, Two's Complement Serial Pipeline Multiplier Chip," *IEEE J. Solid State Circuits*, SC-11:669–678 (Oct. 1976).

[17] P. M. Fenwick, "Binary Multiplication with Overlapped Addition Cycles," *IEEE Trans. Comput.*, C-18(1):71–74 (Jan. 1969).

[18] K. Hwang, *Computer Arithmetic*, Wiley, New York, 1979.

[19] S. F. Anderson et al., "The IBM System 360/Model 91: Floating Point Execution Unit," *IBM J. Res. Dev.*, Jan. 1967, pp. 34–53.

[20] A. D. Booth, "A Signed Binary Multiplication Technique," *Q. J. Mech. Appl. Math.*, 4, Part 2:236–240 (1951).

[21] E. L. Braun, *Digital Computer Design*, Academic Press, New York, 1963.

[22] S. D. Pezaris, "A 40ns 17 Bit by 17 Bit Array Multiplier," *IEEE Trans. Comput.*, C-20(4):442–447 (Apr. 1971).

[23] C. R. Baugh and B. A. Wooley, "A Two's Complement Parallel Array Multiplication Algorithm," *IEEE Trans. Comput.*, C-22(1–2):1045–1047 (Dec. 1973).

[24] T. Brubaker and J. Becker, "Multiplication Using Logarithms Implemented with Read Only Memories," *IEEE Trans. Comput.*, C-24:761–765 (Aug. 1975).

[25] J. N. Michell, "Computer Multiplication and Division Using Binary Logarithms," *IRE Trans. Electron. Comput.*, EC-11:512–517 (Aug. 1962).

[26] W. K. Jenkins, "A Highly Efficient Residue Combinational Architecture for Digital Filters," *Proc. IEEE*, 66:700–702 (June 1978).

[27] J. M. Pollard, "Implementation of Number Theoretic Transforms," *Electron. Lett.*, 12(22):378–379 (July 1976).

[28] F. J. Taylor, "Large Moduli Multipliers for Signal Processing," *IEEE Trans. Circuits Syst.*, CAS-28(7):731–735 (July 1981).

[29] J. S. Walther, "A Unified Algorithm for Elementary Functions," *AFIPS Conf.*, Vol. 38, 1971 SJCC, pp. 379–385.

[30] G. Haviland and A. Tuszynski, "A CORDIC Arithmetic Processor Chip," *IEEE Trans. Comput.*, C-29(2):68–79 (Feb. 1980).

[31] H. Ahmed and M. Morf, "VLSI Array Architectures for Matrix Factorization," *Proc. Workshop Fast Algorithms Linear Syst.*, Aussois, France, Sept. 1981.

[32] H. Ahmed, J. M. Delosme, and M. Morf, "Highly Concurrent Computing Structures for Matrix Arithmetic and Signal Processing," *IEEE Comput.*, 15(1):65–82 (Jan. 1982).

Part III

APPLICATION
OF CONCURRENT
ARRAY PROCESSORS

This section addresses the application of concurrent array processor concepts to signal processing. In the past, major work has been done on mapping various signal processing applications into specific very large scale integration (VLSI) architectures, and vice versa. The insight gained from such hands-on experiences will certainly greatly enhance the understanding of VLSI's real impact on signal processing, with more unified theoretical footing and much closer interactions. Nevertheless, the diversified applicational disciplines can magnify such impact manyfold. To this end, this part includes chapters on the implementation of processor chips and fast Fourier transform (FFT)-type signal processors, and various applications of concurrent array processors. A very important application area of concurrent array processors, emphasized in this part, is image processing, image understanding, and pattern recognition. Moreover, for certain special applications, the architectural solution may be limited in improving processing throughput rates. Therefore, the controversial issue of device speed limits for silicon versus some other (e.g., galium arsenide) integrated circuits also has to be put into perspective.

Chapter 17, by Nudd and Nash, describes application of concurrent array processors to two-dimensional signal processing developed at the Hughes Research Laboratories. Included is the description of a VLSI chip specifically designed to solve a Toeplitz set of simultaneous equations. Such sets of equations frequently arise in the analysis of time- or space-invariant linear systems.

Chapter 18, by Wood, Culler, Greenwood, and Harrison, describes a CMOS VLSI arithmetic processor chip joint developed by CHI Systems and Motorola under DARPA sponsorship. This chip has been designed to efficiently implement general array processing using floating-point arithmetic. By implementing a general-purpose arithmetic chip, many signal processing tasks can be computed with a single architecture.

Chapter 19, by Swartzlander and Hallnor, describes a bipolar chip set of arithmetic processors developed by TRW Defense Systems Group. These chips are specially configured to implement the computationally intensive FFT algorithm. Floating-point arithmetic is used to avoid data scaling and loss of precision in the processor. By specializing the architecture for the computation of FFTs, high throughput rates can be achieved with VLSI technology.

Chapter 20, by Truong, Reed, Yeh, and Shao, discusses a digital filter chip developed by the Jet Propulsion Laboratory and the University of Southern California. The chip is designed to implement number-theoretic transforms directly and is thus capable of exact arithmetic modulo the chosen word size. This architecture is particularly useful for implementing convolutions and correlations corresponding to finite impulse response (FIR) filters.

Chapter 21, by Travassos, describes a systolic signal processor for recursive filtering developed by Integrated Systems. The architecture of the infinite impulse response (IIR) filter is achieved by mapping the Kalman filter equations onto a linearly connected systolic array.

Chapter 22, by Schreiber and Kuekes, considers a design for a systolic linear algebra processor designed by Stanford University and ESL. The architecture is based on the use of unitary transformations to factor matrix representations of the digital signal processing equations. To illustrate these concepts, the application of a two-dimensional systolic array to adaptive beamforming was chosen.

Chapter 23, by Uhr, describes the application of parallel computation to image processing and pattern perception. Architectures using parallel arrays, pipeline arrays, and pyramids are considered.

Chapter 24, by Rosenfeld, surveys the types of parallel algorithms that can be used for image processing and analysis. For fixed-level operations, cellular array architectures are attractive. For image property measurement, tree-connected architectures offer greater efficiency.

Chapter 25, by Fu, Hwang, and Wah, considers both image analysis and image database management. VLSI architectures are proposed for both pattern recognition and image processing. The authors suggest that a VLSI image analysis machine should integrate both the operations of pattern analysis and image database management into a single system.

The last chapter in this section, on the application of architectures, describes work by Gilbert, Kinter, Schwab, Naused, Krueger, and Van Nurden of Mayo Foundation and Zucca of Rockwell International Microelectronics Research and Development Center on signal processing at high data rates. In this chapter the use of both parallel architectures and high-speed circuits is discussed in the context of future circuits developments. The use of gallium arsenide devices for high speed is proposed and the comparison with silicon circuits discussed.

17

Application of Concurrent VLSI Systems to Two-Dimensional Signal Processing

GRAHAM R. NUDD AND J. GREG NASH

Hughes Research Laboratories
Malibu, California

17.1 INTRODUCTION

The advent of high-density metal-oxide-semiconductor (MOS) technologies and VLSI provide an opportunity for significantly increased capability in many areas of signal processing. Many techniques that have until now been considered too expensive or time consuming will soon be performed by real-time "on-board" processors. This is particularly true in the area of two-dimensional signal processing for applications such as radar mapping, beamforming, and image analysis. In these areas the volume of data to be processed typically increases as $(\Delta R)^{-2}$, where ΔR is the resolution. For nonstationary systems with relatively high resolution, input data rates of 10^8 samples/s and processing requirements in excess of 10^3 million operations per second are not uncommon.

The decreased device delay times available with VLSI technology, even using conventional von Neumann-based architectures, can be expected to provide a reduction in machine cycle time of perhaps an order of magnitude. However, for most two-dimensional problems, these improvements will fall far short of the required performance, and it is apparent that parallelism and concurrency must be effectively exploited. This is the area in which VLSI can have its greatest impact. If, for example, within a given processor wafer or chip we can obtain 10, 100, or even 1000 subprocessor elements all working simultaneously, significant throughput increases can be anticipated. However, to achieve this parallel activity several issues must be simultaneously addressed. For example, it is necessary to reevaluate the basic processing functions required by each application and identify new measures to evaluate algorithmic efficiency not based solely on the minimum number of operations or instructions. Increased attention has to be given to developing the optimum paral-

307

lelism and maintaining the highest percentage activity of all elements of any potential processor array. Further, the regularity of the data flow and the needs for global versus local memory become important considerations. These issues are now well recognized and an important science is developing to study VLSI algorithms.

The technology issues, such as VLSI design, processor interconnection, and input/output considerations have to be explored before concurrency can be effectively exploited. For this reason much of our effort to date has been concentrated on developing concurrent architectures which use essentially identical processing elements with minimum global communication. Two classes of architectures are addressed. The first considers the potential of using a large number of identical but relatively small lookup tables to achieve concurrency, and the second addresses the issues of developing an optimum local processing element requiring only local nearest-neighbor interconnection. This second issue is important in that we have already started to identify and build a VLSI multiplication-oriented processor (MOP) which, when interconnected in either a two-dimensional or one-dimensional array, can provide a very wide variety of processing functions. This particular processor presently uses NMOS technology with 5-μm design rules, and can be programmed both at the local processor level and at the array level to perform many of the algorithms required by image analysis and pattern recognition (IAPR). The details of this chip are given in Section 17.3.

17.2 THROUGHPUT REQUIREMENTS FOR IMAGE ANALYSIS

Very significant advances have been made over the past several years in the development of both algorithms and software for image analysis, and as a result, this field has largely emerged from the preprocessing and coding era to more sophisticated approaches typically based on symbolic and noniconic formats [1,2]. These techniques tend to rely more on successive layers of abstraction to represent the data than on the specific two-dimensional topology of the image. As such, the conventional two-dimensional filtering approaches that have been previously employed become less important and emphasis is given to operations such as feature extraction and classification [3,4].

17.2.1 Preprocessing Operations

The concept of a typical modern image analysis system is illustrated in Figure 17.1, where four essential stages of the processing are identified. The first consists of image formation, which for visible imagery can be a trivial problem, but for data bases such as synthetic array radar (SAR), significant throughput may be required. The essential steps include formation of the successive range and cross range cells in real time. For example, an image with a resolution of d in both the directions

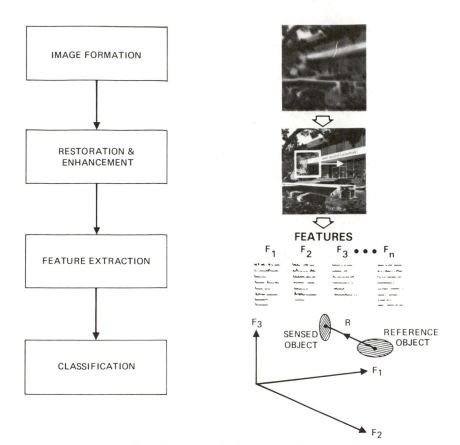

Figure 17.1   Process flow for feature-based image analysis.

requires a throughput of the order of

$$\left(T + \frac{2\,\Delta R}{c} \right) \frac{PRF}{d} \quad \text{operations/s for range calculation}$$

and

$$4v\,\Delta R\,\frac{\ln(R\lambda/2d^2)}{d^2} \quad \text{for cross-range calculation}$$

where T = transmitted pulse duration
ΔR = range window
PRF = pulse repetition frequency
R = radar range
v = vehicle velocity
λ = wavelength of the transmitted signal

For relatively high resolution equivalent to 1 m, this results in a throughput of the order of 100,000,000 operations per second (100 mops).

Once the two-dimensional image is formed, some type of enhancement or restoration might typically be employed to provide a more acceptable image quality for real-time display. Typical of the processing that might be employed at this stage are two-dimensional filtering operations of the form

$$\tilde{I}(x, y) = H(x, y) * I(x, y) + \text{noise} \qquad (17.1)$$

where $\tilde{I}(x, y)$ is the enhanced image at location x, y and $I(x, y)$ the original image, $H(x, y)$ is the filter function, and $*$ represents convolution. In many applications the filter characteristics, $H(x, y)$, are derived in real time from the received image statistics, such as spectral content. For this, a typical operation, in addition to the explicit convolution shown, is an $n \times n$ matrix inversion where n can vary from 3 to 15 pixels.

More sophisticated operations that could be employed at the next stage are based on the maximum entropy (ME) approach [5], for example, which aim at providing an enhanced resolution image. A simple example might be where the resolutions in the two orthogonal directions are unequal and an ME estimation is employed to increase the effective number of pixels on one axis. This might typically involve an inversion of a Toeplitz matrix, for which techniques are described in Section 17.4.

17.2.2 Feature Extraction and Classification

Subsequent processing is concerned with feature extraction and object classification. In this stage each picture element in the image is addressed in conjunction with its local neighbors, so as to form a sliding subwindow, as illustrated in Figure 17.1. The typical window size may vary from 3×3 to 64×64 pixels, resulting in an access rate ranging from 6×10^7 to 3×10^{10} pixels per second for typical television quality imagery. Multiple-feature measurements, such as edge density, variance, histogram, and certain types of moment calculations, are required in real time, resulting in a very computationally intensive processing. Commonly used operations and their required throughputs are listed in Table 17.1. Typical throughput requirements might range from 10^2 to 10^7 mops, as indicated, and as many as 20 different features might be calculated. These data then form the basis of the object classification which, in its simplest form, can consist of nearest-neighbor calculations, based on the clustering of features from both the reference object and the viewed data. An illustration of the process is given in Figure 17.1, where three-dimensional space representing only three features is shown. The length of the vector, R, can be used as a measure of the probability that the viewed object belongs to the same class as the reference object.

This is the essence of the feature-based image classification. Typically, several extensions of this basic technique are employed, including feedback paths and some higher-level symbolic processing to obtain greater levels of abstraction. The

TABLE 17.1 TYPICAL PROCESSING OPERATIONS
USED IN FEATURE-BASED IMAGE ANALYSIS

Processing function	Necessary throughput

Linear operations, $O(N)$
  Spatial filtering
  Convolution
  Edge detection
	10^2–10^5 mops

Second-order operations, $O(N^2)$
  Sorting operations
  Median filtering
  Nearest-neighbor classification
	10^3–10^7 mops

Higher order
  Matrix based
  Spectral processing
  Adaptive operations
	10^4–10^8 mops

throughput requirement for such a process can be calculated by assuming an image quality equivalent to television, with a spatial resolution equivalent to 500 by 500 picture elements and the frame rate of 30 frames a second, resulting in data rate of 10^7 samples per second. This implies that during the feature extraction stage for a subwindow of size 3×3, a memory access rate of the order of 10^8 per second is required. For larger windows, say with 64×64 picture elements, access rates might be of the order of 10^{11} per second. Hence, for even linear operations such as spatial filtering, convolution, and edge detection, the throughput requirement will range from 10^2 to 10^5 mops, as shown in Table 17.1.

17.3 EXISTING IMAGE-ANALYSIS ARCHITECTURES

It is informative to plot the throughput versus the time of introduction of various computing systems, as shown in Figure 17.2. Most commercial machines and presently operating aerospace systems are essentially sequential architectures, perhaps with two or three stages of pipelining [6,7]. Typically, increases in throughput in these architectures have resulted from improved device technology, shorter bus structures, and so on. Over the past several years these developments have provided an increase in throughput from a few mops to as much as 100 mops. However, as can be seen, the rate of increase using these same architectures is not expected to improve dramatically over the next decade. As a result, their performance will fall far short of that required for large image-analysis problems. Even with the increase in technology brought about by the Department of Defense VHSIC program, 10^2 to 10^3 mops might be the near-term limit for these architectures. And it is clear that image-processing and pattern-recognition problems require a throughput which is several orders of magnitude beyond that.

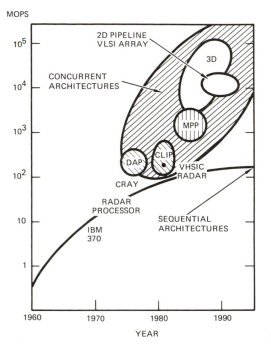

Figure 17.2   Processor throughput using concurrent and sequential architectures.

Those machines which have been successfully developed and achieve throughputs of interest to image analysis have typically used a high degree of parallelism. For example, the Cellular Logic Image Processor [8] (CLIP) from University College London, the Distributed Array Processor [9] (DAP) from ICL, and the Massive Parallel Processor [10] (MPP) of Goodyear and NASA use an N by N array of identical processing elements operating in essentially a single-instruction, multiple-data mode (SIMD), as illustrated in Figure 17.3. The size of these arrays ranges from 64×64 processors for the DAP to 128×128 for the MPP. Since most such configurations are bit serial, a speed increase of the order of N^2/W, where W is the word length, can be achieved when all elements are operating concurrently. This is a very

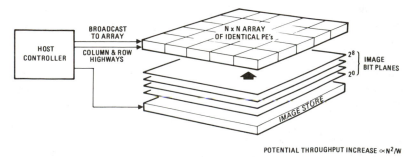

Figure 17.3   Schematic of $N \times N$ SIMD image analysis array.

significant factor. For example, the MPP, when delivered in 1983, will provide a throughput equivalent to 6×10^9 fixed-point additions per second. In a floating-point mode this will reduce to the order of several hundred million floating-point additions per second.

These improvements in throughput will be obtained only when each processor is working at its maximum rate and all picture elements in the frame are essentially being processed identically. This frequently occurs at the lowest levels of the image analysis, but is less common as the processing proceeds. For example, preprocessing functions such as two-dimensional filtering, edge detection, and so on, can easily be programmed to take advantage of this form of parallelism or concurrency, but the high-level operations such as classification and graph matching become more difficult. These difficulties continue through the symbolic analysis, which is presently structured in the form of listed data, for which this type of parallel SIMD architecture is presently considered inappropriate. For this reason it is important to develop alternative VLSI architectures that avoid some of the problems of the SIMD machine and may be configured for both the low- and high-level image analysis.

17.4 CONCURRENT VLSI ARCHITECTURES

An important consideration in any VLSI development is the ease and efficiency of design [11]. As the gate density within the chips increases to hundreds of thousands, the design time and interconnect complexity become overriding issues in the feasibility of production. For this reason a modular design of highly regular structures has been considered an essential element for VLSI implementation. Potentially, the most regular class of structures that have been designed to date occur in high-density memories. For this reason, interest has developed in performing high-speed complex operations by a series of table lookups using random access memory (RAM). The approach has significant advantages for feature extraction and applications, such as edge detection, but if performed directly it can require excessive storage. For example, an image with 12-bit intensity data can require a table lookup of the order of 2^{24}, or 20 million bits. This is clearly not feasible with present or foreseeable VLSI chips, and can be expected to require excessively long access time in any hybrid structure.

Several approaches do exist to reduce the total memory size, including those based on the Ofman–Karatsuba techniques [12] and unconventional number representations. For example, the Ofman–Karatsuba technique involves dividing the required word size, W, within the machine into n parallel channels of size W/n. The result of this for certain operations can be a reduction in the overall memory requirement by 2^n. An interesting example of this approach suggested by Ercegovac [13] is the calculation of image edge density by the Sobel technique [14], where nine pixel intensities, I_{ij}, in a 3 by 3 array, are used to calculate

$$S = (X^2 + Y^2)^{1/2} \qquad (17.2a)$$

where

$$X = (I_{i-1,j-1} + 2I_{i,j-1} + I_{i+1,j-1}) - (I_{i-1,j+1} + 2I_{i,j+1} + I_{i+1,j+1})$$
$$Y = (I_{i-1,j-1} + 2I_{i-1,j} + I_{i-1,j+1}) - (I_{i+1,j-1} + 2I_{i+1,j} + I_{i+1,j+1})$$

(17.2b)

Here S is the two-dimensional edge magnitude at the location X_{ij}. If performed directly, assuming that each pixel intensity, I, is represented by W bits, we require $2^{(2W+2)}$ bits of table lookup to provide full precision in S. If, however, each edge component is expressed as a combination of two words, each being of length $W/2$ and representing the most significant and least significant data, that is,

$$X = X_L 2^{W/2} + X_R$$

(17.3)

then we can write

$$X^2 = X_L^2 2^W + (X_L^2 + X_R^2) - (X_L - X_R)^2 2^{W/2} + X_R^2$$

(17.4)

Since X_L, $(X_L - X_R)$, and X_R have at most $W/2$ nonzero bits, the resulting lookup tables are reduced to the order of 2^W from 2^{2W}. This technique can be used further to divide the word size into n different sections of size W/n and result in an overall memory requirement of $2^{W/n} \times n$. Hughes Research Laboratories are currently developing a processor for edge extraction based on these techniques, which will allow a reduction of the required memory size from 10×10^6 bits to the order of 10^4 bits.

Another approach we are currently investigating for the exploitation of high-density VLSI RAM technology in image analysis is based on the residue arithmetic notation [15]. The technique requires encoding the incoming data by division with prime numbers, as illustrated in Figure 17.4, and then working with only the remainders. In our case, for example, we use four prime numbers (31, 29, 23, and 19), each below 5 bits, and hence the residues or remainders are also below 5 bits. The arithmetic in each channel then proceeds directly, but with the proviso that if any

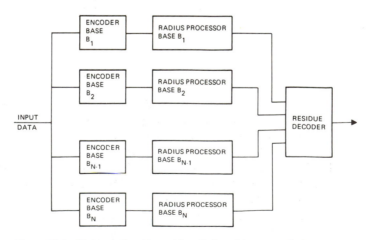

Figure 17.4   Concept of residue arithmetic-based image analysis processor.

internal number exceeds the prime or base, it is reconverted to the remainder (by successive subtraction of the prime). The parallel channels can then be uniquely decoded to binary output data, providing an overall accuracy equivalent to the product of the bases (i.e., 392,863, or approximately 18 bits). The significant advantage of this approach is that all the arithmetic, including the initial encoding into residue notation, the internal calculations, and the final decoding, can be performed with small (5×32)-bit lookup tables. Further, the machine can be made programmable by using high-speed RAMs which are loaded with specific data to perform the various feature extraction functions. The RADIUS [16] machine that was built using these techniques is now operating at our laboratory at essentially real-time rates. It consists of a number of special-purpose custom-built NMOS chips which include the lookup tables and a number of modular arithmetic cells, as shown in Figure 17.5. The full machine performs operations of the form

$$y = \sum f_i(I_{ij}) \tag{17.5}$$

where I_{ij} are the intensity values over a 5×5 element kernel and f_i polynomial functions of a single variable. The effective throughput of the machine in this feature extraction mode is equivalent to 200×10^6 multiplications/s.

An alternative approach to exploit VLSI concurrency, and one that is gaining a great deal of interest, is the two-dimensional pipeline array, illustrated in Figure 17.6. This consists of a number of identical processes, shown here in a rectangular tesselation, wherein the communication is with local neighbors only. This general form of architecture includes both the systolic and wavefront formulations of H. T. Kung [17] and S. Y. Kung [18]. Such an array with a capability to do simple arithmetic operations such as multiplications and additions at each site can be very

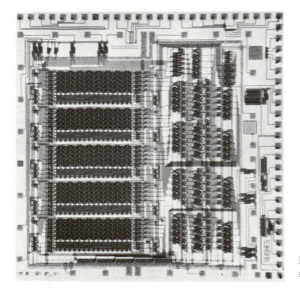

Figure 17.5 Custom NMOS circuit for residue-based processor.

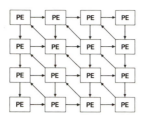

Figure 17.6 General configuration of two-dimensional pipeline array.

effective for image analysis. The square tesselation shown is in many ways ideally suited to image analysis because of the topological correspondence between the processors and the configuration of the picture elements. Our work in this area is aimed at the development of algorithms for such an architecture and the identification of the optimum VLSI processing elements for IAPR.

One class of operation that is well suited to the two-dimensional array is that ascribed to Faddeev [19]. This is useful in solving systems of linear equations of the form

$$AX = B \qquad (17.6)$$

where A is a general $n \times n$ matrix and X and B are vectors of length n. This operation is the kernel of many IAPR algorithms, including most of the pre-processing and enhancement calculations. If, for example, the data are loaded into the array in the form

$$\left. \begin{array}{cccc|c} a_{11} & a_{12} & \cdots & a_{1n} & b_1 \\ a_{21} & a_{22} & \cdots & a_{2n} & b_2 \\ \multicolumn{4}{c}{\cdots\cdots\cdots\cdots} \\ a_{n1} & a_{n2} & \cdots & a_{nn} & b_n \\ \hline -1 & 0 & \cdots & 0 & 0 \end{array} \right. \qquad (17.7)$$

$$\begin{array}{c|c} A & \mathbf{B} \\ \hline -\mathbf{I} & 0 \end{array} \qquad (17.8)$$

and at each point a multiplication, addition, and shift up and to the left is performed, then after n such operations the vector X appears in the left-hand column. This organization is well suited to concurrent VLSI because each processor is operating essentially all the time and the operations (multiplication, addition, and shift) are identical at each node. The technique is essentially a modification of Gaussian elimination but requires no triangular decomposition or back substitution. These results can be generalized so that each of the operations

$$CA^{-1}B + D \qquad (17.9)$$

$$A^{-1} \qquad (17.10)$$

$$A^{-1}B \qquad (17.11)$$

and

$$\mathbf{BC} + \mathbf{D} \tag{17.12}$$

can be performed by the data entries

$$\begin{array}{c|c} A & B \\ \hline -C & D \end{array} \tag{17.13}$$

$$\begin{array}{c|c} A & I \\ \hline -I & 0 \end{array} \tag{17.14}$$

$$\begin{array}{c|c} A & B \\ \hline -I & 0 \end{array} \tag{17.15}$$

$$\begin{array}{c|c} I & B \\ \hline -C & D \end{array} \tag{17.16}$$

respectively.

If the operations are implemented directly, the processors require some global data shifts and communication in that the multiplication at each data site requires data from the edge of both its row and column. This can be considered to violate one of the basic tenents of VLSI: that only local communication is desirable, thereby avoiding the need for long data buses and drivers. For this reason we have developed a pipelined processor, as shown in Figure 17.7, which requires only local communication. The elements of the one-dimensional vectors $\mathbf{C}$ and $\mathbf{B}$ are initially stored in the processors below the double line and to the right of the single line, respectively. The n^2 elements of A are again stored in the upper left corner, and the function of each processor in the array is indicated. Communication between processors is along horizontal, vertical, and diagonal nearest-neighbor paths. At each time step the active processors perform the calculations indicated and immediately transfer data to adjacent processors for their use. This processor organization is of the "wavefront" variety [18], where a single wave of activity proceeds across the entire processor. However, extension of the approach to successive pipeline waves can be made. For larger calculations involving matrices $\mathbf{B}$, $\mathbf{C}$, and $\mathbf{D}$, an augmented processor array would be required.

The issues of timing and control and associated data flow of this two-dimensional processor are complex and necessarily result in some loss of concurrency from the pure broadcast approach.

An alternative approach that requires only a one-dimensional array can be formulated for the solution of Toeplitz equations of the form

$$\begin{bmatrix} a_{11} & a_{12} & a_{13} & \cdots & a_{1n} \\ a_{12} & a_{11} & & & \\ a_{13} & a_{12} & a_{11} & & \\ \vdots & & & \vdots & \\ a_{1n} & & & & a_{nn} \end{bmatrix} \begin{bmatrix} x_1 \\ \cdot \\ \cdot \\ \vdots \\ x_n \end{bmatrix} \begin{bmatrix} b_1 \\ \cdot \\ \cdot \\ \vdots \\ b_n \end{bmatrix} \tag{17.17}$$

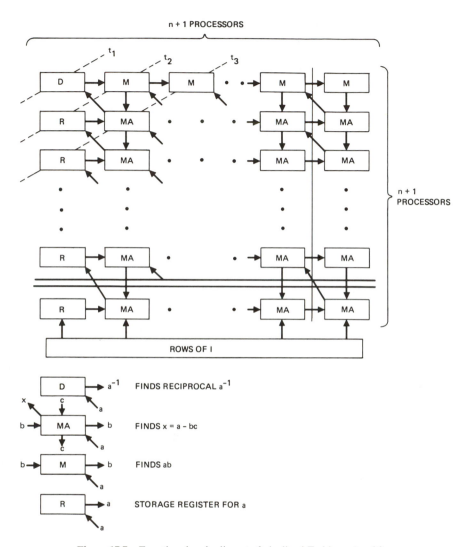

Figure 17.7   Functional embodiment of pipelined Faddeev algorithm.

using the technique described by S. Y. Kung [18]. Essentially, the following four steps are required: decomposition of the A matrix,

$$A = UU^+ \tag{17.18}$$

back substitution, $\phi = [U^+]^{-1}B \tag{17.19}$

multiplication, $\psi = [D]^{-1}\phi \tag{17.20}$

where D is the diagonal matrix $u_{11}^{-1}, u_{22}^{-1}, \ldots, u_{nn}^{-1}$; and a second back-substitution for the final result, $\mathbf{X} = [U]^{-1}\psi$. The technique is based on the Weiner–Levinson

(WL) [20] algorithm, which takes advantage of the symmetry in A and performs the full calculation in $O(n)$ operations. A schematic of the array for this process is shown in Figure 17.8. It consists of n stages of identical sections, each of which contains two multiplication and addition processors, four registers, and a last-in-first-out (LIFO) stack. Two basic sections can be identified: a lattice array of processors that produce the U^+ matrix using the WL algorithm, and a back-substitution section which solves for both the intermediate result, ϕ, and the final result, $\mathbf{X}$. Between the two basic sections is the LIFO stack, which stores the successive elements of U^+ for later output as U.

To generate elements of U^+, the WL lattice array operates recursively on rows of A to produce a vector, which we denote as $\mathbf{A}'$, that is used in conjunction with a second "auxiliary row," $\mathbf{A}''$. At each recursion a "reflection coefficient," K, is generated as the ratio of the first elements of the two rows. This successively multiplies the two rows so that when they are subtracted from each other the new result, $\mathbf{A}'$, is the next row of U^+. A new auxiliary vector, $\mathbf{A}''$, is also produced, and the new reflection coefficient is obtained.

The system solver is initialized by loading both the R_1 and R_2 registers with the elements of the top row of A. The elements of B are loaded into register R_3. When the operation begins, switch S_1 is closed and switches S_2 and S_3 are open. Then R_1 and R_2 supply inputs to the WL array and the results are returned to the same registers. The new values of A correspond to successive rows of U^+. These results are simultaneously stored in the LIFO stack for later use and supplied to the back-substitution section. In each time cycle new values of $\mathbf{A}_i$, $\mathbf{A}'_i$, ϕ_i, and $\mathbf{B}_i$ are calculated and the intermediate results, ψ_i, are stored in a register R_4. When all of the elements of U^+ have been calculated and stored on the LIFO stacks, the shift register containing the intermediate results, ψ_i, will also be full. At this point, switch S_1 is opened and S_2 is closed; switch S_3 is closed long enough to load the R_3

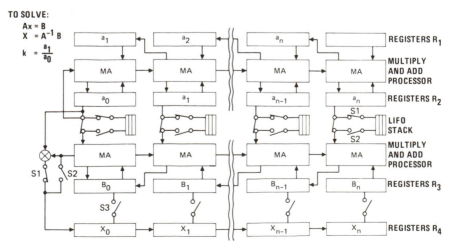

Figure 17.8   Concept of data flow in Toeplitz system solver.

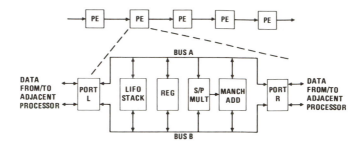

Figure 17.9 VLSI processor module (MOP).

register with the results, ψ_i. The previous operations are repeated with the WL array remaining idle. In each time period a new result, X_i, is calculated and stored in the shift register.

The functional configuration shown in Figure 17.8 can be further reduced to a form that is pipelined and suitable for reduction to hardware. The necessity of pipelining arises from the constraint that the processors be allowed only to communicate with nearest neighbors. This constraint was imposed to minimize the speed and power consuming requirement of a global bus. The basic processing element then appears as shown in Figure 17.9.

We are implementing such a processor in VLSI NMOS technology; the masks are shown in Figure 17.10. The chip itself contains on the order of 15,000 devices

Figure 17.10 Multiplication-oriented processor (MOP) chip built for pipeline array.

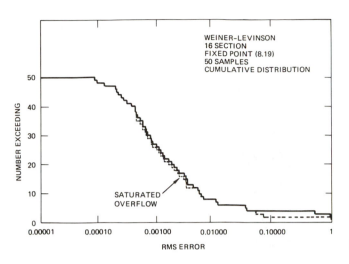

Figure 17.11 APL simulation program.

and provides an overall processing cycle time of approximately 250 ns. Several novel features exist within the chip, including a high-speed carry-save multiplier which provides an addition time essentially independent of the word size. This is implemented in a radix 4 Booth's algorithm, and the internal clock for this is approximately 16 MHz [21]. The present processor uses fixed-point arithmetic and a 28-bit word length. This word size has resulted from an extensive simulation program undertaken using samples of data obtained from specific applications. For this a complete simulation of the processor, including data flow, register lengths, and arithmetic precision, has been completed. The results produced for various internal word sizes have been compared with "infinite" precision processing. From this a cumulative error distribution, as illustrated in Figure 17.11, was calculated and the

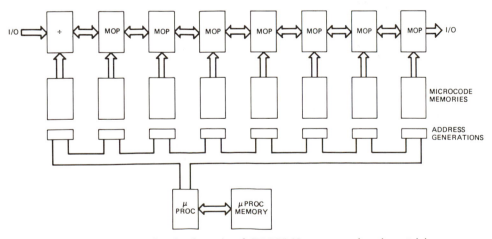

Figure 17.12 Functional schematic of TOPSS-28 processor board containing seven MOPS chips.

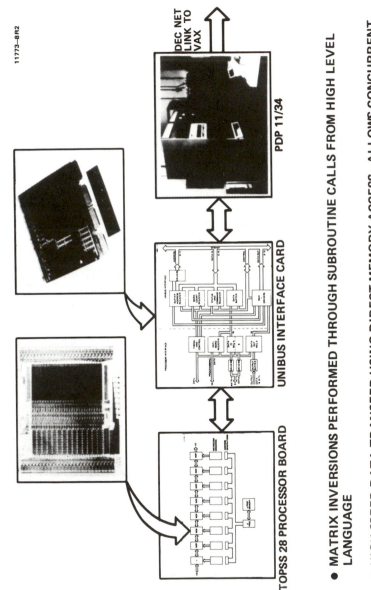

11773–8R2

DEC NET LINK TO VAX

PDP 11/34

UNIBUS INTERFACE CARD

TOPSS 28 PROCESSOR BOARD

- MATRIX INVERSIONS PERFORMED THROUGH SUBROUTINE CALLS FROM HIGH LEVEL LANGUAGE
- HIGH SPEED DATA TRANSFER USING DIRECT MEMORY ACCESS. ALLOWS CONCURRENT OPERATION OF MATRIX PROCESSOR AND HOST
- INVERSION OF 16 x 16 MATRIX PERFORMED IN SAME TIME AS 5 DOUBLE PRECISION MULTIPLICATIONS ON HOST.

Figure 17.13   High-speed interface of TOPSS-28 processor to host computer.

optimum word size determined. In operation the multiplication-oriented processors (MOP) chips will be integrated into a printed circuit board to form the full Toeplitz processing system (TOPSS-28), shown in Figures 17.12 and 17.13. The board will be composed of seven copies of the MOP chip, plus additional circuitry for division, control, and testing. This will give it the capability of processing matrices of sizes up to 16×16 in approximately 175 μs. The circuit has been designed around an internal bus structure so that the MOP chip can be directly interconnected, which simplifies the design and fabrication of the board.

The full system will require an on-line division capability, which is part of the first section, as shown in Figure 17.12. Essentially, the reciprocal of a 28-bit fixed number is required to be calculated in approximately 1 μs. One possibility is to use a special purpose bus-oriented arithmetic chip or a lookup-based approach.

In operation the board controller will generate a 16-bit operation code for each of the 7 MOP chips at the clock rate of 2 to 4 MHz. One approach to this problem is to provide memory with an address generator (counter) for each circuit. The data in the memories can be considered microcoded subroutines. Each of the address generators will then be controlled by a microprocessor which will have its own machine code memory. For certain applications the microcode could be put into a ROM, which would reduce the programming required to that of selecting the subroutines required for each circuit.

The board itself containing a linear array can operate in several modes, corresponding to the several different variations on the Levinson algorithm. These variations represent trade-offs in computation time and memory storage. The fastest algorithm requires $O(n^2)$ storage elements, but solves $AX = B$ in approximation $6n$ time steps. One can also use $O(n)$ storage elements, but computation time increases to approximately $18n$. The present processor uses 5-μm technology and has one processing element on each chip. However, if more sophisticated technology is used, the number of processors per die and the resulting processing speed might increase, as shown in Table 17.2. This might allow a single-chip two-dimensional array of 25×25 elements to be built using submicron design rules.

Several additional capabilities for the processor are also being considered for inclusion in the next iteration. These include a floating-point version and an on-chip square-root and division capability. The design and technology impact of these are now being considered. If, indeed, these options are incorporated, a very wide spec-

TABLE 17.2 PROCESSOR SIZE AND SYSTEM SOLUTION TIME AS A FUNCTION OF TECHNOLOGY FEATURE SIZE

Feature size (μm)	Year available	Processors/ chip	System solution time
6	1976	1	< 0.6 ms
3	1981	4	300.0 μs
1	1985	~36	100.0 μs
0.25	2000	~576	25.0 μs

trum of applications suited to IAPR will be possible, particularly with a two-dimensional array. These include one- and two-dimensional convolutions, enhancement using ME, all of the two-dimensional filtering operations, spline calculations, geometric transforms for model building, and many others at a processing speed that will be sufficient for most real-time applications.

ACKNOWLEDGMENT

Sponsored in part by Contract N0014-81-K-0191 from the Office of Naval Research and Grant ECS 8016581 from the National Science Foundation.

REFERENCES

[1] M. J. Duff and S. Levialdi, *Language and Architectures for Image Processing*, Academic Press, New York, 1981.

[2] T. O. Binford, "Geometric Reasoning and Spatial Understanding," *Proc. Image Understanding Workshop*, Palo Alto, Calif., Sept. 15–16, 1982, pp. 18–20.

[3] G. R. Nudd, "Image Understanding Architectures," *Proc. Natl. Comput. Conf.*, Vol. 49, Anaheim, Calif., May 1980, pp. 370–390. Published by AFIPS Press.

[4] A. Rosenfeld and A. Kak, *Digital Picture Processing*, Academic Press, New York, 1976.

[5] T. J. Ulrych and T. N. Bishop, "Maximum Entropy Spectral Analysis and Autoregressive Decomposition," *Rev. Geophys. Space Phys.*, *13*:183–200 (Feb. 1975).

[6] K. Hwang, *Computer Arithmetic, Principles, Architecture and Design*, Wiley, New York, 1979.

[7] M. J. Flynn, "Some Computer Organizations and Their Effectiveness," *IEEE Trans. Comput.*, *21*:125–140 (Sept. 1972).

[8] M. J. B. Duff, "CLIP 4 A Large Scale Integrated Circuit Array Parallel Processor," *Proc. 3rd Int. Joint Conf. Pattern Recognition*, 1976, pp. 728–733.

[9] S. F. Reddaway, "DAP-A Distributed Processor Array," *Proc. First Annu. Symp. Comput. Architecture*, 1973, pp. 61–65.

[10] K. E. Batcher, "Design of a Massively Parallel Processor," *IEEE Trans. Comput.*, *C-29*(9):836–840 (Sept. 1980).

[11] I. E. Sutherland and C. A. Mead, "Microelectronics and Computer Science," *Sci. Am.*, *237*(3):210–228 (Sept. 1977).

[12] A. Karatsuba and Y. Ofman, "Multiplication of Multi-digit Numbers on Automata," *Sov. Phys.–Dokl.*, *7*(3):595–596 (Jan. 1963).

[13] Milos Ercegovac, private communication, Apr. 1982.

[14] W. K. Pratt, *Digital Image Processing*, Wiley-Interscience, New York, 1978.

[15] N. Szabo and R. Tanaka, *Residue Arithmetic and Its Applications to Computer Technology*, McGraw-Hill, New York, 1967.

[16] S. D. Fouse, G. R. Nudd, and A. D. Cumming, "A VLSI Architecture for Pattern

Recognition Using Residue Arithmetic," *Proc. 6th Int. Conf. VLSI Architecture, Design and Fabrication*, California Institute of Technology, Jan. 1979, pp. 65–90.

[17] H. T. Kung, "Let's Design Algorithms for VLSI Systems," *Proc. Conf. VLSI Architecture, Design and Farbrication*, California Institute of Technology, Jan. 1979, pp. 65–90.

[18] S. Y. Kung, "Impact of VLSI on Modern Signal Processing," *Proc. Workshop VLSI Mod. Signal Process.*, University of Southern California, Nov. 1982.

[19] V. N. Faddeeva, *Computational Methods for Linear Algebra*, translated by Curtis D. Benster, Dover, New York, 1959.

[20] N. Levinson, "The Wiener RMS Error Criterion in Filter Design and Prediction," *J. Math. Phys.* 25(4):261–278 (1947).

[21] M. D. Ercegovac and J. G. Nash, "An Area–Time Efficient VLSI Design of a Radix-4 Multiplier." *Proc. Int. Conf. on Computer Design*, Port Chester, New York, Oct. 1983, pp. 684–687.

18

A VLSI Arithmetic Processor Chip for Array Processing

ROGER WOOD
GLEN CULLER

CHI Systems, Inc.
Goleta, California

ED GREENWOOD
DAVE HARRISON

Motorola, Inc.
Scottsdale, Arizona

18.1 INTRODUCTION

The arithmetic processor unit† (APU) described utilizes Motorola's 3-μm silicon-gate CMOS fabrication technology and CHI Systems, Inc. array processor architecture technology. The architecture of the chip is motivated by the fact that the computational operations involved in combining vectors (arrays) by linear operations can be fully anticipated. Therefore, there exists an architecture that is very efficient for such processing. Optimal efficiency can be achieved in a one multiplier architecture if the number of execution cycles of the process's inner loop can be limited to the number of multiply operations required by the process's inner loop.

Viewed from any application domain (speech, geophysical, radar, tomography, etc.), one thinks of array processors as very special purpose machines. Indeed, machines that are designed to perform a particular process are just that; but the important point is that we have developed an internal architecture that expresses the algebra of bilinear forms. This represents the foundation of array processing and is therefore useful in a broad spectrum of applications.

†The work reported here describes the most recent member of a family of array processing architectures [1,2] which have been developed, in part, under funding from the Defense Advanced Research Projects Agency (DARPA). An earlier member of this architectural family formed the basis for the Floating Point Systems AP-120B [3]. In the current project, the Government Electronics Division of Motorola and CHI Systems, Inc. are working together, under DARPA sponsorship, to produce a VLSI version of the architecture described in [4].

18.2 GENERAL ARRAY PROCESSOR MOTIVATION

To ensure that the resulting architecture would be efficient for general array processing, we used a set of those processes that motivated the earlier architectures and updated them to include additional processes. These test processes were carefully chosen to reveal the inner difficulties concerning numerical analysis, data control, and program control that are usually experienced when ordinary computers are used for general array processing purposes. The test processes used for design verification are listed in Table 18.1. This set of processes was encoded in microcode and used to verify that the architecture meets the efficiency goals.

TABLE 18.1   DESIGN VERIFICATION PROCESSES

Dot product of two arrays
    32-bit accumulation of 16-bit block-floating-point data
    48-bit accumulation of 16-bit block-floating-point data
Floating-point operations:
    Floating multiply, 32-bit mantissa, 16-bit exponent
    Floating add, 32-bit mantissa, 16-bit exponent
LPC algorithms:
    Lattice reduction
    Inverse lattice filter
FFT operations:
    Decimate pass
    General pass
    Pass one
Filter algorithms:
    One-Real-Pole filter
    Two-Real-Pole filter
    Complex-Conjugate Pole–Zero pair with gain

TABLE 18.2   ALGORITHM EXECUTION TIMES

Algorithm	Time[a]
1024-point complex FFT	5632 μs
512-point real FFT	1152 μs
1024-point real vector multiplication	260 μs
1024-point complex multiply	520 μs
DARPA 2400-bit full-duplex voice processing	≤ 5263 μs
32-bit mantissa floating-point addition	1.5–2.5 μs
32-bit mantissa floating-point multiplication	0.75, 1.0, or 1.75 μs
One multiplication and three data additions	250 ns
Complex-conjugate pole–zero pair filter with gain	1.25 μs/point

[a]Based on 4-MHz system clock frequency.

The efficient implementation, that the processes require for the presently known family of LPC algorithms, sets an architectural style for the internal relationship of arithmetic registers and adders, multipliers, and array memories. But when we consider FFTs, complex arithmetic, and matrix inversion, we need more extensive communication and some auxiliary registers to complete the design. The coded test processes result in the application benchmark execution times shown in Table 18.2.

18.3 ARRAY PROCESSOR FEATURES

A simple system, which suffices to demonstrate the array processing system architecture, is shown in Figure 18.1. The system employs the arithmetic processor unit (APU chip), two array memories (X and Y), a partitioned data memory (DR and DL), which could be shared with a host computer, and AP control section, the microprogram memory (P), plus transceivers and pipe registers (DX and DY).

The features implemented on the APU are:

1. Three 16-bit adders and four accumulator registers connected in a fourfold connectivity structure.

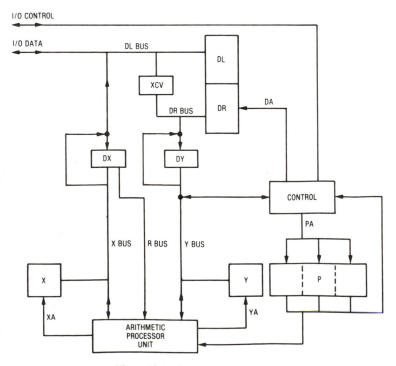

Figure 18.1   Array processor system.

2. A 17-bit × 17-bit pipelined multiplier with a true 17th-bit sign control. The multiplier's output is connected to the three 16-bit adders.

3. Sizing hardware to compute block floating-point vectors rescaling data during the computational process.

4. Float hardware for enhancing floating-point arithmetic.

5. Array addressing by row or column in X and Y memory simultaneously.

6. Direct addressing of a two-dimensional nine-point window, centered at a matrix index.

7. 16-bit hardware bit reversal (decimation).

8. Four auxiliary local registers.

9. The X, Y, and R external buses are preserved inside the APU device.

18.4 APU SYSTEM IMPLICATIONS

Starting with the APU chip, one can design computer systems literally of any type presently known and use no other arithmetic facility than that provided by the APU. That is, it is a true general-purpose part, not just a piece of a special-purpose system on one chip. In this section we indicate some of the ways in which this chip may be used as the major component in systems of very different capabilities and applicational intent.

The small-scale microprocessor-based computers provide user communication that is very successful for program preparation and small-scale problem solving. But when any significant scientific computing is required, the arithmetic capability is simply not adequate. An enormous enhancement of performance will result by using the APU as an accelerator. Keeping with the single address instructions that make compiler writing easy, we suggest that a small mapping ROM be used to build APU microinstructions that specialize the arithmetic operations to those typical of adequate minicomputers. Of course, most of the resulting microinstructions are NOOPs as seen by this type of processor, so we would envision a hard-wired NOOP with field multiplexing that would provide control for individual arithmetic operations.

Special-purpose computers are organized by the concept of performing a particular algorithm with minimum hardware. Because their design must be so tightly coupled to one algorithm, a separate design is required for each case of interest. Consequently, the ease of carrying out such a design is an important issue in the cost-effectiveness of such systems. But with the 16-bit microprocessors presently available, together with the APU chip and wide-word RAMS, such a design will be achieved by just hooking up major components in simple configurations. It is clear that such modular approaches will lead to computer-aided-design programs that use "large module design rules" to generate system designs from user specifications.

Array processors can be considered at various levels of performance. At the low end of the performance scale, we envision systems with a small inventory of supporting memory chips controlled by off-the-shelf sequencers. To keep the memory chip count low, successive reads for microinstruction source will be used as well as successive reads and writes for the vector memories.

In the intermediate range of array processor performance, a full complement of supporting memory chips will be used with parts count determined by how many words are desired in each of the following memory modules:

X, Y	vector memories
PH	microinstruction ROM
PS	microinstruction RAM
D	system memory

High-performance array processors can be achieved by using several APU chips to provide higher-level parallelism of arithmetic operations and to extend arithmetic precision. Of course, even more memory chips are required to keep the memory accesses up to the extended APU performance.

To achieve the superperformance level, we propose micronets of array processors, with each node having general-purpose control, a local operating system, and a micronet communication capability over a multiple connected network. There must, of course, be a global operating system for task and resource management. The actual capability of such a system is limited only by the state of the art in understanding how to organize the design and how to program many processor compilers with variable processor inventories at execution time.

18.5 ARITHMETIC PROCESSING UNIT CHIP

The arithmetic processing unit (APU) chip consists of three sections: (1) a data arithmetic processor, (2) two identical array memory address controllers, and (3) required control logic. The data arithmetic processor does all arithmetic on array elements and has a limited amount of storage for intermediate results and parameters. The address controllers perform the computation of array element addresses associated with the X and Y memories. The control logic provides all clocks and controls for the APU as derived from 20 microinstruction signals, two clock signals, and two special control signals.

The APU chip will have 100 pins and its block diagram is shown in Figure 18.2. The pinouts have the following definitions:

X0–X15	X bus data input/output
Y0–Y15	Y bus data input/ouput
XA0–XA15	X memory address output
YA0–YA15	Y memory address output
I0–I19	Microinstruction inputs; used four times per system clock cycle

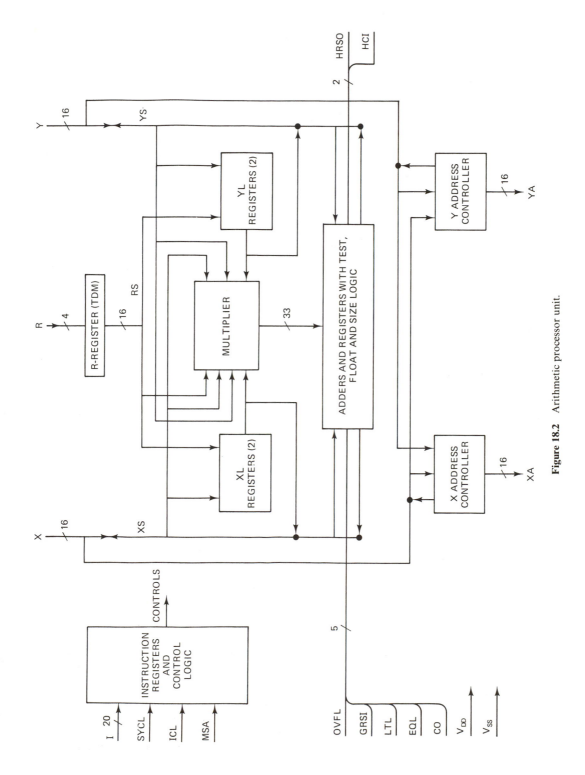

Figure 18.2 Arithmetic processor unit.

SYCL	System clock input, 4 MHz nominal, 5 MHz design goal
ICL	Four-phase clock input for microinstructions and R data, 16 MHz nominal, 20 MHz design goal
RI	R bus data input; use four times per system clock cycle
EQL, LTL	Data test outputs
MSA, OVFL, GRSI, HRSO	Control inputs and outputs for parallel APUs' operation
V_{ss}, V_{dd}	Power supply inputs

The unit will be packaged in a pin grid array package.

The combination of the X and Y address controllers and the data arithmetic unit required isolation of the X bus from the APU X bus (called XS). Similarly, a YS bus exists, isolated from the Y bus. Thus external memory operations (in conjunction with the address controllers) can exist while internal APU data processing uses the extension of the X and Y buses (XS and YS).

18.5.1 General Architecture

The APU architecture is shown in Figure 18.3 and the microoperation codes are shown in Figure 18.4. The principal components of the data arithmetic processor are:

1. A 17-bit × 17-bit pipelined multiplier, with input registers MA and MB, which produces a 33-bit result stored in two registers (called MPL and MPR). MPL contains the 17 most significant bits and MPR contains the 16 least significant bits. This 33-bit register pair is referred to as MP.

2. Three 16-bit adders: F, G, and H. These adders are generally independent, but can be coupled for 32- or 48-bit precision.

3. Two internal buses: XS and YS.

4. Four 16-bit accumulators: T, U, V, and W. U and V can each be loaded from two adders. T and W can be loaded from one adder or from either XS or YS.

5. Four local storage registers: XL0, XL1, YL0, and YL1.

6. An independent input bus (R): it can load the local storage registers or the multiplier.

In addition, the arithmetic unit contains logic to handle shifting, scaling, floating-point normalization, bit reversal, and test logic for the G and H adders. Most elements are interconnected to one or both of the internal XS/YS buses for input and/or output. In addition, there is a series of connections to the multiplier and the adders which do not use the buses. In particular, the multiplier has inputs from the local registers XLi and YLi (i = 0 or 1). The adders have inputs, on the A side, from the multiplier output registers MPL and/or MPR and, on the B side, from a subset of the accumulators; and none of the B side inputs use the XS/YS

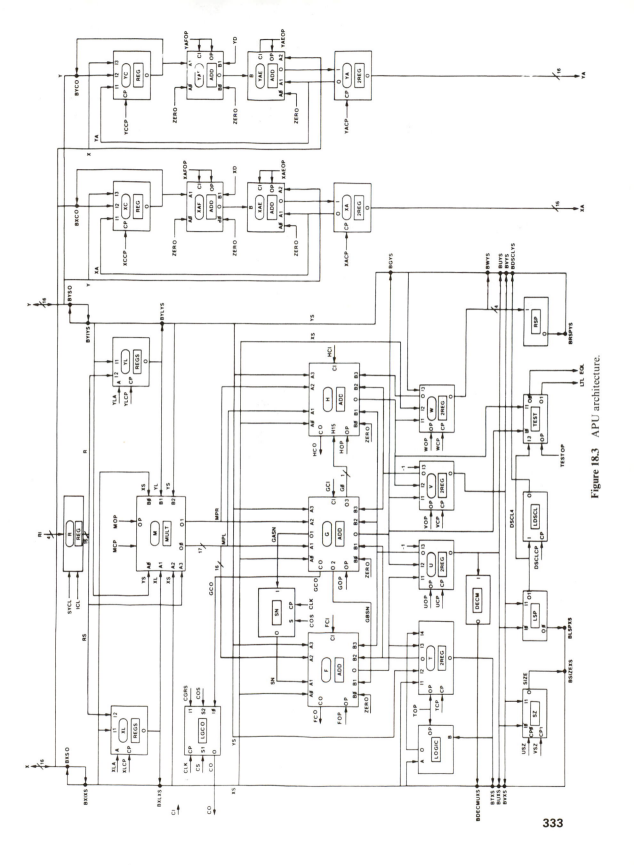

Figure 18.3 APU architecture.

333

CODE	MOP (7)			XSOP (3)	YSOP (3)	SIZE (2)	XLOP (2)		YLOP (2)	
	MULT	A	B				XLA	XL	YLA	YL
0	P2	NOOP	NOOP	XI	YI	NOOP	A0	NOOP	A0	NOOP
1	P2LS	YS/R	XS	XL	YL	U	A1	XS/R →XL	A1	YS/R →YL
2	2P	XL/R	YL	T	W	V				
3	2PLS	XS/R	YS	U	RSP	UV				
4	22			V	U					
5	22LS			SIZE	V					
6	PP			DECM	DSCL					
7	22LSRND			LSP	G					

CODE	FOP (7)				GOP (7)				HOP (8)			
	ADD	A	B	FCI	ADD	A	B	GCI	ADD	A	B	HCI
0	B	XS	0	FOP1	B	XS	0	GOP1	B	XS	0	HOP1
1	A + B	SN	GBSN	CO	A + B	MPL	T	HCO	A + B	MPL	U	CI
2	A − B	MPL	T		A − B	MPR	U		A − B	MPR	V	CO
3	B − A	YS	U		B − A	YS	V		B − A	YS	W	FCO

CODE	REGOP (9)				SHIFT (3)	COS (1)	X0 (2)	Y0 (2)	TEST (2)	ROP (2)
	TOP	UOP	VOP	WOP						
0	NOOP	NOOP	NOOP	NOOP	G, H	GCO	NOOP	NOOP	G	NOOP
1	XS →T	F →U	G →V	H →W	G, RSH	LGCO	NOOP	NOOP	GH	R →MA
2	YS →T	G →U	H →V	XS →W	RSG, H		XS	YS	H	R →XL
3	F →T	−1 →U	−1→V	YS →W	RSG, RSH		XC	YC	DSCL	R →YL
4	XS∧T →T				CSG, H					
5	XS∨T →T				CSG, RSH					
6	XS⊕T →T				CSGH					
7	~XS →T				RSGH					

CODE	XACOP (8)			YACOP (8)		
	XAOP (4)	XCOP (2)	XD (2)	YACOP (4)	YCOP (2)	YD (2)
0	NOOP	NOOP	1	NOOP	NOOP	1
1	XA = Y	XA →XC	2	YA = X	YA →YC	2
2	INC XA (XD)	X →XC	3	INC YA (YD)	Y →YC	3
3	DEC XA (XD)	Y →XC	4	DEC YA (YD)	X →YC	4
4	XA = XC			YA = YC		
5	XA = Y + XC			YA = X + YC		
6	INC XA (XC)			INC YA (YC)		
7	DEC XA (XC)			DEC YA (YC)		
10	XA = XC + XD			YA = YC + YD		
11	XA = Y + (XC + XD)			YA = X + (YC + YD)		
12	INC XA (XC + XD)			INC YA (YC + YD)		
13	DEC XA (XC − XD)			DEC YA (YC − YD)		
14	XA = XC − XD			YA = YC − YD		
15	XA = Y + (XC − XD)			YA = X + (YC − YD)		
16	INC XA (XC − XD)			INC YA (YC − YD)		
17	DEC XA (XC + XD)			DEC YA (YC + YD)		

Figure 18.4   APU microoperation codes.

buses. Each adder can feed two specific accumulators directly; the G adder can also drive YS. The interconnection structure is such that the adders and accumulators can be used to provide an internal pipeline of up to four stages, fed by either bus and unloaded by the other bus. It is also possible to swap the contents of accumulators U and V in one clock via the G and H adders without using either bus or an intermediate register. Two types of shifting are provided in the adders:

1. A 1-bit right shift (RS) of the output of the G adder into U, or the H adder into V, or the GH concatenated adder into UV; the shift will affect all G-adder (and/or H-adder) outputs, including that to the YS bus.

2. A 1-bit conditional right shift (CS) of the G adder into U or the GH adders into UV; if enabled, the shift will be made whenever the adder result reflects an overflow. The shifted result is then the corrected value; the overflow bit indicates that a shift was taken. The LGC0 register output (available on pin CO) will contain the overflow bit whenever conditional shift is requested.

The normal addition can also be performed. Combinations of RS, CS, and the normal addition can be performed with the G and H adder as shown in the SHIFT field of the micro-operation codes of Figure 18.4. Any other shifting in the APU is performed via the multiplier. Shifting with the multiplier can be performed by multiplying the given number by the appropriate power of 2, and saving either MPL (for the right shift) or MPR (for the left shift) or the whole MP (for 32-bit operands). Two operators (RSP and LSP) are provided to determine the appropriate power of 2 (described in the next section).

18.5.2 Specialized Functional Operations

Several specialized logic functions are also provided to enhance algebraic array processing and floating-point arithmetic. Five of these functions are unary operations, called RSP, DSCL, LSP, DECM, and TEST. The remaining operator, called SIZE, employs a SIZE register and allows the microcode designer the capability of deriving the most significant bit position seen in an array of values which have passed through the U register, the V register, or the logical OR of the U and V registers. This function is useful for block-floating-point scaling operations. The TEST logic evaluates the output of the G adder, the H adder, or the concatenated GH adder. Flags specifying whether the data tested were less than zero or equal to zero are output from the chip. This feature allows the conditional branch-on-data operation. The DECM function performs a 16-bit decimate (bit reversal) of the U register onto the XS bus. Certain types of transforms on arrays of data will find this function useful for address calculations. The remaining functions are described and illustrated by example in the following paragraphs.

The LSP operator is based on a left-shift-to-normalize sensor. Its outputs provide the count of how many places must be shifted (DSCL) and the power of 2 required to accomplish this shift (LSP). LSP operates on the 32-bit register pair UV;

if U is negative, then a 1's complement of UV is taken to compute LSP. DSCL is 5 bits wide, and represents delta-scale values between 0 and 31. The LSP result will be determined from the most significant 16 bits following the sign as long as the required shift is less than or equal to 15. Otherwise, LSP will be based on the 16 least significant bits. The user can determine which case occurred by testing DSCL. LSP is gated onto the XS bus; DSCL is gated onto the YS bus and is available one clock after the LSP result.

As an example of the use of this operator, consider the following example.† Suppose that the registers U and V contain one fraction in nonnormalized two's-complement form with value

$$
\begin{array}{cc}
\text{U} & \text{V} \\
00011011 & 01101101
\end{array}
$$

The two zeros following a (positive) sign bit imply that a left shift of two places is required to normalize the result in UV. Hence LSP will be 00000100 and DSCL will be 2. The normalization of U and V is performed by the steps

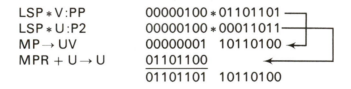

LSP ∗ V:PP	00000100 ∗ 01101101
LSP ∗ U:P2	00000100 ∗ 00011011
MP → UV	00000001 10110100
MPR + U → U	01101100
	01101101 10110100

which is the properly normalized result.

When a full word or longer shift is required, LSP is generated from V instead of U. Suppose that the U and V fractions are given by

$$
\begin{array}{cc}
\text{U} & \text{V} \\
00000000 & 00010101
\end{array}
$$

There are now 10 zeros following the (positive) sign bit, which implies that DSCL = 10 and LSP, generated from V, is given by LSP = 00000100. The steps for normalization are then LSP ∗ V:PP followed (properly delayed) by MPR → U and 0 → V. Thus we obtain

$$
\begin{array}{ll}
\text{LSP} \ast \text{V:PP} & \quad\quad 00010101 \\
& \quad\quad \underline{00000100} \\
& 0000000001010100 \\
& \quad\quad\quad\quad\quad \underbrace{\qquad\qquad} \\
& \quad\quad\quad\quad\quad\quad \text{MPR}
\end{array}
$$

---

†For convenience, the APU will be scaled down by a factor of 2 for all examples; thus the register will be 8 bits long, the multiplier 8 × 8, and so on.

$$MPR \rightarrow U, \ 0 \rightarrow V$$

```
          U                        V
      01010100                 00000000
```

which is the properly normalized result.

All of the previous examples have employed a positive result. Suppose now that U and V are

```
          U                        V
      11110010                 01010101
```

The LSP operator performs the (one's) complementation of U and V, since the sign of U was negative. This yields the result

```
    COMPL:        00001101        10101010
```

and applying the previous algorithm gives DSCL = 3 and LSP = 00001000. If LSP is applied to U and V as the first example, we have

```
LSP * V : PP        00001000 * 01010101
LSP * V : PP        00001000 * 11110010
MP → UV            00000010  10101000
MPR + U → U       + 10010000
                   ─────────────────────
                   10010010  10101000
```

which is the properly normalized result. In all of these cases, DSCL would be used to modify the exponent or to provide a count of the number of places shifted.

The RSP operator is used primarily to align two floating-point numbers for addition. The assumptions are that U and V contain the mantissa and that W contains the positive difference in scale. Then calling for RSP generates a multiplicative operator which will give the desired right shift. For example, if U, V, and W contain

```
          U                 V                 W
      01001001          00100100          00000101
```

the required right shift of UV is five places and RSP = 00001000. The alignment operations are as follows:

```
RSP * U : P2        00001000 * 01001001
RSP * V : PP        00001000 * 00100100
MP → UV            00000010  01001000
MPL + V → V              + 00000001
                   ─────────────────────
                   00000010  01001001
```

which is the properly shifted result. The method works equally well for negative mantissas. However, W is assumed to be positive and to have a value between 0 and 15 (0 to 7 for our reduced scale examples). If W is out of this range, modifications must be made in the microcode to account for this fact.

18.5.3 Conditional Shift Feature

As mentioned earlier, an extra, or 17th, bit is provided on all three adders, on the multiplier (on MPL), and on accumulator U. This provides for testing and correcting, via the conditional shift (CS), a (1-bit) overflow. The extra bit position allows for the correction of overflows which have occurred either on a previous clock cycle or on the current operation. Under overflow conditions, the 16th and 17th bits will have different values; otherwise, they will be identical. When an add with conditional shift is called for, the overflow condition is checked after the add operation and result is right-shifted one bit if an overflow has occurred. The overflow condition is latched into LGCO. The presence of a 1 in LGCO indicates that the shift was performed, a 0 indicates that it was not. Thus account can be taken, on the clock cycle following the CS, of the presence or absence of the shift. Three cases are presented that use the CS. The first is simply to correct for overflow in an addition. For example, in floating-point addition, it is sometimes desirable to have the following code sequence:

$$SNMPL + UV \rightarrow UV$$
$$MP + UV:CS \rightarrow UV$$

where the two multiplier outputs represent, respectively, the right-shifted least significant and most significant portions of the mantissa of the addend. If the mantissa of the augend is assumed to be in UV, an overflow condition can occur on either of the two add operations. However, since the operation is simply the sum of two numbers, one of which has been right-shifted, at most one overflow can occur. Thus, if W contains the scale (exponent) of the larger of the two operands (assuming there to be the augend), the following H-adder operations will correct the scale for the possible occurrence of the overflow and resulting right shift:

$$W + LGCO \rightarrow W$$

(which implies that HCI = LGCO).

A second way in which overflow can occur is during a 1-bit position normalization operation. The code for this is given by

$$UV + UV:CS \rightarrow UV$$

The contents of UV will be added to itself, giving a 1-bit left shift; should this cause overflow, the conditional shift operation will shift it back, leaving the result in normalized form in either case. Hence, if overflow occurred, the scale should remain the same; if there was no overflow, the scale should be decreased by one. This can be performed by the following H-adder operation:

$$W - 0 - 1 + LGCO \rightarrow W$$

which can be achieved by gating W to the A side of the adder, 0 to the B side, CO to HCI, and setting CO to LGCO.

Another way in which overflow can occur is from the multiplication of two numbers which represent negative powers of 2. Such numbers will have a mantissa of 10000000 (8-bit binary representation) which represents −1. The (two's complement) product of two such numbers should be a positive power of 2, but will instead be a negative power of 2 (the multiply will be 22LS). However, the extra (17th) bit on the multiplier will be a true sign bit, calculated as the exclusive-or of the input sign bits. Thus the result will be (in 8-bit example format)

$$010000000\ 00000000$$

and the operation

$$MP + 0:CS \rightarrow UV$$

will correct the overflow by right-shifting and storing

$$01000000\ 00000000$$

in UV. The scale must then be corrected on the next clock by $W + LGCO \rightarrow W$.

18.5.4 Array Memory Address Controllers

Address controllers for the X and Y memory elements are provided on the APU chip. Two identical 16-bit data controllers are used. Two 16-bit adders and two 16-bit registers are used in each controller. The block diagram of the X address controller (XA) is shown in Figure 18.5. This address controller structure provides a base address (Y) number with an index (XC) and an offset variable (XD). The increment/decrement function provides for column displacement for two-

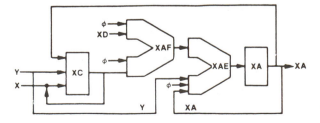

Figure 18.5 *X* address controller.

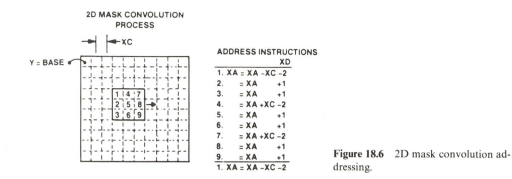

Figure 18.6 2D mask convolution addressing.

dimensional array addressing directly. Thus two-dimensional mask convolution direct addressing is provided, as shown in Figure 18.6. Mask size is limited to twice the maximum value of XD plus one (nine in the APU). The choice of X and Y data as the source for XC allows X memory to be used as a buffer for arrays that have dimensions greater than two. Of course, the Y address controller can be used by the microcode designer for an additional simultaneous address controller.

18.6 RATIONALE FOR CHIP FUNCTIONS

Obviously, the APU chip will implement the basic multiply/add/accumulate functions. But the local multiplier storage (XL/YL), the special logic functions, and the address controllers were designer choices for implementation on the chip. The XL/YL registers were a high priority for implementation, so that bus contention would not limit throughput. We desire to use the multiplier with 100% efficiency in inner processing loops. The only question was the number of registers to be implemented. Two of each was found to be adequate for benchmarks that were reviewed. Similarly, the special logic functions were a high priority since they directly affected inner loop processing throughput.

The address controllers for X and Y memory were certainly an appendage to the arithmetic processor. They were considered for implementation together with on-board program memory (RAM or mask ROM) or X/Y memory RAM or a system controller. This was the primary trade-off area for on-chip versus off-chip functions.

Obviously, the chip contains no RAM or ROM memory, a popular function on newer processor chips. X, Y, R, and program memory were considered for implementation on the chip. However, the rationale for no memory on the chip was as follows:

1. The semiconductor industry will continue to provide denser and faster CMOS memories (PROM, ROM, and RAM).
2. Build a chip that will take advantage of item 1, and do not let the chip be outdated because of item 1.

3. Provide an APU such that the application designer can utilize the type and amount of memory as required by the application.

Thus we envision a system that could use PROM, ROM, or RAM program memory. Obviously, the RAM program memory would have to be down-loaded by a host. Since 80 bits are used for APU instruction, a 96- to 128-bit-wide instruction word is envisioned for systems.

X/Y data memory on-board the chip would be a desirable feature. But if some X/Y memory is on-board and some is off-board, the system application designer would be constrained to a fixed type of RAM. We envision the use of either NX1 or NX8 RAMs (or maybe NX16 in the future) for different applications.

If the system controller (program control, DMA, and I/O control) were added to the chip, the pin count would have been greater than 150 and there would not be area for the address controllers. Since program controllers and DMA controllers were being offered by semiconductor firms, but data controllers were not available, we chose to implement the X/Y data controller function.

18.7 VLSI DESIGN

The silicon implementation of the array processor's arithmetic processor unit (APU) has many degrees of freedom for the designer. The significant decisions and trade-offs pertaining to the VLSI design task will be discussed in the following paragraphs in order to clarify how the IC became what it is. A brief discussion of the technology choice, logic/circuit design, layout approach, and layout implementation will be presented.

18.7.1 Process Technology

The process technology used on the APU is the MICRO-CMOS (μ-CMOS) process. This process was the most advanced silicon-gate CMOS available for use at the time of this design. It has some very important characteristics that make it an excellent choice for the implementation of the APU. It is a 3-μm polysilicon gate with platinum silicide process.

The utilization of platinum silicide gates in the μ-CMOS process makes for two important characteristics that are very desirable for the design of complex, high-speed, bus-oriented circuits. First, it allows both P- and N-doped polysilicon to be used, without the standard metal shorting bars required to tie the two different gate polys together. The shorting of this polysilicon PN diode is accomplished by the platinum silicide layer, which is placed on the patterned polysilicon. By using this procedure, the advantages of a P- and N-type polysilicon process are retained. Normally, this allows contacts (buried) from polysilicon to substrate. This device flexibility saves chip real estate by allowing the typical interconnection of transistor drains to the next gates to be routed completely in polysilicon (without using metal).

Another solution to the PN diode gate problem is to use a highly doped single (typically N-type) polysilicon process. This serves the purpose of eliminating the required shorting bars on the gates. However, it allows for buried contacts on only the N-type device's source or drain and therefore leads to a less efficient layout than that obtained with platinum silicide.

The second feature of platinum silicide is that the resistivity of the polysilicon is very low. This desirable advantage of low-resistance polysilicon shows up in two ways when the VLSI is being planned and designed. The 2- to 5-Ω per square resistivity when compared to the 20 to 30 Ω per square for the single-doped N-type polysilicon indicates a typical interconnection (RC) delay advantage of 7 : 1. Even though only a small portion of the overall delay within a device is attributable to these RC delays, the designer can circumvent this limitation at the expense of additional chip area when using non-platinum silicide processes. This leads to the second advantage of low-resistivity polysilicon, which is the flexibility given layout designers because they no longer have to use extra chip area to interconnect the higher-speed signals on the device.

The μ-CMOS process has 3-μm gates and feature sizes that support 2- to 3-ns gate delays. This circuit speed is fast enough to support greater than 4-MHz APU device operations with the 16-MHz multiplexed instruction port and R-bus port. The process is a 5-V CMOS process and the VLSI I/O's are T^2L compatible. The chip power will be less than 200 mW.

18.7.2 Logic and Circuit Design

The circuit design used standard CMOS logic design techniques. The basic circuits have been design proven over several technologies. For example, the Motorola-patented fast array adder [5] was used throughout the design. Using proven circuits minimized the risk of design errors that lead to costly redesigns and schedule delays. The circuit design effort was a continual trade-off between the design time (and cost) and performance. That is, when the device's performance goal was reached with suitable margin, the design efforts to enhance the circuit's performance were stopped. The performance was analyzed using an MOS modeling and transient analysis circuit simulator which is constantly correlated with measurements of test devices. Designing this general-purpose device to meet the stated goals is the design approach used. However, higher-speed performance is always desired. There is always a system that can use the better performance. It must be stated that by using this design approach, the final design will not yield the best performance that is possible with this MOS technology, but the design time and cost was kept to a minimum acceptable level.

18.7.3 Layout Approach

The layout methodology was defined by knowing the limitations on layout time, circuit complexity, and circuit speed. To design the chip for optimum performance and density a full custom VLSI would be required. However, since the AP program was to show feasibility of integrating an array processor, only minimal funds and time were allocated for the design. With the imposition of this constraint, the best results could be derived with a semicustom layout approach utilizing as much redundancy of layout as could be used in several places on the IC. The best example of repeatability of use was demonstrated by the fast carry adder cell which was used in nearly every section of the chip. The layout form factor of this cell was certainly not optimum for every section in which it was used. But the time saving and risk reduction derived by not having to re-layout the full adder each time was considered to be a good trade-off for the loss in chip area.

In the layout process for a section of the device, the approach was to pack the cells to their best density within the constraints imposed by the chip floor plan. In this process the cells of a section, once interconnected, typically yielded a section layout with a form factor different from that predicted and desired by the chip floor plan. In a design where the density of the layout is a premium design factor, not layout time (and cost), several iterations of a section layout are used to optimize the overall chip layout. But for the VLSI array processor program the original section layouts of the design were used in the chip layout even where they fit awkwardly and at the sacrifice of chip area.

With this approach, the design time was 17 person-months and the layout time was 24 person-months. Table 18.3 illustrates that 30,500 or 30.5-k transistors are on the device. Using Fitzpatrick et al.'s [6] definition of "regularization factor," no ROM devices are on the chip, 1207 transistors were drawn, and thus the regularization factor was 25.3. Of course, Fitzpatrick et al.'s comparison data are for NMOS devices; our design was low in design cost similar to their Gold RISC I device. From Table 18.3, other comparisons can be made to Fitzpatrick et al.'s design metrics.

With the general methodology defined by the trade-offs just described, the layout of the VLSI was started by examining the desired architecture. The obvious

TABLE 18.3 ALLOCATION OF CHIP AREA

Function	Area (mils^2)	Number of transistors	Density (transistors/mils^2)	Percent of chip area
Multiplier	15,484	10,349	0.668	17
Bit slice	18,849	10,000	0.531	21
XAC	2,584	1,952	0.755	3
YAC	2,584	1,952	0.755	3
XL, YL	2,553	1,824	0.714	3
Control	10,269	3,506	0.341	11
I/O (bond pads)	18,038	978	0.054	20
Data buses	15,022	0	0.000	16*
Control buses	2,444	0	0.000	3*
Available area	2,770	—	—	3
Total	90,597	30,561	0.337	100

*Does not include buses within each section.

conclusion from examination of the architecture and op-code tables is how essentially 20 separate 16-bit data buses, 184 control lines, power, and ground can be interconnected on a single IC while maintaining the 250-ns cycle time. Many chip plans and interconnect structures were evaluated before the structure of Figure 18.7 was chosen.

The architecture of the APU was divided into five functional sections: (1)

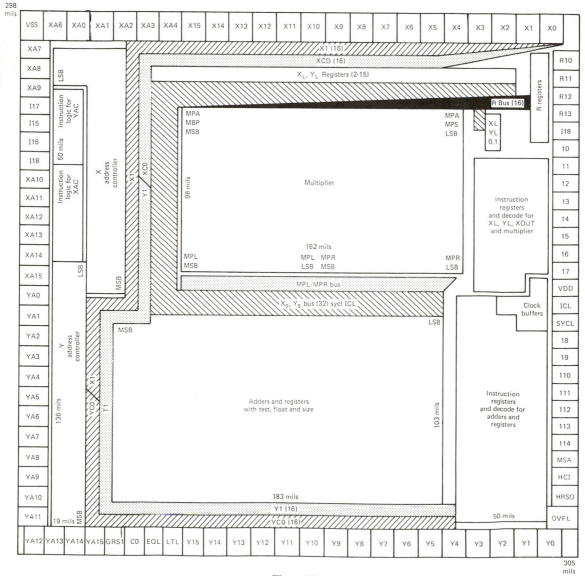

Figure 18.7   APU chip plan.

multiplier; (2) adders, accumulators with special logic functions; (3) XL and YL general-purpose registers; (4) address controllers (to be used twice); and (5) instruction registers and control. Each of the sections is identified on the chip plan and their relative size can be compared.

Each section except the control used the same fundamental layout and design scheme. The layouts of each section were constructed with a hierarchical approach starting with the basic logic cells stacked together vertically to form a single bit slice of the function. Each bit slice utilized as much redundancy of cells as possible. Each of the bit slices were then stacked side by side to form an N-bit function. This bit-slice approach saves significant chip real estate in that only the data buses which enter and exit the function typically had to parallel the section as a wide bus. As an example, the XS and YS buses entering the adder/accumulator section require that each of these buses travels the length of each bit-slice section (0 through 16), even though it never is required as an interconnection in bit slice 16. Conversely, the buses T, F, G, U, V, H, and W are embedded within the adder/accumulator/logic section and communicate only along a vertical slice from cell to cell. This technique saves not only chip area but the parasitics associated with a less efficient layout approach, as would have occurred if the adders were designed as stand-alone cells.

As implied above, the data paths for each section were routed vertically through the section for each bit slice and they were interconnected with polysilicon. This use of very long polysilicon runs was possible because of the low resistivity of the platinum silicide process. Otherwise, the interconnection delays associated with typical higher-resistance poly would have caused excessive delay and the design would not meet the device's speed goals.

Vertical routing of the data flow paths allowed the power, ground, and high-speed control paths to be interconnected horizontally across the bit slices in metal. This approach takes advantage of the μ-CMOS process that allowed the use of buried contacts. The internal node interconnection of the logic cells could be connected on polysilicon, thereby allowing the control and power lines to be routed directly over the active devices.

The only section of the design that was laid out differently from this slice approach was the control and decode logic. This section contains the 4 : 1 shift register and op-code pipe delay registers as well as logic decode of individual control signals from the operation code inputs. This logic decoded the 78-bit microcode word into 184 control lines. The 78 bits are derived from the 20 instruction pins that are input at four times (i.e., ≥ 16 MHz) the system clock rate. This decode was performed with gate-level logic. The number of cells required to do the decode was held to a minimum and were essentially NAND gates and two different-size buffer drivers, all of which were designed to have short delay times. Short delay times required matching the characteristics of the instruction register's output to the high capacitances of the long control lines that transverse the layout of each functional section.

The cells for the control were designed to be easily interconnected and the approach was to build cell rows and interconnect the cells via metal wiring between the rows. The summary of device layout metrics in Table 18.3 bears out the fact that this approach is not very efficient regarding layout density. However, with the constraints of a short schedule and high-speed performance, the results are acceptable. Other decode methods were considered (i.e., PLA, ROM, etc.) and rejected on the basis that high performance was more important than the layout savings.

The next area of layout considered was the power interconnections for the VLSI. The approach isolated the power and ground paths for the I/O devices from the power and ground of the internal logic. One hundred T^2L compatible I/Os result in substantial chip real estate being used for power interconnection. This isolation of power systems was implemented to avoid having any peripheral devices that may cause voltage spikes on the I/O power system from affecting the chip's internal logic. The chip has over 60 T^2L drivers and this requires the power buses for the I/O system to be >2 mils wide to avoid metal migration and voltage crop problems.

18.7.4 Device Design Metrics

Table 18.3 is a summary of the chip layout design, showing the number of devices, layout density, and percentage of chip area on a per function basis. The overall layout density of the VLSI is not as dense as the 3-μm μ-CMOS process could allow. But considering the magnitude of the layout, the limitations on design time, and the structure of the architecture, this layout density (i.e., chip size) is considered acceptable from the standpoint of chip cost per design function performed.

The chip design and layout was completed in 14 months. This time included the iterative design phase between the system architecture and the VLSI designer. The design methodology described here allowed the task to be performed with minimal personnel. The job required a senior VLSI designer, a VLSI draftsman, and a computer-aided graphics operator. Also, another engineer was required for 5 months to perform the logic simulation. System architecture design was provided by additional engineers. The detailed layout procedure was to design the basic cell complete with a form factor consistent with the chip plan and analyze the cell for timing. The cell was then drawn at 1000 times actual size by the VLSI draftsman. The cell artwork was then captured on the CALMA GDS II graphics data system by the operator.

Once the cells were all drafted for a section, the operator would interactively interconnect the cells into slices and then interconnect the slices into a complete section on the graphics data system. As each section was completed, it was interactively integrated into the final chip layout. This system of interactively connecting the cells, slices, sections, and overall chip was an effective approach of mask artwork generation. In most custom or semicustom designs, the interconnection of these pieces is hand drawn and then digitized, requiring more layout time. The interactive interconnection of the design was possible because of the meticulous planning of the interconnection at the chip planning phase.

The design required approximately 50 cells to be drafted and another 50 cells to be created from the original set. The second 50 cells were derived from the original by minor edits by the graphics system operator and were required because of layout usage from section to section. The CALMA GDS II was also used for the artwork design rule verification, the plot preparation for the electrostatic plotter, and the electron-beam pattern-generator tape preparation.

18.8 CONCLUDING REMARKS

As a result of this work, it has been demonstrated that a state-of-the-art arithmetic processor (for an array processor) could be built in a 3-μm CMOS technology. We believe that this device is representative of future VLSI processors with regard to the

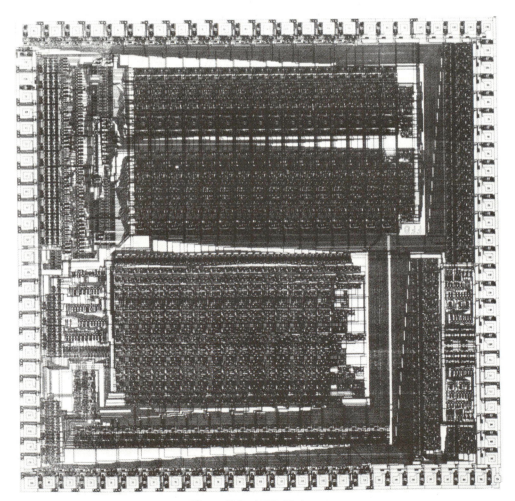

Figure 18.8   APU chip layout.

following parameters:

1. The device has a high percentage of bus area.
2. The device has a high pin count.

These parameters illustrate the need for denser interconnect technology, such as double-layer metal and smaller contacts. It also shows the need for low-cost, high-pin-count semiconductor packaging schemes.

This effort shows the feasibility of even more powerful parallel processors being built on single semiconductor devices as the VLSI technology improves.

Review of the layout (after completion) has allowed us to better critique our design approach. A photo of the chip layout is shown in Figure 18.8. We have estimated that a fully custom layout (with the same design rules) would have resulted in a chip with 80% of the present area. However, this would have required at least three times the design hours. Had we used standard cells with computer placement and routing tools (that are known to us today), the chip size would not have been feasible to build. Future tools being discussed and their utilization on a chip of this type have yet to be proven and are good for present-day coffee-break soliloquies.

REFERENCES

[1] J. B. Bruckner, *CHI-5 Micro-Programming Reference Manual*, CHI Systems, Inc., Goleta, Calif., May 1982.

[2] G. J. Culler, "Array Processor with Parallel Operations per Instruction," U.S. Pat. 4,287,566, Sept. 1, 1981.

[3] A. E. Charlesworth, "An Approach to Scientific Array Processing: The Architectural Design of the PP-120B/FPS-164 Family," *Computer*, 14(9):18–27 (Sept. 1981).

[4] Ed Greenwood et al., "Array Processor Architecture Report," Rep. N00014-80-C-693-I, Feb. 15, 1982.

[5] B. Fette and L. Hazelet, "Electronic Digital Adder," U.S. Pat. 3,843,876, Oct. 22, 1974.

[6] D. T. Fitzpatrick et al., "A RISCy Approach to VLSI," *VLSI Des.*, 4th Quarter 1981, pp. 14–20.

19

Frequency-Domain Digital Filtering with VLSI

EARL E. SWARTZLANDER, JR.

TRW Defense Systems Group
Redondo Beach, California

GEORGE HALLNOR

TRW Defense Systems Group
McLean, Virginia

19.1 INTRODUCTION

There is a strong synergy between VLSI and signal processing. This is because digital signal processing traditionally has led the demand for high levels of computational throughput. Most signal processing applications have the advantage (from an implementation viewpoint) of requiring much computation with little data-dependent operation. Accordingly, they are amenable to implementation with pipeline and parallel architectures.

An important aspect in exploiting technology to the fullest is clear identification of the optimization criterion. This varies depending on application (i.e., minimizing the parts count, power, size, cost, number of part types, etc.). For this chapter, the goal is to minimize the parts count, which requires maximizing the computation per part. The computation per part is estimated by the product of the number of gates times their clock rate. A rough estimate of the computation per part as a function of clock rate for commercially available parts is shown in Figure 19.1. Currently, there is a peak at clock rates of about 10 MHz. Below that speed, the logic density is limited by design complexity and grows relatively slowly with decreasing speed. Above 10 MHz the logic density decreases quite rapidly, due in part to the prevalence of TTL interfaces, which are not amenable to operation at rates above 10 MHz and the high power consumption of current high-speed logic families.

Another way of viewing the technology choice is to examine the attainable level of complexity. For most functions, there is a "logical critical mass" required

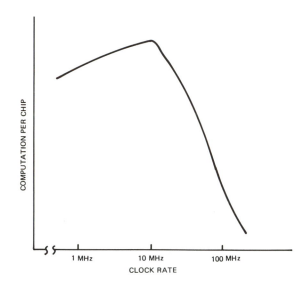

Figure 19.1 Computation per chip as a function of clock rate.

for implementation. At lower logic levels the function cannot be implemented on a single chip and must be partitioned for either multiple-chip implementation (which involves extra I/O drivers, package pins, etc.) or for sequential implementation (which requires additional control logic, temporary variable storage, etc.). In either event the implementation efficiency is greatly improved if the complete function is realized on a single chip.

A high-performance digital filter was selected as a design example that typifies demanding applications. This filter requires processing data at input rates of 40 megasamples per second (MSPS) with a filter resolution of 10 kHz. The filter is realized by computation of the spectrum of 4096 input data, multiplication of the spectrum by a selected frequency response, and inverse transformation to produce the filtered time-domain signal. This processing flow is repeated for data skewed by 50% of the transform length to eliminate edge effects due to the data windowing operation. This system requires computation of the equivalent of approximately 1 billion radix-2 butterfly operations per second (240 million butterflies/s each for the two forward and the two inverse transforms).

The design approach is to convert the input 40-MSPS data stream into four 10-MSPS data streams. These four streams are processed using a radix-4 Cooley–Tukey FFT pipeline processor according to the basic design of McClellan and Purdy of the MIT Lincoln Laboratory [1]. The Cooley–Tukey FFT algorithm was selected because both the forward and inverse transforms are computed by networks built from two distinct basic computational elements. The McClellan and Purdy design was developed nearly a decade ago for implementation using small-scale ECL circuits by General Electric Company for use at Kwajelein Missile Range. The GE implementation uses 27,000 integrated circuits to compute 960 million radix-2 butterflies/s [1]. The design described here is comparable in performance to its predecessor but is implemented with a factor of 5 fewer circuits. It also uses a 22-bit floating-point number to represent each component of the complex data instead of the 11-bit fixed-point numbers with a 6-bit exponent for each complex number used in the GE implementation.

19.2 FREQUENCY-DOMAIN FILTER REQUIREMENTS

The basic algorithm flow is shown in Figure 19.2. Input data flow (along the upper path) through a data acquisition element into a 4096-point complex FFT. Data windowing is performed in the time domain in the data acquisition element. The transformed data are filtered by multiplication in the frequency domain followed by inverse FFT processing. The magnitude of the unfiltered spectrum is computed to determine the power spectral density, which is used to develop the filter kernel. A similar process is performed on data skewed in arrival time (by half of the transform length) through the overlap channel. Use of the two overlapped channels prevents the loss of data that normally occurs at data block boundaries due to windowing.

Examination of this flow indicates the need for the following modules.

1. *Data acquisition:*   provides input (data source) interface, rate conversion from a single 40-MSPS channel to four 10-MSPS channels, and windowing with any prestored time domain window (see [2] for a comparison of windows).

2. *4096-point complex FFT:*   accepts data in time sequential order and produces the complex FFT. The transform element is programmable to accommodate transforms of lengths that are powers of 4 (i.e., 4, 16, 64, 256, 1024, 4096).

3. *Frequency-domain filter:*   multiplies the FFT spectrum by a spectral filter. This is a simple point-by-point multiplication.

4. *Power spectral density:*   computes the magnitude of the spectrum, orders the data in frequency order, and averages using an exponential weight.

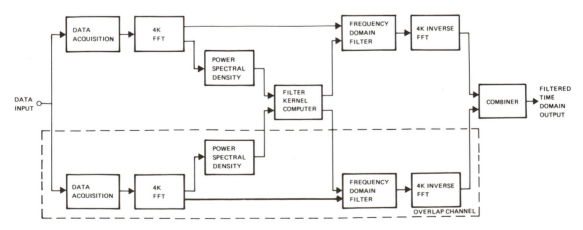

Figure 19.2   Frequency-domain filter.

5. *Filter kernel:*   a general-purpose computer develops an adaptive filter kernel based on the observed power spectral density.

6. *4096-point complex inverse FFT:*   converts the filtered spectrum back to the time domain. The inverse FFT is programmable like the FFT module to accommodate transforms of various binary lengths.

Thus six major modules are required. It should be emphasized that these modules are not of the same level of complexity; the FFT and inverse FFT modules are much more complex than the other modules.

After the initial processing in the data acquisition module, all computation is implemented with 22-bit floating-point arithmetic (6-bit exponent, 16-bit fraction) to maintain precision and avoid the need for adaptive data scaling.

For this processor four parallel channels are implemented using pipeline processing with a cascade architecture [3]. This choice was made to achieve high speed (to accommodate 40-MSPS data rates) while using readily available VLSI technology (as noted in the introduction, current VLSI devices easily achieve 10-MHz clock rates; higher speeds would require specialized circuits with much lower levels of functional integration). A wide variety of VLSI components are currently available in several technologies to support 10-MHz clock rates. Examples include high-speed 16K-bit static random access memory (RAM), 22-bit floating-point adders and multipliers [4], and so on.

19.3 COMPUTATIONAL MODULE DESIGN

The design of each of the basic modules will be summarized based on the assumption of single-chip functional elements for the arithmetic operations.

Data acquisition module.   This module provides the data source interface, reorders the data from a single data stream to four channels, and performs the time-domain windowing. Its implementation is shown in Figure 19.3. Data arriving in complex form are clocked at a 40-MHz rate into eight shift registers, four for each of the two complex components of each data word. The contents of the shift registers are then transferred into eight parallel latches at a 10-MHz rate. The shift register/latch combination converts the single serial complex channel to four parallel complex channels. The four parallel data channels are weighted with multiplicative data windows that are implemented using a 1024-element window [2] stored in a read-only memory. The ROM is implemented with four parallel memories each containing 1024 weights since four channels are weighted simultaneously.

On this module the shift registers and latches are implemented with ECL devices in order to achieve the 40-MSPS rates. Since ECL devices are quite low in density, only four register or latch stages are packaged in a single chip. The multiplier and ROM chips are VLSI, so that of the 120 chips required to implement this module, all but 15 are ECL chips.

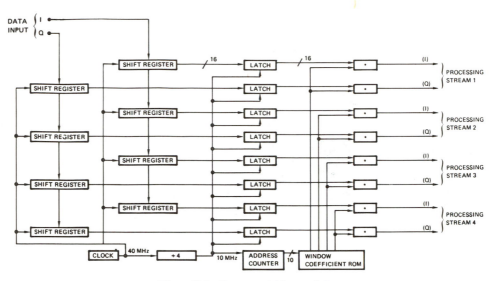

Figure 19.3   Data acquisition module.

The input data are assumed to be 16-bit fixed-point words. Since most high-speed A/D converters are lower in precision (i.e., 8 bits, 10 bits, or occasionally as high as 12 bits), it may be possible to use smaller data words in this unit. Because of the low level of integration of the ECL components, reduction to 12- or 8-bit input data word sizes reduces the complexity by 25 to 50%, respectively. For this sizing, the conservative assumption of 16-bit data was selected.

FFT module.   As noted earlier, the FFT module is considerably more complex than the other modules. As a result, for implementation purposes, it is partitioned into functional elements of two types. This partitioning is a compelling advantage of the Cooley–Tukey pipeline FFT algorithm.

The structure of the FFT processor is shown in Figure 19.4. It is constructed with computational elements (CEs) and delay/commutators (D/Cs). In this design the computational element is a radix-4 butterfly and the delay/commutators perform the data reordering required for the Cooley–Tukey FFT algorithm [5]. The basic pipeline processing concept was developed at Raytheon in the late 1960s [6], and extended to radix-4 implementation by the MIT Lincoln Laboratory [1]. A radix-2 FFT is currently being implemented in VLSI with a similar partitioning onto two chip types by NEC [7].

The radix-4 butterfly computational element realizes a four-point discrete Fourier transform. It is implemented with three complex multipliers, eight complex adders (four perform addition and four perform subtraction), sin–cos tables (stored in ROM), and miscellaneous addressing logic. The computational element is implemented with approximately 80 integrated circuits. This low level of complexity is a direct result of the availability of single-chip arithmetic components.

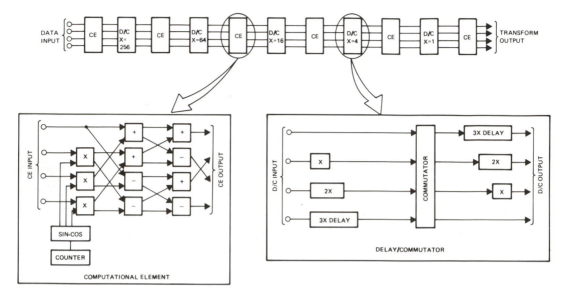

Figure 19.4   Radix-4 pipeline Cooley–Tukey FFT.

The other element of the pipeline FFT is the delay/commutator that is used to perform the interstage data reordering required by the Cooley–Tukey algorithm. It consists of input delay lines of graduated length, a commutator switch, and output delay lines of graduated length. The length of the unit delay is 1 at the final stage of the FFT and increases by a factor of 4 at each preceding stage. The module requires about 180 integrated circuits to implement transform lengths up to 4096 points. The delays are implemented as shown in Figure 19.5. A RAM is used with data written in sequential order (using an address generated by the counter plus the offset value) and read (using the counter value alone as the address). Thus data are written in sequence through the memory (as though it were a sequential access drum memory) and the delayed data are read in sequence from the memory.

In its basic form the radix-4 pipeline FFT computes transforms of sequence lengths that are powers of 4. Addition of a radix-2 computation element (a subset of the radix-4 CE) and a special delay/commutator should allow computation of transforms of lengths that are powers of 2.

Frequency-domain filter module.   The frequency-domain filtering operation involves multiplying the 4096-point complex spectrum by a filter kernel on a point-by-point basis. The module is implemented with four identical complex channels, as shown in Figure 19.6. The filter kernel is stored in a RAM for recall as input data arrive. This module is implemented with approximately 70 integrated circuits.

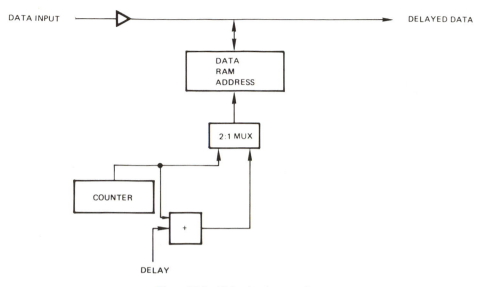

Figure 19.5 Delay implementation.

Power spectral density module.   This module is used to compute an estimate of the power spectral density, PSD (by summing the squares of the real and complex components of each frequency of the spectrum) and averaging successive PSD using a recursive averaging scheme. The design of the element is shown in Figure 19.7. The magnitude of the complex input data is computed followed by the

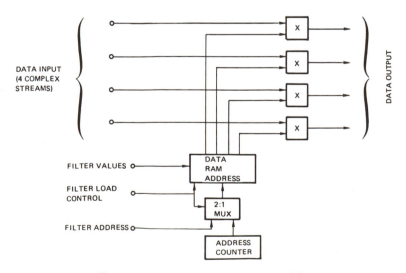

Figure 19.6 Frequency-domain filter module.

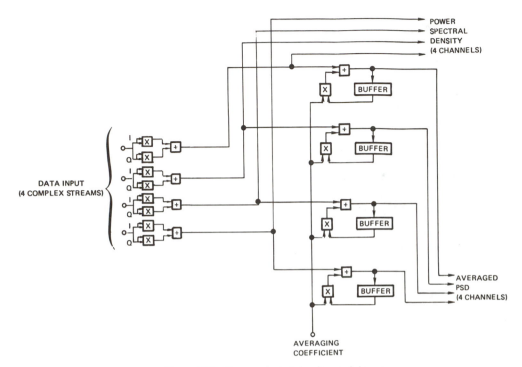

Figure 19.7   Power spectral density module.

averaging operation. The averaging involves summing the current spectrum with a fraction of the previously accumulated spectrum on a point-by-point basis. The complexity of the power spectral density module is approximately 90 integrated circuits.

Filter kernel module.   The filter kernel is computed by a general-purpose mini- or microcomputer. The filter response can be fixed or made adaptive, depending on the system application. The general-purpose computer is not included in the filter complexity estimates since most systems have a host computer that will be used to provide this function.

Combiner module.   The inverse FFT and overlapped inverse FFT outputs are multiplexed together in the combiner as shown in Figure 19.8. All input data are converted to ECL data levels. The data are then multiplexed onto a single output channel through a two-level 8 : 1 multiplexer. It is realized with a 4 : 1 multiplexer followed by a 2 : 1 multiplexer. The combiner complexity is 143 integrated circuits.

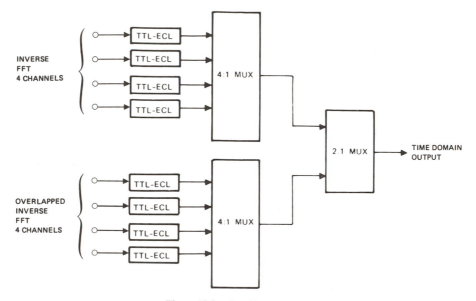

Figure 19.8   Combiner module.

19.4 COMPLETE FREQUENCY-DOMAIN FILTER

The total parts complement for the complete frequency-domain filter processor, including the two overlap channels and two inverse FFT modules, is 49 circuit cards containing approximately 6100 integrated circuits and dissipating under 6 kW, as detailed in Table 19.1. The unit is packaged on circuit cards containing 78 to 179 integrated circuits and dissipating 100 to 150 W. Each circuit card measures 9 by 16 in., which allows the use of commercial card cages and (for breadboard purposes) commercial wirewrap cards.

TABLE 19.1   FAST TRANSFORM PROCESSOR COMPLEXITY

Module type	Module count	Total chips	Total power (W)
Data acquisition and weighting	2	240	250
FFT			
Computational element	12	960	1500
Delay/commutator	10	1790	1000
Frequency-domain filter	1	140	150
Power spectral density	1	78	100
Combiner	1	143	150
Inverse FFT (same as FFT)	22	2750	2500
Total	49	6101	5650

The circuit cards are packaged in four drawers of a standard equipment rack. Each of the four transform processors (i.e., forward, overlapped forward, inverse, and overlapped inverse) are packaged in a separate drawer. The total volume for a breadboard implementation of the complete frequency domain filter is about 10 ft^3. It accepts data at an input rate of 40 MSPS (equivalent to 1.28 gigabits/s if the inputs are complex 16-bit words). The filtered data are output at the same 40-MSPS rate.

After the initial processing on the data acquisition and weighting module, all computation is implemented with 22-bit floating-point (6-bit exponent, 16-bit fraction) arithmetic to maintain precision and avoid the need for adaptive data scaling.

19.5 CONCLUDING REMARKS

This chapter illustrates the potential for achieving throughputs of 1 billion equivalent radix-2 butterflies per second with current VLSI technology. Table 19.2 summarizes the characteristics of the complete filter. Two significant figures of merit are the computational rate per unit volume (960 million butterflies per second in 10 ft^3 for approximately 100 million butterflies per second per cubic foot) and the computational rate per watt (approximately 170 thousand butterflies per second per watt). These figures of merit and the computational rate may be viewed in perspective by noting that the SPS-1000 with four pipes achieves about 42 million butterflies per second [8]. The SPS-1000 computational rate per unit volume is on the order of 50 to 100 million butterflies per second per cubic foot. Another advanced processor achieves a computational rate (with 16 computational elements) of 71 million butterflies per second [9]. Thus the VLSI design attains an order-of-magnitude higher total throughput than either of the referenced commercial systems.

This study certainly does not indicate an upper bound on processor throughput. Indeed, it is reasonable to speculate that computational rates on the order of 100 billion butterfly operations per second may be achieved with advanced ECL or GaAs technology in the near future.

TABLE 19.2 FREQUENCY-DOMAIN FILTER
CHARACTERISTICS

Technology:	Off-the-shelf 1982
	Logic: bipolar
	Memory: nMOS
Size:	10 ft^3 (using wirewrap construction)
Power:	5.6 kW
Throughput:	960 million radix-2 butterflies per second
Word size:	Input: 16-bit fixed point
	Intermediate: 22-bit floating point
	(6-bit exponent, 16-bit fraction)
Figures of merit:	96 million butterflies/s/ft^3
	170,000 butterflies/s/W

REFERENCES

[1] J. H. McClellan and R. J. Purdy, "Applications of Digital Signal Processing to Radar," in A. V. Oppenheim, ed., *Applications of Digital Signal Processing*, Prentice-Hall, Englewood Cliffs, N.J., 1978, Chap. 5.

[2] F. J. Harris, "On the Use of Windows for Harmonic Analysis with the Discrete Fourier Transform," *Proc. IEEE, 66*:51–83 (1978).

[3] C. D. Bergland, "Fast Fourier Transform Hardware Implementations—An Overview," *IEEE Trans. Audio Electroacoust., AU-17*:104–108 (1969).

[4] J. A. Eldon and C. Robertson, "A Floating Point Format for Signal Processing," *Proc. ICASSP-82*, 1982, pp. 717–720.

[5] J. W. Cooley and J. W. Tukey, "An Algorithm for the Machine Calculation of Complex Fourier Series," *Math. Comp., 19*:297–301 (1965).

[6] H. L. Groginsky and G. A. Works, "A Pipeline Fast Fourier Transform," *IEEE Trans. Comput., C-19*:1015–1019 (1970).

[7] A. Kanemasa, R. Maruta, K. Nakayama, Y. Sakamura, and S. Tanaka, "An LSI Chip Set for DSP Hardware Implementation," *Proc. ICASSP-81*, 1981, pp. 644–647.

[8] R. A. Collesidis, T. A. Dutton, and J. R. Fisher, "An Ultra-High Speed FFT Processor," *Proc. ICASSP-80*, 1980, pp. 784–787.

[9] C. S. Joshi, J. F. McDonald, and R. H. Stinvorth, "A Video Rate Two Dimensional FFT Processor," *Proc. ICASSP-80*, 1980, pp. 774–778.

20

A Parallel VLSI Architecture for a Digital Filter Using a Number-Theoretic Transform

T. K. Truong

California Institute of Technology
Pasadena, California

I. S. Reed, C.-S. Yeh, J. J. Chang, and H. M. Shao

University of Southern California
Los Angeles, California

20.1 INTRODUCTION

Many applications of digital signal processing require cyclic convolutions [1]. Both the fast Fourier transform (FFT) and the number-theoretic transform (NTT) were developed to efficiently compute cyclic convolutions [2–7]. The cyclic convolution of two sequences of integers is obtained by taking the inverse NTT of the product of the NTTs of the two sequences. The FNT can be applied to digital filtering [4,5], image processing [8,9], x-ray reconstruction [10], and to the encoding and decoding of certain Reed–Solomon codes [11,12].

Some advantages of the NTT over the more usual FFT are: (1) for some NTTs, no multiplications are required; (2) integer arithmetic is used in the NTT; (3) and no round-off error is introduced. Some disadvantages of the NTT are: (1) there is a conflict between the long transform lengths and a large dynamic range under a reasonable word size, which often limits the applicability of the NTT; and (2) a modulus operation is required in the arithmetic of the transform.

An important NTT is the Fermat number transform (FNT). The FNT has the advantage over most NTTs in that the transform length is a power of 2. As a consequence, the FFT structure can be applied to compute the FNT; but 2 is a possible root of unity in the particular finite field or ring. With an FNT that has 2 as a root of unity, multiplications by powers of 2, modulo the Fermat number, are implemented easily by bit rotations in binary arithmetic. On the other hand, the

360

FFT requires multiplications by the complex roots of unity of the form $e^{2\pi inK/N}$ Therefore, in the case of the FFT form, multiplications by both sine and cosine coefficients are required.

McClelland [13] designed a hardware system to realize a 64-point 17-bit FNT that used commercially available ECL IC chips. For this FNT a nonstandard binary number representation was developed and used to perform the binary arithmetic operations, modulo a Fermat number. Leibowitz [14] also developed a simpler binary number representation for the same purpose.

In this chapter a generalized overlap-save method is developed to solve the conflict between a long transform length and a wide dynamic range. Using the generalized overlap-save technique with FNTs, a parallel architecture is designed to realize a FIR digital filter of arbitrary length. In Section 20.2 the FNT, Leibowitz's binary number representation, and the required binary operations, modulo a Fermat number, are briefly reviewed.

In Section 20.3 a pipeline architecture suitable for large-scale-integration (LSI) circuitry is designed to compute a 128-point FNT over F_5. Only additions and bit rotations are required in this structure. The bit rotations needed for multiplication by a power of 2 can be realized by a modification of a standard barrel shifter circuit [15].

In Section 20.4 the generalized overlap-save method is developed in detail for computing the linear convolution of two sequences of arbitrary length. Finally, a parallel systolic-type architecture is designed to realize the generalized overlap-save method using the FNT and inverse FNT circuits of 128 points. This architecture is regular, expansible, and therefore suitable for VLSI implementation.

20.2 THE FERMAT NUMBER TRANSFORM

Let $F_t = 2^{2^t} + 1$ be the tth Fermat number, wnere $t \geq 0$. F_t is a prime number for $0 \leq t \leq 4$. Let $\{x_n\}$ be an N-point sequence of integer numbers, where $0 \leq x_n \leq F_t - 1$, $0 \leq n \leq N - 1$, and N is a power of 2. The Fermat number transform $\{X_k\}$ of $\{x_n\}$ over F_t is defined as follows:

$$X_k \equiv \sum_{n=0}^{N-1} x_n \alpha^{nk} \pmod{F_t} \qquad 0 \leq k \leq N - 1 \qquad (20.1)$$

where $0 \leq X_k \leq F_t - 1$ and α is an Nth root of unity. That is, N is the least positive integer such that $\alpha^N \equiv 1 \pmod{F_t}$. The corresponding inverse FNT is the following:

$$x_n \equiv \frac{1}{N} \sum_{k=0}^{N-1} X_k \alpha^{-nk} \pmod{F_t} \qquad 0 \leq n \leq N - 1 \qquad (20.2)$$

where $1/N$ is the inverse of N modulo F_t. $1/N$ is also an integer. In equations (20.1) and (20.2), exponents are assumed to be taken modulo N.

A cyclic convolution is computed with the FNT pair in (20.1) and (20.2). The transform length N depends on F_t and the α chosen [4,5]. More details of the FNT can be found in [4,5].

In order to obtain a convolution without arithmetic overflow with an FNT over F_t, one must restrict the operating ranges of the variables. To compute an unambiguous cyclic convolution $\{y_k\}$ of two N-point integer sequences $\{x_n\}$ and $\{h_m\}$, that is,

$$y_k = \sum_{n=0}^{N-1} x_n h_{(k-n)} \tag{20.3}$$

the final convolution should lie within a range determined by the magnitude of F_t where $(k - n)$ is the residue of $(k - n)$ modulo N. This can be achieved if

$$|y_k| = \left| \sum_{n=0}^{N-1} x_n h_{(k-n)} \right| \leq \sum_{n=0}^{N-1} |x_n||h_{(k-n)}| \leq \frac{F_t - 1}{2} \tag{20.4}$$

Let max $|x_n| =$ max $|h_m| = A$ be the maximum variation of the variables x_n and h_m. A is called the dynamic range of x_n and h_m. In order to restrict $|y_k|$ to satisfy (20.4), it suffices to let

$$A = \left[\sqrt{\frac{F_t - 1}{2N}} \right] \tag{20.5}$$

where $[x]$ denotes the greatest integer less than x. However, for many applications this value of A is quite pessimistic.

In this chapter F_t, α, and N are selected specifically to be $F_5 = 2^{32} + 1$, $\sqrt{2}$, and 128, respectively. That is, the data of this FNT are integers between 0 and 2^{32}. Hence 33 bits are required to represent a number. The transform length of this FNT is 128.

In an FNT over F_t, the quantity $\sqrt{2}$ represents the integer $2^{2^{t-2}}(2^{2^{t-1}} - 1)$ [4,5]. For $t = 5$, since $2^{32} \equiv -1 \pmod{F_5}$, $\sqrt{2} = 2^{24} - 2^8 = 2^{24} + 2^{40}$. A conservative value of the dynamic range [12] is $\sqrt{2^{32}/(2 \cdot 128)} = 2^{12}$. This value is sufficiently large for a number of applications.

TABLE 20.1 CORRESPONDENCE AMONG DECIMAL NUMBERS, THEIR VALUES IN THE NORMAL BINARY REPRESENTATION, AND IN THE DIMINISHED-ONE REPRESENTATION

Decimal number	Normal binary representation							Diminished-one representation						
	a_{32}	a_{31}	a_{30}	$\cdots$	a_2	a_1	a_0	a_{32}	a_{31}	a_{30}	$\cdots$	a_2	a_1	a_0
0	0	0	0	$\cdots$	0	0	0	1	0	0	$\cdots$	0	0	0
1	0	0	0	$\cdots$	0	0	1	0	0	0	$\cdots$	0	0	0
2	0	0	0	$\cdots$	0	1	0	0	0	0	$\cdots$	0	0	1
$\vdots$				$\vdots$							$\vdots$			
$2^{32} - 2$	0	1	1	$\cdots$	1	1	0	0	1	1	$\cdots$	1	0	1
$2^{32} - 1$	0	1	1	$\cdots$	1	1	1	0	1	1	$\cdots$	1	1	0
2^{32}	1	0	0	$\cdots$	0	0	0	0	1	1	$\cdots$	1	1	1

Two binary number representations [13,14] have been proposed to achieve an easy implementation of the binary arithmetic operations needed to realize an FNT circuit. The diminished-one representation proposed by Leibowitz [14] is used in the following design. Let A be represented by $[a_{32} \ a_{31} \ \cdots \ a_1 \ a_0]$, where $0 \leq A \leq 2^{32}$ and a_i is the ith bit of A. Table 20.1 shows the correspondence between decimal numbers in a normal binary representation and their values in the diminished-one representation. In the diminished-one representation the most significant bit (MSB) a_{32} is 1 if and only if the integer represents 0. Thus a_{32} can be viewed as a zero-detection bit.

Two basic binary arithmetic operations needed to define an FNT are addition and multiplication by a power of 2, both modulo F_t. Other operations can be expressed in terms of these two operations. In the following these operations are described briefly. More details can be found in [14].

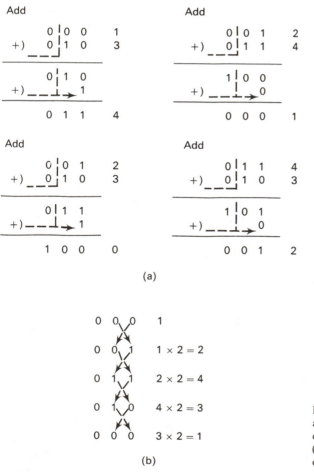

(a)

(b)

Figure 20.1 Examples of binary operations modulo $F_1 = 2^2 + 1$ in the diminished-one binary representation: (a) addition; (b) multiplication by a power of 2.

1. *Addition:* Let $S = A + B$. (a) If $A = 0$, then $S = B$. (b) If $B = 0$, then $S = A$. (c) If neither A nor B equals 0, add $[a_{31} \quad a_{30} \quad \cdots \quad a_1 \quad a_0]$ and $[b_{31} \quad b_{30} \quad \cdots \quad b_1 \quad b_0]$. Then complement the carry and add it to the previous sum. This yields S.

2. *Multiplication by a power of 2:* Let $B = A \cdot 2^C$. (a) If $A = 0$, then $B = 0$, (b) If $A \neq 0$, left rotate $[a_{31} \quad a_{30} \quad \cdots \quad a_1 \quad a_0]$ C bit positions, but complement the value of the 31st bit when it is rotated to bit position 0. Finally, set $b_{32} = 0$.

3. *Negation:* Since $2^{32} \equiv -1 \pmod{F_5}$, $-A = A \cdot 2^{32}$. Hence, if $A \neq 0$, $-A = [a_{32} \quad \bar{a}_{31} \quad \bar{a}_{30} \quad \cdots \quad \bar{a}_1 \quad \bar{a}_0]$, where $\bar{a}_i$ denotes the complement of a_i. If $A = 0$, then $-A = 0$.

4. *Multiplication by $\sqrt{2}$:* Since $\sqrt{2} = 2^{24} + 2^{40}$, $A \cdot \sqrt{2} = A \cdot 2^{24} + A \cdot 2^{40}$.

5. *Multiplication by a power of $\sqrt{2}$:* Let $B = A \cdot (\sqrt{2})^C$. If C is even, then $B = A \cdot (2)^{C/2}$. If C is odd, then $B = (A \cdot \sqrt{2}) \cdot 2^{(C-1)/2}$.

Several examples are given in Figure 20.1 to illustrate addition and multiplication by a power of 2, modulo F_t. For simplicity, F_t is chosen to be $F_1 = 2^2 + 1$ in Figure 20.1. In this case, 3 bits are required to represent an integer. The diminished-one representation is used in Figure 20.1.

20.3 A PIPELINE ARCHITECTURE FOR COMPUTING A 128-POINT FNT

By virtue of (20.1) the FNT has a mathematical algorithm very similar to the FFT. Hence an almost exact analog of an FFT-type structure can be used to perform a fast FNT. Figure 20.2 shows an FFT-type pipeline structure [1] for computing a 128-point FNT over F_5. This pipeline structure can be viewed as a one-dimensional systolic array. The basic elements of the structure are delays, switches, and FNT butterflies. These basic elements are arranged alternately in the structure. The radix-2 decimation-in-time (DIT) technique is used in this structure. The structure for performing an inverse FNT is the mirror image of the circuit shown in Figure 20.2, where a radix-2 decimation-in-frequency (DIF) technique is used.

In Figure 20.2 z^{-j} denotes a j-step delay element. This delay element can be realized by 33 sets of shift registers of j 1-bit stages. SW_i in Figure 20.2 is a switch controlled by the control signal S_i for $1 \leq i \leq 6$. The operations of the SW_i are shown in Figure 20.3. The S_i's can be implemented simply by a six-stage up-counter if buffer registers are not used in the FNT butterflies [1]. If buffer registers are used in the butterflies, delay elements are needed at the outputs of the counter, as shown in Figure 20.4, for the purpose of synchronization.

In Figure 20.5 are shown the symbolic diagram and operations of a DIT FNT butterfly. A similar DIF FNT butterfly was designed in [13]. A block diagram of a DIT FNT butterfly is shown in Figure 20.6. In this design, A, B, D, and E are 33-bit

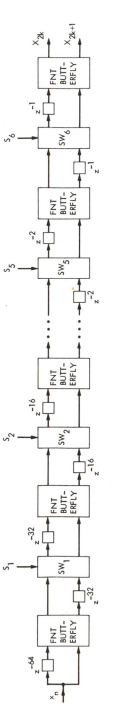

Figure 20.2 Pipelined structure for computing a 128-point Fermat number transform.

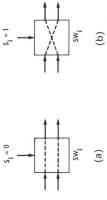

Figure 20.3 Shuffle-exchange switch: (a) direct connection; (b) crossed connection.

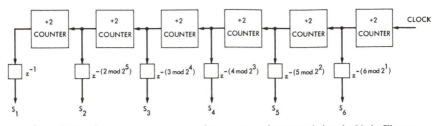

Figure 20.4 Six-stage up-counter used to generate the control signals S_i's in Figure 20.1.

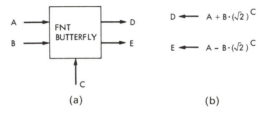

(a)   (b)

Figure 20.5 (a) Symbolic diagram; (b) operations of a DIT FNT butterfly.

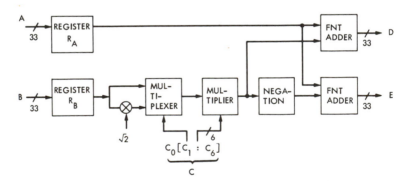

Figure 20.6 Block diagram of a DIT FNT butterfly.

data, and C is the 7-bit exponent of α in (20.1), that is, nk mod N. Two realizations of an FNT adder can be found in [13]. One of them is shown in Figure 20.7. The multiplier in Figure 20.6 is used to multiply a number by a power of 2 modulo F_5. Figure 20.8 shows a block diagram of this multiplier. The shifter in Figure 20.8 is a modification of a barrel shifter [15] for performing bit rotation operations.

For purposes of illustration, consider the simple FNT over $F_1 = 2^2 + 1$. In such an FNT butterfly the functional table and circuit of a modified barrel shifter are shown in Figure 20.9, where the inputs are $[b_1 \quad b_0]$ and $[s_3 \quad s_2 \quad s_1 \quad s_0]$, and the outputs are $[b_1^* \quad b_0^*]$.

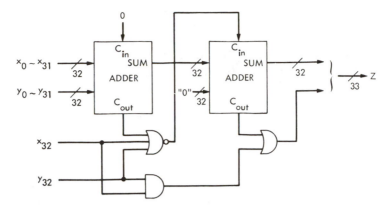

Figure 20.7 Block diagram of an FNT adder to perform $Z = X + Y \pmod{F_5}$.

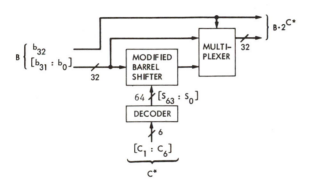

Figure 20.8 Circuit to perform $B \cdot 2^{C*}$.

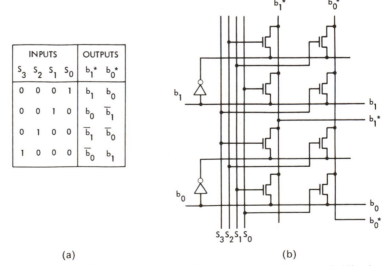

INPUTS		OUTPUTS	
S_3	S_2	S_1	S_0	b_1^*	b_0^*
0	0	0	1	b_1	b_0
0	0	1	0	b_0	$\overline{b_1}$
0	1	0	0	$\overline{b_1}$	$\overline{b_0}$
1	0	0	0	$\overline{b_0}$	b_1

(a)

(b)

Figure 20.9 (a) Functional table, and (b) circuit of a modified barrel shifter in an FNT over $F_1 = 2^2 + 1$.

20.4 DIGITAL FILTER ARCHITECTURE OF ARBITRARY LENGTH USING THE FNT

In the preceding section, F_t, α, and N are chosen to be F_5, $\sqrt{2}$, and 128, respectively. $N = 128$ is the maximum transform length over F_5 [2,3], and 2^{12} is the dynamic range. One could increase the transform length by choosing F_t for $t \geq 6$. In so doing, however, at least $2^6 + 1 = 65$ bits are required to represent a number. Alternatively, one could use a specific α, where α is not a power of $\sqrt{2}$, over F_3 or F_4 to increase the transform length. In such a case complete multiplication is necessary. In addition, the dynamic range may not be sufficient. To remedy this difficulty, the overlap-save method [1] is generalized to compute the linear convolution of a digital filter of arbitrary input data and filter lengths. Finally, a parallel architecture is developed to realize this generalized overlap-save method using the 128-point FNT structure designed in the preceding section.

Let $\{x_n\}$ and $\{h_m\}$ be the input and filter sequences of a digital filter, respectively, where $0 \leq n \leq N - 1$ and $0 \leq m \leq M - 1$. The output sequence $\{y_k\}$ of the filter is the linear convolution of $\{x_n\}$ and $\{h_m\}$, where $0 \leq k \leq N + M - 1$ [13]. For purposes of exposition it is assumed that $N \geq 128$ and $M = 256$ in the following discussion.

In order to use 128-point FNTs to compute $\{y_k\}$, four 128-point subfilters $\{h_m^1\}$, $\{h_m^2\}$, $\{h_m^3\}$, and $\{h_m^4\}$ are formed by partitioning $\{h_m\}$ as follows:

$$h_m^i = \begin{cases} h_m & \text{for } 64(i-1) \leq m \leq 64i - 1 \\ 0 & \text{otherwise} \end{cases} \qquad (20.6)$$

for $1 \leq i \leq 4$.

Next, the linear convolution $\{y_k^i\}$ of $\{x_n\}$ and $\{h_m^i\}$ is computed by using cyclic convolution technique. The standard overlap-save method [1] and the FNT are used to compute $\{y_k^i\}$. To accomplish this $\{x_n\}$ is sectioned into 128-point subsequences $\{x_n^j\}$ each with an overlap of 64 points of $\{x_n\}$ between any two consecutive subsequences, where $j \geq 0$. That is,

$$x_n^j = \begin{cases} x_n & \text{for } 64(j-1) \leq n \leq 64(j+1) - 1 \\ 0 & \text{otherwise} \end{cases} \qquad (20.7)$$

Next a 128-point cyclic convolution $\{z_k^j\}$ of $\{x_n^j\}$ and $\{h_m^i\}$ is computed by using the FNT. Sequence $\{y_k^i\}$ for $1 \leq i \leq 4$ is obtained by discarding the first half and saving the second half of each cyclic convolution $\{z_k^j\}$ computed above.

Finally, the desired output sequence $\{y_k\}$ equals the standard arithmetic sum of $\{y_k^i\}$ for $1 \le i \le 4$. In summary, the linear convolution of $\{x_n\}$ and $\{h_m\}$ is computed by the following algorithm.

1. Partition $\{h_m\}$ into subsequences $\{h_m^i\}$ according to (20.6).
2. Compute the linear convolution of $\{h_m^i\}$ and $\{x_n\}$ using the standard overlap-save method as follows.
    a. Section $\{x_n\}$ into subsequences $\{x_n^j\}$ as shown in (20.7).
    b. Compute the cyclic convolution $\{z_k^j\}$ of $\{x_n^j\}$ and $\{h_m^i\}$ by using the FNT.
    c. $\{y_k^i\}$ is obtained by discarding the first half and saving the second half of each convolution $\{z_k^j\}$ computed in step (b).
3. Finally, $\{y_k\}$ equals the arithmetic sum of $\{y_k^i\}$.

Figure 20.10 illustrates this example of the generalized overlap-save method.

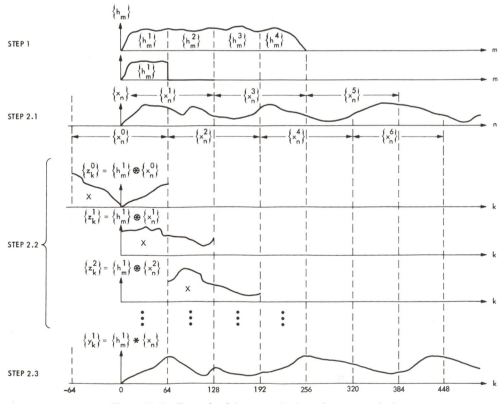

Figure 20.10   Example of the generalized overlap-save method.

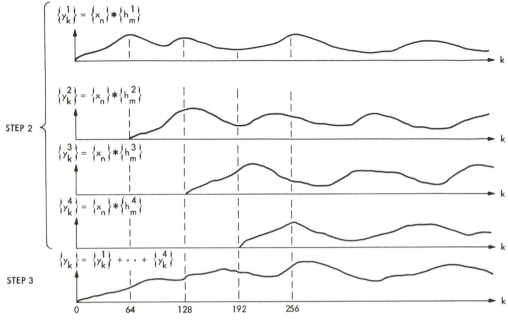

Figure 20.10 (*Continued*)

For simplicity, the drawing in Figure 20.10 shows the data as if they were continuous rather than digital. Also, the results of the convolutions are not drawn with complete accuracy. Other cases of the generalized overlap-save method are constructed in a similar manner.

A simulation program was written to verify the correctness of the generalized overlap-save method. The flowchart of the program is shown in Figure 20.11. In Figure 20.11, t, M, and N denote the subscript of F_t, the filter length, and the length of data sequence, respectively. The transform length L is 2^{t+2} if $\alpha = \sqrt{2}$ for an FNT over F_t. $L = 2^{t+1}$ if $\alpha = 2$ for an FNT over F_t. Several examples are computed by the program. The convolutions obtained using the generalized overlap-save method are exactly the same as the convolutions obtained by direct brute-force computation. In Figure 20.12 is shown the results of the convolution of two sequences, each of unity amplitude, obtained by the overlap-save simulation program for the FNT. The lengths of the two sequences are 64 and 128, respectively.

In Figure 20.13 is shown a parallel architecture to realize the digital filter, described above using the generalized overlap-save method. The sequence $\{H_k^i\}$ in Figure 20.13 is the FNT of $\{h_m^i\}$ multiplied by $1/N$, where $1 \le i \le 4$. The multipliers in Figure 20.13 perform multiplications, modulo F_5. From step 2c of the algorithm above, one of two outputs of an inverse FNT is not needed. Hence the last FNT butterfly in the inverse FNT circuit is a degenerative one, and the delay elements associated with this butterfly are not required. The adders in Figure 20.13 perform normal binary addition, not addition modulo F_t.

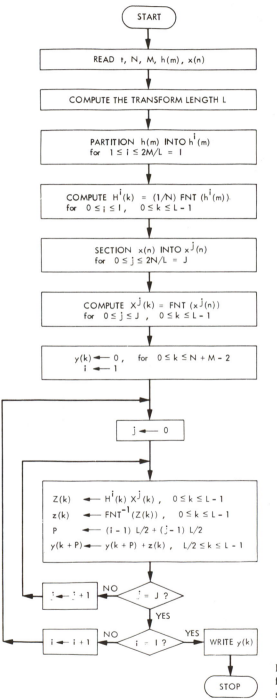

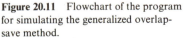

Figure 20.11   Flowchart of the program for simulating the generalized overlap-save method.

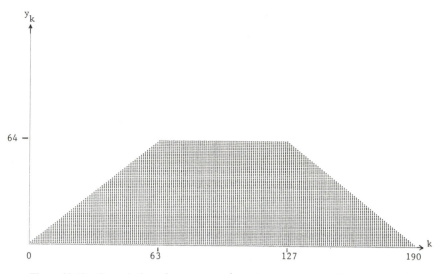

Figure 20.12 Convolution of two rectangular pulses over a ring of integers modulo a Fermat number using the generalized overlap-save method.

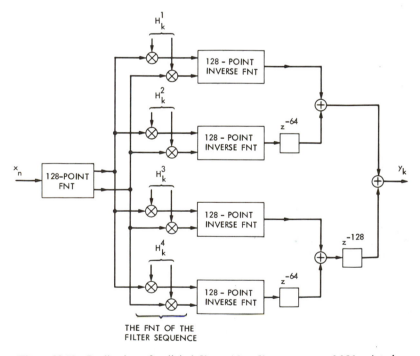

Figure 20.13 Realization of a digital filter with a filter sequence of 256 points by using the generalized overlap-save method and the FNT technique.

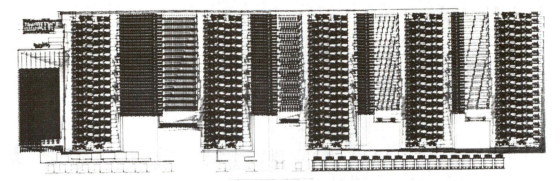

Figure 20.14   Layout of an FNT butterfly for a 32-point FNT over F_4.

20.5 CONCLUDING REMARKS

A pipeline structure is developed to compute a 128-point Fermat number transform. In this 128-point FNT, only additions and bit rotations are required. A barrel shifter circuit is modified to perform the multiplication of an integer by a power of 2 modulo a Fermat number. The layout of an FNT butterfly for 32-point FNT over F_4 is shown in Figure 20.14. The overlap-save method is generalized to compute the linear convolution of a digital filter with arbitrary input data and filter lengths. An architecture is developed to realize this generalized overlap-save method by a simple combination of one 128-point FNT and several inverse FNT structures.

The advantages of realizing a digital filter using the generalized overlap-save method and the FNT over the more usual FFT are the following: (1) no multiplications are required; (2) the generalized overlap-save method solves the conflict between long transform lengths and a wide dynamic range; (3) the word size used in the FNT is practical for many applications, as large word sizes are needed only at the last step of the algorithm; (4) the lengths of the input data and filter sequences can be arbitrary and different; and (5) the architecture is simple, regular, expansible, and hence suitable for VLSI implementation.

ACKNOWLEDGMENTS

The authors would like to thank the referee for several useful suggestions and J. M. Liao for his help in layout design.

This work was supported in part by the U.S. Air Force Office of Scientific Research, under Grant APOSR-80-0151, and also in part by NASA, under Contract NAS7-100.

REFERENCES

[1] L. R. Rabiner and B. Gold, *Theory and Application of Digital Signal Processing*, Prentice-Hall, Englewood Cliffs, N.J., 1975.

[2] J. H. McClellan and C. M. Rader, *Number Theory in Digital Signal Processing*, Prentice-Hall, Englewood Cliffs, N.J., 1979.

[3] C. M. Rader, "Discrete Convolutions via Mersenne Transforms," *IEEE Trans. Comput.*, *C-21*(12):1269–1273 (Dec. 1972).

[4] R. C. Agarwal and C. S. Burrus, "Fast Convolution Using Fermat Number Transforms with Applications to Digital Filtering," *IEEE Trans. Acoust. Speech Signal Process.*, *ASSP-22*(2):87–97 (Apr. 1974a).

[5] R. C. Agarwal and C. S. Burrus, "Number Theoretical Transforms to Implement Fast Digital Convolution," *Proc. IEEE*, 63(4):550–560 (Apr. 1975).

[6] I. S. Reed and T. K. Truong, "The Use of Finite Field to Compute Convolutions," *IEEE Trans. Inf. Theory*, *IT-21*(2):208–213 (Mar. 1975).

[7] J. M. Pollard, "The Fast Fourier Transform in a Finite Field," *Math. Comp.*, 25:365–374 (Apr. 1971).

[8] I. S. Reed, T. K. Truong, Y. S. Kwoh, and E. L. Hall, "Image Processing by Transforms over a Finite Field," *IEEE Trans. Comput.*, *C-26*(9):874–881 (Sept. 1977).

[9] C. M. Radar, "On the Application of the Number Theoretic Methods of High Speed Convolution to Two-Dimensional Filtering," *IEEE Trans. Circuits Syst.*, *CAS-22*(6):575 (June 1975).

[10] I. S. Reed, Y. S. Kwoh, T. K. Truong, and E. L. Hall, "X-Ray Reconstruction by Finite Field Transforms," *IEEE Trans. Nuclear Sci.*, *NS-24*(1):843–849 (Feb. 1977).

[11] I. S. Reed, T. K. Truong, and L. R. Welch, "The Fast Decoding of Reed-Solomon Codes Using Fermat Number Transforms," *IEEE Trans. Inf. Theory*, *IT-24*(4):497–499 (July 1978).

[12] H. F. A. Roots and M. R. Best, "Concatenated Coding on a Spacecraft-to-Ground Telemetry Channel Performance," *Process. ICC 81*, Denver, Colo., 1981.

[13] J. H. McClellan, "Hardware Realization of a Fermat Number Transform," *IEEE Trans. Acoust. Speech Signal Process.*, *ASSP-24*(3):216–225 (June 1976).

[14] L. M. Leibowitz, "A Simplified Binary Arithmetic for the Fermat Number Transform," *IEEE Trans. Acoust. Speech Signal Process.*, *ASSP-24*(5):356–359 (Oct. 1976).

[15] C. A. Mead and L. A. Conway, *Introduction to VLSI Systems*, Addison-Wesley, Reading, Mass., 1980.

21

Application of Systolic Array Technology to Recursive Filtering

RICHARD H. TRAVASSOS

Systolic Systems, Inc.
San Jose, California

21.1 INTRODUCTION

Estimating the frequency and angle of arrival of a point emitter can be viewed as an application of Kalman filtering. The signals received at the linear sensor array, shown in Figure 21.1, carry both angle-of-arrival and frequency information. The estimator in Figure 21.1 may be viewed as a space-time sampling system where the number of sensors, m, and the number of memory elements per sensor, n, determine the size of the two-dimensional window in the signal field.

Linear-prediction techniques have been applied to two-dimensional signals [1,2]. In these methods, least-squares prediction is performed on the observed data and the results are used to extrapolate the data beyond the observation interval. The two-dimensional fast Fourier transform is then applied to the extended data. Although these techniques yield high-resolution estimates of both frequency and angle of arrival, the estimates are biased when the observed data are corrupted by additive white noise.

In the one-dimensional case, Pisarenko [3] has proposed a harmonic retrieval method for detecting sinusoids in additive white noise that removes the bias problem while preserving high resolution. Pisarenko's method, however, is not computationally efficient and must be applied off-line. Thompson [4] derived an adaptive version of Pisarenko's method, but its convergence rate is slow. Reddy et al. [5] derived a least-squares-type adaptive version of Pisarenko's method that is similar to the γ-LMS algorithm proposed by Frost [6] for noise power cancellation of sinusoids in noise.

A performance comparison of the adaptive Pisarenko method and a recursive maximum-likelihood method was reported in [7]. The results indicated that the recursive maximum-likelihood method gives higher-resolution estimates.

375

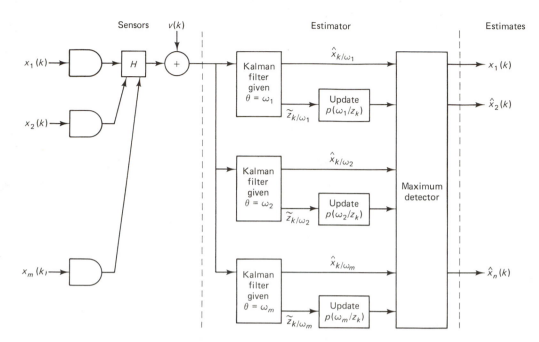

Figure 21.1   Maximum likelihood estimation of sinusoidal signals.

In this chapter systolic array concepts are used to develop efficient architectures for implementing the bank of Kalman filters needed by the recursive maximum-likelihood estimator.

21.2 PROBLEM FORMULATION

Consider the problem of estimating the frequency and angle of arrival of multiple sinusoids corrupted by wideband noise. Let $x_i(t_k)$ denote the kth time sample at the ith sensor in the linear array of Figure 21.1 and suppose that the sinusoidal signals are modeled by

$$x_i(t_k) = A_i \cos \left[\omega_i(t_k - i\gamma_i) + \phi_i \right] \qquad (21.1)$$

where $\omega_i = 2\pi f_i$, A_i, and ϕ_i are the frequency, amplitude, and phase of the sinusoid as received by the ith sensor. The wave number γ_i of the ith sinusoid is defined as

$$\omega_i \gamma_i = \omega_i \frac{d}{c} \sin \theta_i$$

where d is the spacing between the sensors, c the propagation velocity of the plane wave, and θ_i the angle of arrival with respect to the broadside direction. The phase angles, ϕ_i, of the sinusoids are assumed to be uniformly distributed over $[0, 2\pi]$. In the model $t_k = kT$ denotes the kth sampling time and T is the sampling interval.

Differentiating x_i with respect to time gives

$$\dot{x}_i = -A_i \omega_i \sin \left[\omega_i(t - i\gamma_i) + \phi_i\right] \tag{21.2}$$

and

$$\ddot{x}_i = -A_i \omega_i^2 \cos \left[\omega_i(t - i\gamma_i) + \phi_i\right] \tag{21.3}$$

Introducing the state variables

$$x_{1i} = A_i \cos \left[\omega_i(t - i\gamma_i) + \phi_i\right]$$

and

$$x_{2i} = A_1 \sin \left[\omega_i(t - i\gamma_i) + \phi_i\right]$$

then $\dot{x}_{1i} = -\omega_i x_{2i}$ and $\dot{x}_{2i} = \omega_i x_{1i}$. The problem can be formulated in state-variable form as follows:

$$
\begin{bmatrix} \dot{x}_{11} \\ \dot{x}_{21} \\ \dot{x}_{12} \\ \dot{x}_{22} \\ \vdots \\ \dot{x}_{1m} \\ \dot{x}_{2m} \end{bmatrix}
=
\begin{bmatrix}
0 & -\omega_1 & & & & & \\
\omega_1 & 0 & & & & & \\
& & 0 & -\omega_2 & & & \\
& & \omega_2 & 0 & & & \\
& & & & \ddots & & \\
& & & & & 0 & -\omega_m \\
& & & & & \omega_m & 0
\end{bmatrix}
\begin{bmatrix} x_{11} \\ x_{21} \\ x_{12} \\ x_{22} \\ \vdots \\ x_{1m} \\ x_{2m} \end{bmatrix}
\tag{21.4}
$$

or

$$\dot{x} = Fx \tag{21.5}$$

The observation model for the received signals may be written in the following state-variable form:

$$
y =
\begin{bmatrix}
h_{11} & h_{12} & \cdots & h_{1m} \\
h_{21} & h_{22} & \cdots & h_{2m} \\
\vdots & & & \\
h_{n1} & h_{n2} & \cdots & h_{nm}
\end{bmatrix}
\begin{bmatrix} x_{11} \\ x_{12} \\ \vdots \\ x_{1m} \\ x_{2m} \end{bmatrix}
+ v
\tag{21.6}
$$

or

$$y = Hx + v \tag{21.7}$$

where v is a measurement noise process. When sinusoidal signals are transmitted, the frequencies of the signals, ω_i, are usually held constant. Thus the m blocks in the signal model above may be uncoupled and processed simultaneously.

The Kalman filter for each block in Figure 21.1 is of the form

$$\hat{x}_{k+1} = \Phi \hat{x}_k + K[y_k - H\hat{x}_k] \tag{21.8}$$

or equivalently,

$$\hat{x}_{k+1} = \underbrace{[\Phi - KH \mid K]}_{\substack{\text{filter design}\\\text{matrix}}}\begin{bmatrix}\hat{x}_k\\ \hline y_k\end{bmatrix} \tag{21.9}$$

where Φ is the discrete-time state transition matrix associated with the F matrix in equation (21.4) and K is the discrete-time Kalman gain. In this formulation, the elements of the filter design matrix are constant. Φ and K, however, are dependent on the specific values of ω_i, $i = 1, 2, \ldots, m$. The structure of the filter indicates that it can be implemented as a single matrix-vector multiply. The elements of the filter design matrix can be stored in memory to cover the anticipated range of ω_i. The bank of Kalman filters shown in Figure 21.1 can be "tuned" by using the appropriate filter design matrix stored in memory for a given set of signals.

The conditional probabilities, $p(\omega_i \mid Z_k)$, can be computed using Bayes's rule as follows [8]:

$$p(\omega_i \mid Z_k) = \frac{p(Z_k \mid \omega_i)}{p(Z_k)}$$

$$= \frac{p(z_k \mid Z_{k-1}, \omega_i)p(\omega_i \mid Z_{k-1})}{\sum\limits_{i=1}^{m} p(z_k \mid Z_{k-1}, \omega_i)p(\omega_i \mid Z_{k-1})} \tag{21.10}$$

where Z_k denotes a sequence of measurements $z_0, z_1, \ldots, z_k$. The evaluation of $p(\omega_i \mid Z_k)$ is dependent on the distribution of the measurement noise. For example, for Gaussian noise $p(\omega_i \mid Z_k)$ can be calculated recursively from the following formula:

$$p(\omega_i \mid Z_k) = c \det (R^{-1})^{-1/2} \exp \left(\tfrac{1}{2}\tilde{z}_{k\mid\omega_i}^T R^{-1}\tilde{z}_{k\mid\omega_i}^T\right) \cdot p(\omega_i \mid Z_k) \tag{21.11}$$

where c is a normalizing constant, independent of ω_i, chosen to ensure that $\sum p(\omega_i \mid Z_k) = 1$. The innovations, $\tilde{z}_{k\mid\omega_i} = z_k - H\hat{x}_{k\mid\omega_i}$, and the covariance of the innovations, R, needed to evaluate $p(\omega_i \mid Z_k)$ can be obtained from the bank of Kalman filters. Note that once $p(\omega_i \mid Z_k)$, $i = 1, 2, \ldots, m$, have been evaluated, the value of ω_i that maximizes $p(\omega_i \mid Z_k)$ may be computed via successive comparisons to obtain

$$\hat{\omega}_{\mathrm{ML}} = \max_{\omega_i \in \Omega} p(\omega_i \mid Z_k) \tag{21.12}$$

The maximum-likelihood estimates of the sinusoidal signals are determined by evaluating any of the blocks in (21.4) with $\omega = \hat{\omega}_{\mathrm{ML}}$.

Since $p(\omega_i \mid Z_k)$ needs to be evaluated rapidly, $p(\omega_i \mid Z_k)$ may be stored in memory in a table-lookup fashion. Thus the key to implementing the overall scheme illustrated in Figure 21.1 is to develop a high-speed processor for evaluating the

Kalman filter equations. Due to the structure of the filter equations, systolic array architectures may be used to effectively speed up computations.

21.3 PARALLEL KALMAN FILTERING

The Kalman filter has been successfully applied to many signal processing applications, including target prediction, target tracking, radar signal processing, onboard calibration of inertial systems, and in-flight estimation of aircraft stability and control derivatives. The applicability of the Kalman filter to real-time processing problems is generally limited, however, by the filter's relative computational complexity. In particular, the number of arithmetic operations required for implementing the Kalman filter with n state variables grows as $O(n^2)$ for the time update and as $O(n^3)$ for the covariance update. In general, real-time filtering cannot be performed on large-scale problems using a uniprocessor architecture because serious processing lags can result.

The Kalman filter can be extended to a much greater class of problems by using parallel processing concepts. Full utilization of parallelism can be obtained through insight in the structure of the problem and decoupling of arithmetic processes to permit concurrent processing.

To date, relatively little research has been conducted on restructuring the Kalman filter for parallel processing. Three approaches that have been considered include: (1) *vectorizing* the standard Kalman filtering equations by running the filter on a vector (or array) processor [9], (2) uncorrelating the measurement data to the filter so that each measurement can be *pipelined* into each processor simultaneously [10], and (3) decoupling the predictor and corrector equations in the filter so that these computations can be evaluated simultaneously on separate processors using *multiprocessing* [11,12]. The first approach has been studied previously using an array processor to process the filter equations [8,13]. Although this approach can speed up computations considerably over conventional techniques (speed-up factors of 6 to 10 have been realized), the computational throughput was limited primarily by the architecture of the array processor. This occurred because the array processor architecture was optimized for fast Fourier transform (FFT) computations, not Kalman filter computations.

The architectures discussed in this chapter, however, are based on mapping the Kalman filter equations directly onto a linearly connected systolic array. Thus the systolic architectures discussed in this chapter exploit the structure of the filter to improve the overall throughput rate.

To speed up Kalman filter computations, parallel processing is performed at two levels: (1) the predictor and corrector equations of the Kalman filter are decoupled so that the predictor and corrector can be computed on separate processors, and (2) the measurement data are pipelined into each processor. Therefore, both multiprocessing and pipelining are considered to achieve large improvements in computational speed.

21.3.1 Decoupling the Time and Measurement Updates

For the sinusoidal signal model defined by (21.4) and (21.5) the standard Kalman filtering equations are given by

$$\text{Predictor} \quad \begin{cases} \hat{x}_k(-) = \Phi_{k-1}\hat{x}_{k-1}(+) & (21.13) \\ P_k(-) = \Phi_{k-1}P_k(+)\Phi_{k-1}^T & (21.14) \end{cases}$$

$$\text{Corrector} \quad \begin{cases} \hat{x}_k(+) = \hat{x}_k(-) + K_k[y_k - H_k\hat{x}_k(-)] & (21.15) \\ P_k(+) = [I - K_k H_k]P_k(-) & (21.16) \end{cases}$$

where the Kalman gain is

$$K_k = P_k(-)H_k^T[H_k P_k(-)H_k^T + R_k]^{-1} \qquad (21.17)$$

Note that (21.13) to (21.17) are inherently sequential since the temporal updates (predictor equations) must be evaluated before the observation updates (corrector equations). From a computational point of view, this is not desirable since to evaluate the corrector, a uniprocessor must wait until the predictor has been evaluated. To avoid this difficulty, the predictor-corrector equations can be decoupled to obtain a parallel Kalman filter (PKF).

Decoupling the state update. The decoupling of the state predictor and corrector is achieved by forcing the corrector to lag the predictor by one time step, as follows:

$$\text{Predictor:} \quad \hat{x}_{k+1}(-) = \Phi_k \Phi_{k-1}\hat{x}_{k-1}(+) \qquad (21.18)$$

$$\text{Corrector:} \quad \hat{x}_k(+) = \hat{x}_k(-) + K_k[y_k - H_k\hat{x}_k(-)] \qquad (21.19)$$

Decoupling the covariance update. Let the covariance of the estimation error before and after a measurement update be denoted by

$$P_{k+1}(-) = E\tilde{x}_{k+1}(-)\tilde{x}_{k+1}^T(-) \qquad (21.20)$$

$$P_k(+) = E\tilde{x}_k(+)\tilde{x}_k^T(+) \qquad (21.21)$$

where

$$\tilde{x}_{k+1}(-) \triangleq \hat{x}_{k+1}(-) - x_{k+1} \qquad (21.22)$$

$$\tilde{x}_k(+) \triangleq \hat{x}_k(+) - x_k \qquad (21.23)$$

By direct computation, it can be shown that the covariance of the estimation error before the update is given by [11,12]

$$P_{k+1}(-) = \Phi_k \Phi_{k-1}(+)\Phi_{k-1}^T\Phi_k^T \qquad (21.24)$$

Because the form of (21.19) is the same as (21.15), the covariance of the estimation error after a measurement update in the PKF is given by

$$P_k(+) = (I - K_k H_k)P_k(-) \qquad (21.25)$$

where

$$K_k = P_k(-)H_k^T[H_k P_k(-)H_k^T + R_k]^{-1} \qquad (21.26)$$

is the Kalman gain.

Summary of the decoupled PKF equations.   Because the predictor and corrector equations in the Kalman filter can be decoupled, computations can be performed simultaneously on two separate processors, one processor for the predictor equations and one processor for the corrector equations. In summary, the parallel Kalman filter (PKF) equations are

$$\text{Predictor} \begin{cases} \hat{x}_{k+1}(-) = \Phi_k \Phi_{k-1} \hat{x}_{k-1}(+) & (21.27) \\ P_{k+1}(-) = \Phi_k \Phi_{k-1} P_{k-1}(+) \Phi_{k-1}^T \Phi_k^T & (21.28) \end{cases}$$

$$\text{Corrector} \begin{cases} \hat{x}_k(+) = \hat{x}_k(-) + K_k[y_k - H_k \hat{x}_k(-)] & (21.29) \\ P_k(+) = [I - K_k H_k]P_k(-) & (21.30) \end{cases}$$

$$\begin{matrix} \text{Kalman} \\ \text{Gain} \end{matrix} \begin{cases} K_k = P_k(-)H_k[H_k P_k(-)H_k^T + R_k]^{-1} & (21.31) \end{cases}$$

The decoupling concept can be used to further partition the predictor and corrector equations for larger reductions in computation time. Note that when $\Phi_k = \Phi_{k-1}$ (e.g., Φ is constant) the computational complexity of the PKF is the same as in the standard formulation. Once the decoupling has been performed, systolic array architectures based on the parallel Kalman filter equations may be developed.

21.3.2 Systolic Kalman Filter Processor Architecture

The structure of the parallel Kalman filter algorithm described by (21.27) to (21.31) suggests that the equations can be implemented on the linearly connected systolic array initially reported by H. T. Kung [14]. The architecture shown in Figure 21.2 illustrates how a linearly connected array may be used to evaluate (21.29), for example, written in compact form:

$$\hat{x}_k(+) = [F_{11} \mid F_{12}] \left[\frac{\hat{x}_k(-)}{y_k} \right]$$

where $F_{11} = I - K_k H_k$ and $F_{12} = K_k$. To illustrate the computations, let f_{ij} denote the ij element of $[F_{11} \mid F_{12}]$ and s_j denote the jth element of $[x_k(-) \mid y_k]^T$. The filter design matrix coefficients, f_{ij}, are stored in memory for a given range of ω. The coefficients are passed to the inner product processors so that the state estimates in

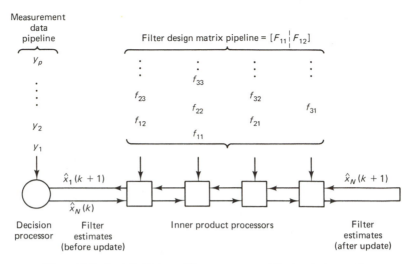

Figure 21.2   Systolic Kalman filter processor architecture ($n + m = 4$).

the systolic Kalman filter can be computed (see Figure 21.2). The following recurrence is computed by the processors to update the state estimates.

$$
\begin{aligned}
&\text{FOR} \quad i = 1 \text{ to } m \\
&\qquad \hat{x}_i^{(1)} = 0 \\
&\qquad \hat{x}_i^{(j+1)} = \hat{x}_i^{(j)} + f_{ij}s_j \qquad j = 1, 2, \ldots, n + m \\
&\qquad \hat{x}_i = \hat{x}_i^{n+m+1} \\
&\text{NEXT} \quad i
\end{aligned}
$$

Since the filter design matrix is $n \times (n + m)$, the $n + m$ linearly connected processors are needed to evaluate the recurrence above.

The data flow of the pipelining algorithm can be viewed as follows. The $\hat{x}_i$, which are initially zero, move to the left while the s_j are moving to the right and the f_{ij} are moving downward. All the data movements are synchronized. Each inner product step processor accumulates all its terms, namely, $f_{i, i-2}s_{i-2}$, $f_{i, i-1}s_{i-1}$, and $f_{i, i+1}s_{i+1}$ before the data leave the network. The decision processor passes either measurement data or state estimates to the inner-step processors according to the recursions above. The procedure is repeated until all the rows of the matrix $[F_{11} \vdots F_{12}]$ have been processed.

Note that in the systolic Kalman filter data flows throughout the processors to eliminate separate loading and unloading of data. Each processor pumps data in, performs a prespecified inner product computation, and moves data out in a regular fashion. In addition, note that the interconnection of the inner product processors is simple and that data movement is only between adjacent processors. This observation greatly simplifies the implementation of the systolic Kalman filter.

21.4 PROCESSOR DESIGN CONSIDERATIONS

The operational characteristics of the systolic Kalman filter architecture were evaluated via simulation. This aspect of the design cycle is particularly important since several operations are performed simultaneously in the systolic Kalman filter architecture. The simulation was carried out on a VAX 11/750 to address the following implementation issues: (1) finite word-length effects on the inner product processors, (2) stability of the Kalman filter equations, (3) sampling rate selection, (4) round-off error propagation, and (5) estimator sensitivity.

21.4.1 Stability Analysis

Consider the Kalman filter implementation equation

$$\hat{\mathbf{x}}_{k+1} = \boldsymbol{F}_{11}\hat{\mathbf{x}}_k + \boldsymbol{F}_{12}\mathbf{y}_k \tag{21.32}$$

where $\boldsymbol{F}_{11} = \boldsymbol{\Phi} - \boldsymbol{KH}$ and $\boldsymbol{F}_{12} = \boldsymbol{K}$. Taking the z transform of (21.32) it can be shown that the characteristic (21.32) is given by

$$\Delta(z) = \det\,(z\boldsymbol{I} - \boldsymbol{F}_{11}) \tag{21.33}$$

For the parallel Kalman filter computation to be stable, the roots of $\Delta(z)$ must lie within the unit circle [15]. Since the roots of $\Delta(z)$ are equal to the eigenvalues, λ_i, of $\boldsymbol{F}_{11}$, the parallel Kalman filter computations will be stable if and only if $|\lambda_i| \leq 1$.

Because of modeling uncertainties, it is important to examine the margin of stability associated with the Kalman filter computations. Figure 21.3 gives a geometric interpretation of a stability margin in the z plane. $|z|$ represents the magnitude of the roots of $\Delta(z)$ and $|\Delta z|$ is the distance from $|z|$ to the edge of the unit circle. For stability, it is necessary that $|z| + |\Delta z| \leq 1$. Hence $|\Delta z|$ can be viewed as a stability margin. The magnitude of $|\Delta z|$ is dependent on a number of factors, such

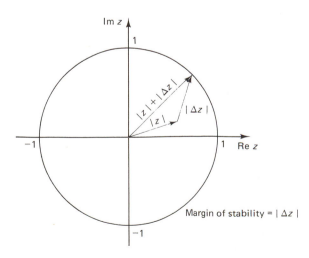

Figure 21.3 Stability in the z plane.

as modeling uncertainties and finite word-length effects. From an analysis viewpoint, it is convenient to lump these considerations into a perturbation of F_{11} and analyze the stability of the eigenvalues of $F_{11} + \Delta F_{11}$.

If F_{11} is perturbed to $F_{11} + \Delta F_{11}$, the corresponding change in the eigenvalues of F_{11} can be estimated from the following formula (Faddeeva [16]):

$$\Delta\lambda_i = \mathbf{v}_i^T \Delta F_{11} \mathbf{u}_i \qquad i = 1, 2, \ldots, n \qquad (21.34)$$

where the vectors $\mathbf{u}_i$ and $\mathbf{v}_i$ satisfy

$$F_{11}\mathbf{u}_i - \lambda_i \mathbf{u}_i = 0 \qquad |\mathbf{u}_i| = 1 \qquad (21.35)$$

$$F_{11}^T\mathbf{v}_i - \lambda_i \mathbf{v}_i = 0 \qquad \mathbf{v}_i^T \mathbf{u}_i = 1 \qquad (21.36)$$

If $\lambda_i + \Delta\lambda_i$ are the roots of $\Delta(z)$ when F_{11} is perturbed to $F_{11} + \Delta F_{11}$, then $|\lambda_i + \Delta\lambda_i| \leq |\lambda_i| + |\Delta\lambda_i|$ must lie within the unit circle for the Kalman filter computations to remain stable. Thus $|\Delta z| = \max_{1 \leq i \leq n} |\Delta\lambda_i|$ represents a stability margin for the Kalman filter computations.

21.4.2 Wordlength Considerations

The stability analysis of the preceding section can be used to specify the minimum number of bits required to ensure stability of the Kalman filter computations. To show this, let the perturbation in each component of $F_n \in R^{n \times n}$ be of the order 2^{-b}, where b is the number of bits. Hence $\|\Delta F_{11}\|_\infty \leq (n/2)2^{-b}$ since the elements of F_{11} are less than 1 due to the discretization process. Now by (21.34), the eigenvalues of the perturbed system are given by

$$\lambda_i + \Delta\lambda_i = \lambda_i + \mathbf{v}_i^T \Delta F_{11} \mathbf{u}_i \qquad (21.37)$$

For stability,

$$|\lambda_i + \mathbf{v}_i^T \Delta F_{11} \mathbf{u}_i| \leq |\lambda_i| + |\mathbf{v}_i^T \Delta F_{11} \mathbf{u}_i| \leq 1 \qquad (21.38)$$

Taking logarithms of both sides of (21.38) and rearranging terms give the desired relationship:

$$b = \lceil -(1 + \log_2((1 - \lambda_{max})/n)) \rceil \geq 0 \qquad (21.39)$$

where $\lambda_{max} = \max_{1 \leq i \leq n} |\lambda_i|$ and $\lceil (\cdot) \rceil$ is a ceiling function that denotes the largest integer nearest to but greater than $(\cdot)$.

The wordlength selection formula given by (21.39) represents an effective method for rapidly determining the minimum number of bits required to implement a Kalman filter on the linearly connected systolic array architecture. The magnitude of the maximum eigenvalue of F_{11} is the critical parameter which influences the margin of stability $= |1 - \lambda_{max}| = |\Delta\lambda| = |\Delta z|$.

Equation (21.39) was evaluated for different values of λ_{max} and n. The results are shown in Figure 21.4. Note that the number of bits required for stability in-

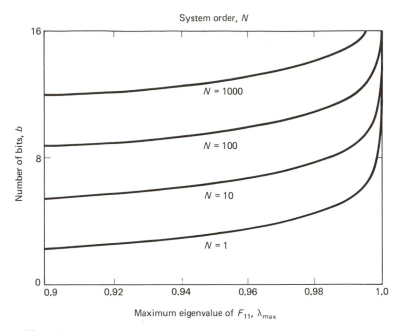

Figure 21.4   Minimum number of bits for stable Kalman filter computations.

creases substantially as $\lambda_{max} \to 1$ and as n increases. Because λ_{max} is directly related to the sampling rate $(1/\Delta t)$, Figure 21.4 indicates that at high sampling rates, a large number of bits are needed to keep the parallel Kalman filter computations stable. Since the improvement in stability is small beyond 16 bits, this analysis suggests that a 16-bit arithmetic processor may be used in the systolic Kalman filter architecture for low-frequency applications.

21.4.3 Sampling Rate Selection

The sampling rate may be selected by studying the sensitivity of the Kalman filter computations as a function of the sampling period. For example, referring to the Kalman filter (21.32), the discrete time $\boldsymbol{\Phi}$ and $\boldsymbol{K}$ depend explicitly on the sampling period, Δt. By expanding $\Phi = \exp{(F \, \Delta t)}$ and K in a Taylor series, Φ and K can be approximated as follows:

$$\boldsymbol{\Phi} \doteq \boldsymbol{I} + \boldsymbol{F} \, \Delta t + \tfrac{1}{2} \boldsymbol{F}^2 \, \Delta t^2 + O(\Delta t^3) \tag{21.40}$$

and

$$\boldsymbol{K} \doteq \tilde{\boldsymbol{K}} \, \Delta t + \boldsymbol{F} \tilde{\boldsymbol{K}} \, \Delta t^2 + O(\Delta t^3) \tag{21.41}$$

where $\boldsymbol{F}$ is defined in the continuous model given by (21.5) and $\tilde{\boldsymbol{K}}$ is the continuous

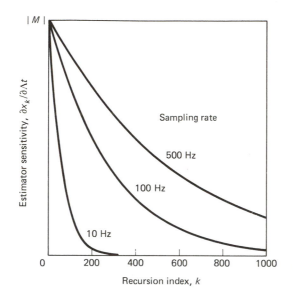

Figure 21.5 Estimator sensitivity parameterized by sampling rate.

Kalman gain associated with the continuous model. Substituting the approximations above for Φ and K into (32) gives

$$\hat{\mathbf{x}}_{k+1} = [\boldsymbol{\Phi} - \boldsymbol{KH}]\hat{\mathbf{x}}_k + \boldsymbol{Ky}_k$$

$$= [\boldsymbol{I} + (\boldsymbol{F} - \tilde{\boldsymbol{K}}H)\,\Delta t + \tfrac{1}{2}(\boldsymbol{F}^2 - 2F\tilde{\boldsymbol{K}}H)t^2]\mathbf{x}_k$$

$$+ (\tilde{\boldsymbol{K}}\,\Delta t + F\tilde{\boldsymbol{K}}\,\Delta t^2)\mathbf{y}_k + O(\Delta t^3) \qquad (21.42)$$

The sensitivity of the signal estimate, $\hat{\mathbf{x}}_k$, to changes in the sampling period, Δt, is

$$\frac{\partial \mathbf{x}_{k+1}}{\partial\,\Delta t} = \boldsymbol{M}\hat{\mathbf{x}}_k \qquad (21.43)$$

where

$$\boldsymbol{M} = (\boldsymbol{F} - \tilde{\boldsymbol{K}}H) + (\boldsymbol{F}^2 - 2F\tilde{\boldsymbol{K}}H)\,\Delta t \qquad (21.44)$$

The magnitude of the signal estimator sensitivity to changes in the sampling period is illustrated in Figure 21.5 for a typical Kalman filter design with sampling rates well above the Nyquist rate. Figure 21.5 indicates that at high sampling rates $\hat{\mathbf{x}}_k$ is more sensitive to changes in the sampling period than at low sampling rates. This follows because the eigenvalues of the Kalman filter move closer to the unit circle, and hence the filter is more susceptible to round-off error and instability.

21.4.4 Analysis of Round-off Error Propagation

In addition to the sensitivity analysis and stability analysis previously discussed, the propagation of errors due to round-off in the Kalman filter was analyzed. The propagation of errors due to round-off was studied by examining the fundamental processor equations:

$$\hat{\mathbf{x}}_{k+1} + \Delta\hat{\mathbf{x}}_{k+1} = (\boldsymbol{F}_{11} + \Delta\boldsymbol{F}_{11})(\hat{\mathbf{x}}_k + \Delta\hat{\mathbf{x}}_k) + (\boldsymbol{F}_{12} + \Delta\boldsymbol{F}_{12})(\mathbf{y}_k + \Delta\mathbf{y}_k) \qquad (21.45)$$

ΔF_{11} and ΔF_{12} result from representing F_{11} and F_{12} in a finite-word-size data memory. Δy represents a quantization error due to sampling. Keeping only the significant terms in (21.45) gives

$$\hat{\mathbf{x}}_{k+1} + \Delta\hat{\mathbf{x}}_{k+1} = \boldsymbol{F}_{11}\hat{\mathbf{x}}_k + \boldsymbol{F}_{12}\, y_k + \boldsymbol{\varepsilon}_k \qquad (21.46)$$

where

$$\boldsymbol{\varepsilon}_k = \boldsymbol{F}_{11}\,\Delta\hat{\mathbf{x}}_k + \Delta\boldsymbol{F}_{11}\hat{\mathbf{x}}_k + \Delta\boldsymbol{F}_{12}\,y_k + \Delta\boldsymbol{F}_{12}\,\mathbf{y}_k \qquad (21.47)$$

Now suppose that the round-off error ε_k is zero mean and is uniformly distributed between $\pm\varepsilon_{\max}$, where $|\varepsilon_k| \le \varepsilon_{\max}$. Then it can be shown that the error propagation, $\mathbf{e}_k = \hat{\mathbf{x}}_k - \mathbf{x}_k$, is [12]

$$\mathbf{e}_{k+1} = \boldsymbol{F}_{11}\mathbf{e}_k + \boldsymbol{\varepsilon}_k, \qquad e_0 = 0 \qquad (21.48)$$

The covariance of the errors can be determined from the following equations [12]:

$$\boldsymbol{P}_{k+1} = \boldsymbol{F}_{11}\boldsymbol{P}_k\boldsymbol{F}_{11}^T + \boldsymbol{R}_k \qquad (21.49)$$

where $\boldsymbol{R}_k = E\boldsymbol{\varepsilon}_k\boldsymbol{\varepsilon}_k^T$. By appropriately scaling $\boldsymbol{F}_{11}$ to be stable in the Kalman filter, round-off error effects can be minimized.

21.5 CONCLUDING REMARKS

A systolic Kalman filter processor architecture was developed based on mapping the Kalman filter recursions directly onto a linearly connected systolic array. An analysis of the recursive equations was performed to specify analytically the word length, sampling rate, and memory requirements of the processor cells. The analysis showed that 16-bit processor cells are adequate to ensure stability of the recursive equations for low-frequency applications provided that the Kalman gains have been precomputed off-line. Although the results in this paper were derived for the Kalman filter recursions, the results are generally applicable to all linear recursive update procedures.

ACKNOWLEDGMENT

This work was supported in part by the Office of Naval Research under Contract N00014-81-K-0191 and the U.S. Air Force Flight Dynamics Laboratory under Contract F33615-82-C-3604 while the author was with Integrated Systems, Inc., Palo Alto, California.

REFERENCES

[1] O. L. Frost and T. M. Sullivan, "High-Resolution Two-Dimensional Spectral Analysis," *Proc. 1979 IEEE Int. Conf. Acoust. Speech Signal Process.*, Apr. 1979, pp. 673–676.

[2] O. L. Frost, "High-Resolution 2-D Spectral Analysis at Low SNR," *Proc. 1980 IEEE Int. Conf. Acoust. Speech Signal Process.*, Apr. 1980, pp. 580–583.

[3] V. F. Pisarenko, "The Retrieval of Harmonics from Covariance Functions," *Feof. J. R. Astron. Soc.*, 1973, pp. 347–366.

[4] P. A. Thompson, "An Adaptive Spectral Analysis Technique for Unbiased Frequency Estimation in the Presence of White Noise," *Proc. 13th Asilomar Conf. Circuits, Syst. Comp.*, New Mexico, 1979, pp. 529–533.

[5] V. U. Reddy, B. Egardt, and T. Kailath, "Least-Squares Type Algorithm for Adaptive Implementation of Pisarenko's Harmonic Retrieval Method," *IEEE Trans. Acoust. Speech Signal Process.*, June 1982, pp. 399–405.

[6] O. L. Frost, "Resolution Improvement in AR Spectral Analysis by Noise Power Cancellation," *EASCON*, 1977.

[7] V. U. Reddy, R. H. Travassos, and T. Kailath, "A Comparison of Nonlinear Spectral Estimation Techniques," ISI Tech. Memo 5016-05, Jan. 1982.

[8] B. D. O. Anderson and J. B. Moore, *Optimal Filtering*, Prentice-Hall, Englewood Cliffs, N.J., 1979, pp. 267–274.

[9] R. S. Bucy and K. D. Senne, "Nonlinear Filtering Algorithms for Vector Processing Machines," *Compl. Math. Appl.*, 6(3):317–338 (Mar. 1980).

[10] A. Andrews, "Parallel Processing of the Kalman Filter," *Int. Conf. Parallel Process.*, Aug. 1981, pp. 216–220.

[11] R. H. Travassos, "Parallel Kalman Filtering," ISI-04 Tech. Rep., Oct. 1981.

[12] R. H. Travassos and A. Andrews, "VLSI Implementation of Parallel Kalman Filters," *AIAA Guid. Cont. Conf.*, Advanced Avionics Session, San Diego, Aug. 1982.

[13] E. C. Dudzinski, "Software Optimization of a Kalman Filter for an AP-120B Array Processor," *Natl. Aerosp. Electron. Conf.*, May 1982, pp. 221–227.

[14] H. T. Kung, "Communication and Concurrency in Conventional Computers," in C. Mead and L. Conway, *Introduction to VLSI Design*, Addison-Wesley, Reading, Mass., 1980, pp. 264–270.

[15] B. C. Kuo, *Discrete-Data Control Systems*, Prentice-Hall, Englewood Cliffs, N.J., 1980.

[16] P. K. Faddeeva and V. N. Faddeeva, *Computational Methods of Linear Algebra*, Freeman, San Francisco, 1963, p. 288.

22

Systolic Linear Algebra Machines in Digital Signal Processing

ROBERT SCHREIBER†
PHILIP J. KUEKES

ESL, Incorporated
Sunnyvale, California

22.1 INTRODUCTION

Several recent contributions to the literature in signal processing, computer architecture, and VLSI design showed that systolic arrays are extremely useful for designing special-purpose high-performance devices to solve problems in numerical linear algebra [1–5]. But no attention has been paid to the problem of integrating these designs into any computing or signal processing environment. The purpose of this chapter is to examine systolic arrays in a specific context. We have chosen an adaptive beamforming problem as that context.

The adaptive beamforming problem chosen is simple, yet typical of those encountered in sonar signal processing. The signals of a large array of *m identical* sensors are sampled, stored, and Fourier transformed in time. The result is a set $x(\omega, i)$ of complex values depending on frequency (ω) and sensor (i). Then, for each resulting frequency ω, an array output function $g(\omega, \vartheta)$, depending on a bearing angle ϑ, is to be produced by

$$g(\omega, \vartheta) = \sum_{i=1}^{m} x(\omega, i)\bar{w}(i, \omega, \vartheta)$$

where the overbar denotes complex conjugate. The vector **w** determines the characteristics of the beamformer. For the minimum-variance distortionless response beamformer, **w** is chosen to minimize the output power, the expected value of $|g|^2$, subject to a signal-protection constraint

$$\sum_{i=1}^{m} c(i, \omega, \vartheta)\bar{w}(i, \omega, \vartheta) = 1$$

†Permanent address: Department of Computer Science, Stanford University, Stanford, CA 94305.

where $c(i, \omega, \vartheta)$ is the output of sensor i at frequency ω given no signal other than that due to a source at bearing angle ϑ.

The solution is to choose the weight vector

$$\mathbf{w}(\omega, \vartheta) \equiv (w(1, \omega, \vartheta), \ldots, w(m, \omega, \vartheta))^T$$

as

$$\mathbf{w} = \frac{R_{xx}^{-1} \mathbf{c}}{\mathbf{c}^* R_{xx}^{-1} \mathbf{c}}$$

where $\mathbf{c}$ is the desired signal vector and R_{xx} is the covariance matrix of the signal at frequency ω:

$$R_{xx}(\omega) = E\{\mathbf{x}(\omega)\mathbf{x}^*(\omega)\}$$

In practice, for every interesting frequency, an $n \times m$ matrix of samples of the signal

$$X^*(\omega) \equiv \begin{bmatrix} {}^1\mathbf{x}^*(\omega) \\ {}^2\mathbf{x}^*(\omega) \\ \vdots \\ {}^n\mathbf{x}^*(\omega) \end{bmatrix}$$

would be obtained, and R estimated by

$$R \approx XX^*$$

(Here $*$ denotes the conjugate transpose.) Possibly, different weights would be given to the rows of $X^*(\omega)$.

The following algorithm gives the solution.

1. Factor

$$X^* = QU' \tag{22.1}$$

   where Q is an $n \times n$ unitary matrix, and U' is the $n \times m$ matrix

$$U' \equiv \begin{bmatrix} U \\ 0 \end{bmatrix}$$

2. For each bearing angle ϑ,
   a. Forward solve:

$$U^*\mathbf{a}(\vartheta) = \mathbf{c}(\vartheta) \tag{22.2a}$$

   b. Back solve:

$$U\mathbf{w}(\vartheta) = \mathbf{a}(\vartheta)(\mathbf{a}^*\mathbf{a})^{-1} \tag{22.2b}$$

[Here $\mathbf{a} = (U^*)^{-1}\mathbf{c}$. Therefore, $\mathbf{a}^*\mathbf{a} = \mathbf{c}^*U^{-1}(U^*)^{-1}\mathbf{c} = \mathbf{c}^*R_{xx}^{-1}\mathbf{c}$.]

For the remainder of this chapter we shall concentrate on the design of an adaptive weight-selection processor that performs the two major steps of the algorithm. Two systolic arrays will be used. One, a variant of the design of Gentleman and Kung for QU factorizations [2], performs step 1. The second does the triangular solves of step 2 and is new.

22.2 THE QU-FACTORIZATION PROCESSOR

This section is an extension of previous results of Gentleman and Kung in several directions: complex matrices, unloading the result U from the array, control and synchronization, fabrication of the cells, and simulation of a large array by a physically smaller array through decomposition of the problem.

QU factorization of A is performed by finding an orthogonal matrix, Q^*, such that $Q^*A = U$ is upper triangular. Q^* can be a product of simple orthogonal matrices (Givens rotations) each chosen to zero one element of A below the diagonal. To zero a_{ij}, the ith and $(i-1)$th rows of A are replaced by

$$\begin{pmatrix} a_{i-1,k} \\ a_{i,k} \end{pmatrix} = \begin{pmatrix} .c & s \\ -s & c \end{pmatrix} \begin{pmatrix} a_{i-1,k} \\ a_{i,k} \end{pmatrix} \qquad k = 1, 2, \ldots, m$$

where (c, s) are chosen so that $a_{i,j}$ becomes zero and the matrix shown is orthogonal:

$$c = \frac{a_{i-1,j}}{\sqrt{a_{i-j,j}^2 + a_{i,j}^2}}$$

$$s = \frac{a_{i,j}}{\sqrt{a_{i-1,j}^2 + a_{i,j}^2}}$$

The elements can be zeroed column by column from the bottom up; elements are zeroed in the sequence

$$(n, 1), (n-1, 1), \ldots, (2, 1); (n, 2), \ldots, (3, 2); \ldots; (n, m), \ldots, (m+1, m)$$

Note that the rotation zeroing $a_{i,j}$ needs to be applied only to columns $j, j+1, \ldots, m$ since, at the time it is applied, the elements $a_{i-j,k}$ and $a_{i,k}$ for $k < j$ are already zero.

The Gentleman–Kung (GK) array, an $m \times m$ triangular array of two cell types that computes the QU factorization (22.1) of an $n \times m$ input matrix A, is shown in Figure 22.1. We call the circles "boundary" cells. Figure 22.2 shows the cell's functions. We shall now explain the GK array's operation. First, note that the cells each have a single memory. These initially are zero. The diagonal boundary cells compute rotation parameters (c, s). Cell (j, j) computes the rotations that zero elements of column j of A. These rotations then move right, along the rows of the array. The square "internal" cells apply these rotations to the other columns.

The matrix A enters the array at the top in the pattern shown in Figure 22.1. To the upper left of each cell is the time that the first element of A arrives. Suppose that $a_{n,1}$ enters cell $(1, 1)$ at time $t = 1$. The first element of U, $u_{1,1}$ is computed in

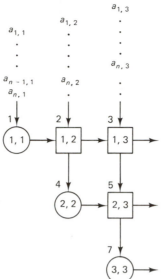

Figure 22.1 Gentleman–Kung array for QU.

cell $(1, 1)$ at time $t = n$. By time $t = n + 2(m - 1)$, the last element, $u_{m, m}$, has been computed. U now resides in the array. The rotations defining Q^* will have emerged from the right edge.

22.2.1 The Trapezoidal Subarray

We would like to solve beamforming problems of various sizes using one physical array, so we must consider how to simulate a full $m \times m$ array using a smaller piece. Suppose that we have a $p \times q$ trapezoid, as shown in Figure 22.3. Let B be a matrix having n rows and q columns. If B is presented at the array top, we compute the factorization

$$Q^*B = \begin{bmatrix} U_{11} & U_{12} \\ 0 & B_2 \end{bmatrix} \tag{22.3}$$

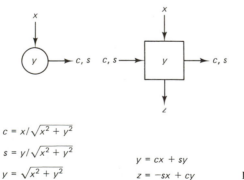

$c = x/\sqrt{x^2 + y^2}$

$s = y/\sqrt{x^2 + y^2}$

$y = \sqrt{x^2 + y^2}$

$y = cx + sy$

$z = -sx + cy$

Figure 22.2 Cells in the QU array.

$p = 2$

$q = 4$

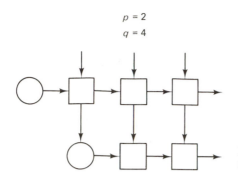

Figure 22.3   Tiling with subarrays: trapezoidal subarray.

where $\mathbf{0}$ denotes the zero matrix order $(n - p) \times p$, $\boldsymbol{U}_{11}$ is $p \times p$ upper triangular, and $\boldsymbol{U}_{12}$ and $\boldsymbol{B}_2$ are full $(q - p)$-column matrices. $\boldsymbol{B}_2$ emerges from the bottom of the array; $\boldsymbol{U}_{11}$ and $\boldsymbol{U}_{12}$ are stored in it.

Can we obtain a complete $\boldsymbol{QU}$ factorization with the trapezoid? If q is not less than m, the number of columns of $\boldsymbol{A}$, we can; we would zero $\boldsymbol{A}$ in groups of p columns, from left to right, using $\lceil m/p \rceil$ passes through the array. The details are obvious. But if $m > q$, we cannot. For suppose that we zero the first p columns of $\boldsymbol{A}$ below the diagonal by passing columns $1, 2, \ldots, q$ through the array. We want next to zero columns $p + 1, \ldots, 2p$, by passing $p + 1, \ldots, p + q$ through. But until we have applied the rotations from the first pass to columns $q + 1, \ldots, q + p$, we may not apply those of the second pass.

To allow factorization of matrices with more than q columns, the array must provide a second function: the ability to apply previously computed (and stored) rotations to a set of input columns. We can give the array this ability by turning off the cells in the leftmost $p \times p$ triangular part, keeping a $p \times (q - p)$ rectangle active. We also allow rotation parameters to come in via the left edge. A set of $q - p$ columns can enter the active rectangle at the top. The result—the input rotations applied to the input columns—emerges from the bottom, except for the first p rows, which are stored in the cells of the active rectangle.

22.2.2 Simulating the Full Array

Here we show how the $p \times q$ trapezoidal array, supported by an appropriate memory system for partial results, can be controlled to generate a $\boldsymbol{QU}$ factorization when $m > q$. Imagine that the set of work to be done is represented by an $m \times m$ triangular array of pairs,

$$(i, j): \quad 1 \le i \le j \le m$$

where the pair (i, j) represents the task of applying to column j the rotations used to zero elements of column i. The array can be used to perform "generate" passes, where columns are actually zeroed, and "apply" passes where stored rotations are applied. A generate pass performs a $p \times q$ trapezoidal piece of the set of task pairs; an apply pass performs a $p \times (q - p)$ rectangular piece. Sequencing the passes to

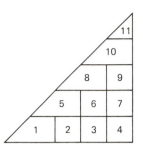

Figure 22.4   Tiling with subarrays:
schedule of passes for QU factorization.

perform the entire job is analogous to covering a triangle by trapezoidal and rectangular "tiles" following these rules:

1. Trapezoidal tiles must be placed at the triangle's diagonal edge.
2. No tile may be placed unless the diagonal edge to its left has been tiled.
3. No tile may be placed if any space directly below it is untiled.

There are many legal tilings; Figure 22.4 shows one.

In generating the QU factorization by multiple passes, temporary results are produced. These must be stored and reentered into the array later. Figure 22.5 shows a suitable memory design. The important features are these. There is a

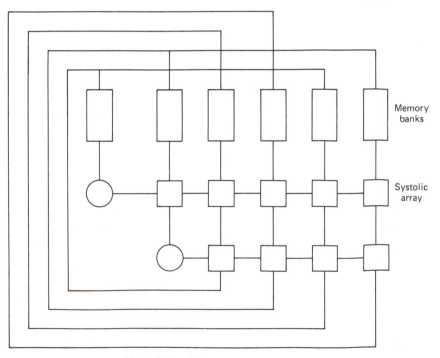

Figure 22.5 QU memory support system.

separate, independent memory bank for each array column. A matrix column is stored in the bank above the cell it first enters. A column (of temporary results) that emerges from the bottom is sent to the bank above the cell it will next enter. When the passes are sequenced as in Figure 22.4, this destination is for some columns uniquely determined, and for others is one of two possibilities. Thus, the extra-array interconnections are very simple.

Control of the memories is also simple, since the pattern of access to the data, (Figure 22.7) is so regular. Memory addresses could be generated once and passed from one bank to the next.

22.2.3 Alternative Scheme

There is another possibility. We could also work with tiles of these shapes:

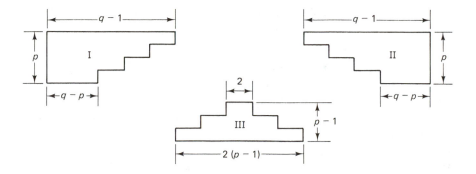

The array implements tile II by using all internal cells actively, tile I in the same way, but with columns brought in reverse order, and tile III by shutting off certain internal cells. Tile III can be realized only if it fits into the array—if $2(p - 1) \le q - p$.

Let use compare the two possibilities. The first has an average tile size of $p(q - p)$, while the second has $2p(q - 1)/3$. The tiles are used to cover rectangles like this:

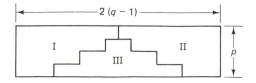

Thus the first scheme is more efficient if $p(q - p) > 2p(q - 1)/3$, that is, if $q > 3p - 2$. Since the second scheme *requires* that q be at least $3p - 2$, it can never be more efficient than the first.

22.2.4 Choosing the Array Shape

We suppose that some constraint, cost for example, limits the size, $pq - p(p-1)/2$, of the array. What is the best shape possible? Two conflicting factors influence the decision. The I/O bandwidth to support the array is least for a "well-rounded" array ($p \approx q$) since only the cells at the array edge communicate with the surrounding systems. But the number of passes needed to solve the given problem will usually not be minimized by taking $p = q$. In typical cases [say $m = 100$, $pq - p(p-1)/2 = 31$] the minimizing shape may be quite narrow ($p = 2$, $q = 16$ in this example).

22.2.5 Unloading the Cell Memories

The QU trapezoid implements the matrix factorization (22.3). But how can we remove the elements of U_{11} and U_{12}, which are stored in the cells of the array? Here we shall develop a scheme with these properties:

1. The outward flow of data is entirely uniform.
2. Control signals are applied only at the boundary cells.
3. No specially controlled functions are required of the internal cells.
4. The array can finish the factorization of a matrix, unload its cell memories, and begin the factorization of another matrix with no delay whatever.

    The key to the unloading scheme is the way an internal cell behaves when given the "identity" rotation ($c = 1$, $s = 0$). In this case it acts as a unit delay:

    To begin, suppose that the input matrix A is followed by a matrix B. The data will be presented to the array in this format:

$$
\begin{array}{cccccc}
 & & & & b_{nm} & \\
 & & & \ddots & 0 & \\
 & & b_{n2} & \ddots & a_{1m} & \\
 & b_{n1} & 0 & \ddots & & \\
 & 0 & a_{12} & & & \\
 & a_{11} & a_{22} & \cdots & a_{mm} &
\end{array}
\tag{22.4}
$$

so the two matrices are separated by a line of zeros. The scheme will, in effect, push out U as it is created and fill each cell with a zero just before a B element reaches it. It therefore "looks" to the new matrix B as if the array initially contained only zeros.

When a B element first arrives at a boundary cell it meets a zero (we show this later). The boundary cell's normal function (see Figure 22.2) is to store this element's absolute value in its memory and output the identity rotation, if the element was nonnegative, or -1 times the identity rotation otherwise. As this rotation moves to the right it meets cells containing zeros in their memories and pushes these zeros out, loading instead elements (possibly negated) of row n of B. Thus the zeros continue to lead the columns of B down through the array.

We now show how elements of U are unloaded. Let time $t = 0$ be the time $a_{1,1}$ enters cell $(1, 1)$. Then cell (i, j) accepts its last A element, and computes its U element, thereby finishing its work, at time $t = i + j - 2$. By our assumed sequence of inputs [equation (22.4)] the datum zero appears at the input to cell $(1, j)$ at time j, immediately after it has computed $u_{i, j}$. To make the scheme work, we want an identity rotation to get there at the same time, knocking out the computed element $u_{1, j}$ and loading the zero. This will be made to happen by a special boundary cell function.

Let cell $(1, 1)$ do this at time 1:

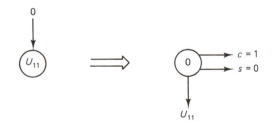

The rotation so generated will reach cell $(1, j)$ at time j, as required. Now, at time 3, let cell $(2, 2)$ do the same thing:

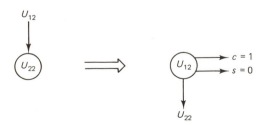

(The datum $u_{1,2}$, has been forced out of the first row, as described above.) This rotation will proceed to knock elements $u_{2,k}$ out of cells $(2, k)$ and load $u_{1,k}$ in, $k = 3, 4, \ldots, q$. At time 4, this special function is required at cell $(2, 2)$ again:

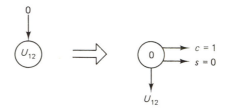

This rotation follows the earlier one down the row, removing elements $u_{1,k}$ and loading zeros. In general, cell (k, k) performs the special function k times, at $t = 2k - 1, \ldots, 3k - 2$. Then k identity rotations flow down row k, pushing out rows $k, k - 1, \ldots, 1$ of U, and finally loading the zero that precedes the next matrix.

   To make the array output uniform, we would actually add some cells at the lower left to make the array a rectangle:

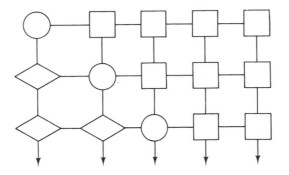

The only purpose of these diamond-shaped cells is to delay final output of elements of U, which now leave the array in the same format as elements of A—element i, j leaves at relative time $m - i + j$.

22.3 THE BACK-SOLVE ARRAY

To solve the triangular system $U^T Y = B$ (Y and B are $m \times n$) we can use a triangular array shown in Figure 22.6. The details of this array are quite straightforward and are omitted. We note that it consists of a triangular array of cells each containing a single element of U exactly as does the GK array, so that it might, in some applications, be useful to build a common realization of both these arrays.

   In the present context, n is often quite large. We intend to solve two systems, $U^T Y = B$, then $UX = Y$; we are not otherwise interested in elements of Y. If X is stored, we can store Y in its place. Suppose, however, that X will not be stored. We

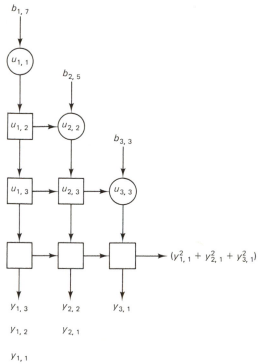

Figure 22.6 $U^T Y = B$ array.

may want to minimize temporary storage for Y. This can be reduced to $O(m^2)$ locations in two ways. We could use a second array to solve $UX = Y$ and stream the first array's output into this second array. An interface of $3m(m-1)/2$ delay cells is needed, as Figure 22.7 shows. Another possibility is to use one array to solve both

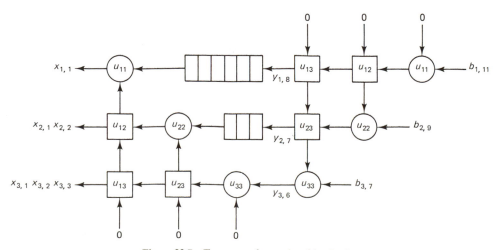

Figure 22.7 Two-array forward and backsolve.

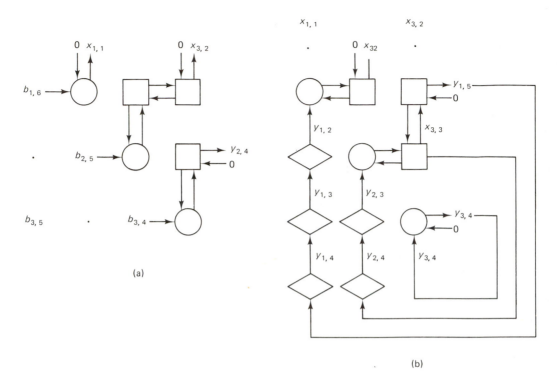

Figure 22.8 Single-array for forward and back solve: (a) cycle 11; (b) cycle 12.

systems at the same time. Figure 22.8 shows two successive cycles of such a device. At a given instant, every second diagonal is working on one problem, the other diagonals on the other problem. An array of $(3m^2 - 2m)/4$ delay cells is needed in this case, about half as many as in the two-array design.

22.3.1 Computing the Scale Factors

The computation (22.2b) requires the scale factor $\mathbf{a}^*(\vartheta)\mathbf{a}(\vartheta)$, that is, the dot product of the solution of the system (22.2a) with itself. Since the solution vectors $\mathbf{a}$ are produced one element per cycle by successive array columns, we can accumulate these dot products by attaching a row of cells at the bottom of the array; see Figure 22.6, for example.

22.4 COMPLEX QU FACTORIZATION

In fact, we must deal with complex matrices A. Of course, one can solve a complex $n \times m$ least-squares problem by means of a real QU factorization of the $2n \times 2m$

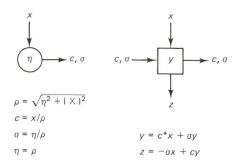

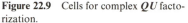

$$\rho = \sqrt{\eta^2 + |X|^2}$$
$$c = x/\rho$$
$$\sigma = \eta/\rho \qquad\qquad\qquad y = c^*x + \sigma y$$
$$\eta = \rho \qquad\qquad\qquad\qquad z = -\sigma x + cy$$

Figure 22.9 Cells for complex QU factorization.

matrix

$$\begin{bmatrix} A_R & -A_I \\ A_I & A_R \end{bmatrix}$$

where $A = A_R + iA_I$ is the decomposition of A into its real and imaginary parts. But when Givens rotations are used, this requires $\frac{8}{3}$ times as many real multiplications as a direct complex QU factorization of A.

The QU factorization is unique only up to scaling of the rows of U by factors of unit modulus: $A = (QD)(D^{-1}U)$ is also a QU factorization for any unitary diagonal matrix D. Thus we may require that the diagonal elements of U be positive real. This is the (unique) factorization computed by our array.

Let us change the cell definitions of Figure 22.1 to those of Figure 22.9. We shall now employ this convention: lowercase Roman letters are complex, Greek are real, and, for example,

$$a = \alpha + j\alpha', \qquad x = \xi + j\xi'$$

where $j = \sqrt{-1}$. These cells are, of course, more difficult to implement than the real Givens cells.

With this there is little else that changes. Now, when a leading element of an input matrix hits a boundary cell, the effect is this:

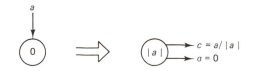

Thus, instead of the identity produced in the real case, a unitary diagonal rotation that simply rescales a row is produced.

22.5 VLSI IMPLEMENTATION

In this section we discuss how the internal cells of the real and complex QU arrays and the complex back-solve array might be fabricated using a systolic internal chip (SIC) that is being developed at ESL. What is particularly noteworthy is that these different cells can be obtained using a common VLSI building block, without the use of additional chips. Moreover, we have used the SIC to design other compound cells (for real and complex *LU* factorization of dense and band matrices and for band-*QU* factorization). We expect that systolic arrays for most of the standard algorithms of numerical linear algebra can be generated using this chip and one other that we will also mention.

First we give a rough description of the SIC. It is being designed in a TRW 2μ-CMOS technology. Its function is this:

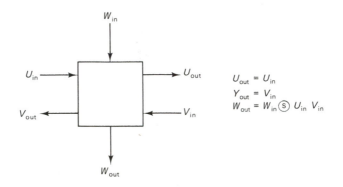

$$U_{out} = U_{in}$$
$$Y_{out} = V_{in}$$
$$W_{out} = W_{in} \, Ⓢ \, U_{in} \, V_{in}$$

where $s \in \{+, -, +-,$ nothing over $+\}$ denotes the sign used in the addition. This can be $+$, $-$, or can alternate between the two. Operands are floating point. Substantial effort has gone into providing switching functions and internal registers to increase the SIC's flexibility.

Compound cells for complex arithmetic can be built. Here are two of Kung's designs: a complex $c + ab$ cell using two SICs:

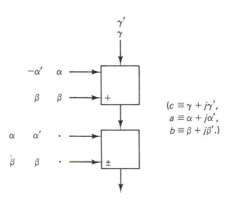

$(c \equiv \gamma + j\gamma',$
$a \equiv \alpha + j\alpha',$
$b \equiv \beta + j\beta'.)$

and another, using four chips, in which one operand is stationary:

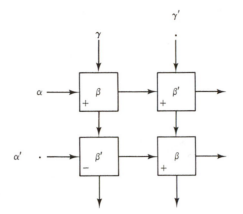

This is the internal cell for the complex back-solve array.

Figure 22.10 gives the cell layout for a complex Givens cell that uses six SICs.

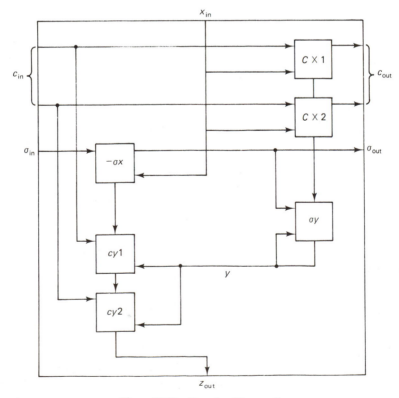

Figure 22.10   Complex Givens cell.

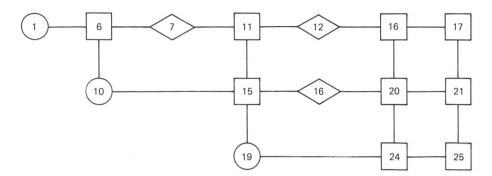

Figure 22.11   Effect of boundary cell latency.

There are two two-chip complex multiply-add cells, one for $c*x$ and another for cy, and two one-chip cells for real*complex multiply-add, one for σx and another for σy. The complex quantities are represented using one of the formats discussed above. New operands can enter the cell every second clock. There is a three-clock delay on the x-z path, but only a one-clock delay on the c_{in}-c_{out} and σ_{in}-σ_{out} paths.

We lack the space to discuss fully the boundary cell's implementation. Another chip is necessary. A chip using either faster logic or more internal parallelism could provide divide, square root, and reciprocal square root in one SIC cycle. This second chip could, with the SIC, be used to design a pipelined boundary cell for QU, back-solve, or LU operations with sufficient throughput to keep up with the array. The boundary cells will usually have more latency than the internal cells.

For a trapezoidal QU array, this extra boundary cell latency throws the array out of synchronization unless appropriate intercell delays are provided. Fortunately, these delays are needed only in the left-hand triangular part of the array. Figure 22.11 gives an example. Here the boundary cell's latency is 5 and the internal cell's is 4, although its left-right latency is just 1. In each cell the time an operand first arrives is shown. Delays are shown as diamonds; they delay the data four cycles. Their effect is to equalize the latency of alternate paths through the array. In a $p \times q$ array, $p(p - 1)/2$ delay cells would be required. These delays have an effect on the overall latency of the array and on the pattern of input and output (it is not a parallelogram any more). But there is no reduction in throughput.

ACKNOWLEDGMENTS

We thank Dr. Theo Kooij of DARPA for encouraging this work and suggesting and explaining to us the beamforming problem. We also thank our colleagues at ESL, Geoffrey Frank, Dragan Milojkovic, and Larry Ruane, for their helpful comments.

REFERENCES

[1] A. Bojanczyk, R. P. Brent, and H. T. Kung, "Numerically Stable Solution of Dense Systems of Linear Equations Using Mesh-connected Processors," *Siam Jour. on Scientific and Statistical Computing*, 5:1, pp. 95–104, 1984.

[2] W. M. Gentleman and H. T. Kung, "Matrix Triangularization by Systolic Arrays," *Real Time Signal Processing IV*, SPIE, Vol. 298, Society of Photo-optical Instrumentation Engineers, Bellingham, Wash., 1981.

[3] D. E. Heller and I. C. F. Ipsen, "Systolic Networks for Orthogonal Decompositions." *Siam Jour. on Scientific and Statistical Computing*, 4:2, pp. 261–269, 1983.

[4] R. Schreiber, "Systolic Arrays for Eigenvalue Computation," *Real Time Signal Processing V*, SPIE Vol. 341, Society of Photo-optical Instrumentation Engineers, Bellingham, Wash., 1982.

[5] H. J. Whitehouse and J. M. Speiser, "Sonar Applications of Systolic Array Technology," *IEEE EASCON Proc.*, 1981.

23

Multicomputer Parallel Arrays, Pipelines, and Pyramids for Pattern Perception

Leonard Uhr

University of Wisconsin
Madison, Wisconsin

23.1 INTRODUCTION: THE ENORMOUS INCREASES IN POWER AND SPEED FROM VLSI

The basic transistor-equivalent devices that are fabricated onto tiny VLSI (very large scale integration) silicon chips (each chip roughly 4 mm^2) are now so small and cheap that we are reaching the moment when extremely large numbers of computers can be combined into a single network. In 1978, 100,000 devices were successfully fabricated onto the 64K RAM. In 1981, Hewlett-Packard announced their 450,000-device computer chip. We have witnessed a doubling in packing densities roughly every 12 to 18 months for 10 or 20 years, and there is a high probability that this will continue for the next 10 to 20 years. Therefore, "medium-sized" multicomputers with several thousand chips, each chip with a million devices, or even with 10 million devices, will soon be feasible. This means that a single computer can have $10^4 \times 10^7 = 10^{11}$ devices.

These multicomputer systems can be used very effectively to attack enormously large problems, for example, the modeling of three-dimensional masses of atmosphere to predict the weather, the modeling of three-dimensional sections of the earth's crust to study such phenomena as earthquakes, the modeling of the large network of neurons that make up a human brain, and the very large set of transformations needed to perceive complex scenes of patterned objects.

Image processing and pattern perception will be among the major applications for such computers, since they pose extremely large information-processing problems that demand *very large and fast highly parallel micromodular computers*. The very large raw image input and digitized into a 512^2, 1024^2, or even larger raster array of picture elements ("pixels") can very effectively be processed with local operations that look at each pixel and also a judiciously chosen set of its neighbors.

Such *local "window" operations* can be carried out with great efficiency by a suitably designed parallel computer. Orders-of-magnitude speed-ups over conventional serial computers can be obtained. This for the first time will make possible image-processing and pattern-perception programs able to recognize complex objects in real-world images as they move about in real time, that is, within 30 ms or less per image frame.

23.2 TRADITIONAL SINGLE-CPU SERIAL COMPUTERS AND MULTICOMPUTER NETWORKS

A conventional "general-purpose serial computer" (see Figure 23.1) is, basically, built by linking a large high-speed memory to a single CPU (central processing unit) that fetches program statements from this memory, decodes each statement, and (as specified by these statements) fetches data stored in the indicated memory locations, executes the specified operations on these data, and stores the results in the specified locations. In addition, input devices and output devices must be linked to the system. Figure 23.1 indicates, in several variations, how we can picture and represent such a system.

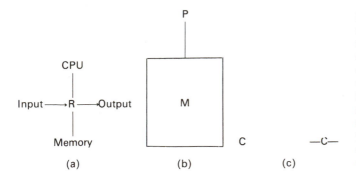

(a)   (b)   (c)

Figure 23.1   Traditional 1-CPU serial computer. (a) Memory is linked to a CPU (containing both controller and processor). These in turn are linked to input and output devices (often, as indicated in this figure, via R, registers). (b) Processor is linked to a relatively large main high-speed memory. (Input and output are assumed, but not shown.) (c) Single computer is indicated, with input and output assumed, and with input and output indicated by arrows.

23.3 ASSEMBLY-LINE PIPELINES OF COMPUTERS (OR PROCESSORS)

Several *pipelines* of processors or computers have been built, much in the spirit of an assembly line of workers. Each processor in the pipe repeatedly executes the same instruction on a sequence of data flowing through that pipe. This means that if the same sequence of instructions is to be executed on a larger number of different pieces of data (the case whenever the data stored in an array are to be operated on in parallel as by a nested FOR loop, as in the problem examples just mentioned), a pipeline as long as the sequence of instructions can be built, and data (e.g., information stored in the cells of an array) flowed through the pipeline's processors (see Figure 23.2).

Figure 23.2   Pipeline of N computers
through which data flow. At time 1, C1
will execute the first instruction on the
first piece of data; at time 2, C2 will execute the second instruction on the 1st
piece of data, and C1 will execute the first
instruction on the second piece of data;
and so on.

Input device ———→C1 →C2 →C3 →C4 → · · · CN ——→output device

If the pipeline has N processors, the program will execute up to approximately dN times as fast as a one-processor computer. (d is the often appreciable saving from not having to fetch and decode the next instruction, since each processor fetches only once, and keeps executing the same instruction.)

The fastest of today's "super-computers," for example, Seymour Cray's Cray-1 and the CDC-255 (see [11]), use such pipelines of a dozen or so very powerful and expensive processors to execute vector operations on arrays of data for "number-crunching" purposes.

Bjorn Kruse's [12] *PICAP* uses a pipeline of processors specially designed to effect local 3×3 window functions that compute any logical or arithmetic operation whose operands are the center cell of the window and also the eight neighbor cells that the programmer chooses to code.

The longest pipeline built to date is in Stanley Sternberg's [22] *Cytocomputer*, a multicomputer specialized to execute image-processing operations. The Cytocomputer has two types of processors, one that computes Boolean (i.e., 1-bit "true" or "false") functions over the center cell and its eight neighbors, and a second that computes arithmetic functions over 3-bit gray-scale values stored in the nine cells of the window. Each processor is much simpler and smaller than those found in the Cray-1, but the present Cytocomputer has a total of 113 processors. Even more are planned in projected future systems to be built using new VLSI chips (one processor to each chip) that can be linked together into (in theory) arbitrarily long pipelines [23].

23.4 ARRAYS OF VERY LARGE NUMBERS OF EACH SIMPLE PROCESSORS

Three very large two-dimensional arrays have been built in recent years, or are now being completed. These include the 64×64 *DAP* (distributed array processor) designed by Stewart Reddaway [16] at ICL; the 96×96 *CLIP*4 (cellular logic image processor), designed by Michael Duff [3] at University College London; and the 128×128 *MPP* (massively parallel processor), designed by Kenneth Batcher [1] together with David Schaeffer and others at Goodyear-Aerospace and NASA Goddard, and delivered in May 1983.

Each of these systems' thousands of computers executes the same instruction, but on a different set of data. The data to be processed are input to a large array, ideally the same size as the array of computers, so that each computer has one subset of those data in its own memory, for example, one pixel from the total image. Then each computer operates on data stored in its own memory, and also on near-neighbor data (see Figure 23.3).

CLIP4 is built with CMOS custom chips designed to be driven by a 2.5-MHz clock. Each chip has a 2×4 array of eight computers, each consisting of the processor, additional logic for the parallel fetch and processing of the eight neighbor values needed in a window operation, and 32 bits of memory (plus several 1-bit registers). The DAPs that are actually running have been built from off-the-shelf TTL ICs driven by a 10-MHz clock (each processor has a 4K RAM). But a chip with four processors plus registers has been checked out, and it is expected that future DAPs will use that chip, plus commercial RAMs for memory [17]. The MPP uses NMOS custom chips designed to be driven by a 40-MHz clock. Each chip has eight processors, each with a 32-bit register, each linked to a 1K 80-ns off-chip RAM.

A basic CLIP4 machine-language instruction has (in 11 μs) every computer fetch (in parallel) and operate on information from its own memory and also from any or all of the eight immediately surrounding near-neighbor computers' memories—that is, from the 3×3 window surrounding it. This instruction is quite similar to the window operations executed by PICAP and the Cytocomputer. But

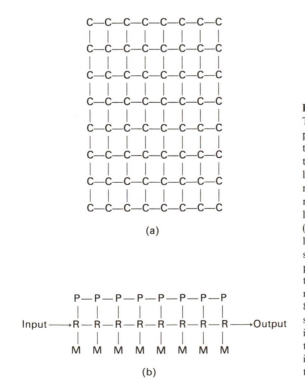

(a)

(b)

Figure 23.3 Arrays of computers. (a) Two-dimensional 8×8 array of computers, viewed from above. Each is linked to its four square neighbors. (Alternately, the four diagonal neighbors might also be linked to directly; or each computer might be linked to its six hexagonal neighbors.) Each computer's processor links directly to that computer's memory (not shown in this view); and also, via the links shown, to the memories of its four square neighbors. (b) 1×8 array, each processor linked to its own memory, and to input and output, via registers. (Alternately, this can be viewed as the above 8×8 array looked at from one side, showing how each computer's processor in a 1×8 one-dimensional subarray of that 8×8 two-dimensional array links to its memory and to input/output via registers.)

the actual hardware embodiment is radically different, since CLIP executes the entire window operation everywhere in the array in one instruction, whereas the pipeline systems must serially flow each pixel and its surrounding window through the processor that executes this instruction.

The DAP and MPP do not have built-in window operations. Their basic instructions consist of "fetch-add," "fetch-or," or some other sequence of one fetch followed by an arithmetic or logic operation on that (1-bit) piece of data and a second piece of data presently in an accumulator. This means they must use a sequence of up to 12 instructions (including two to shift the values in the window's diagonal cells to square directly linked cells) to execute a single window operation. This basic instruction takes 250 ns on the DAP, and is expected to take 100 ns on the MPP.

Today's large arrays have from 4000 to 16,000 computers. But well within current technology and current enonomics is the capability to build much larger arrays, with 256×256, 512×512, 1024×1024, or even more computers. For example, CLIP4 with 10,000 computers cost roughly $100,000 to build in 1980, or $10 per computer. This is the cost of the finished multicomputer; CLIP4 has eight computers on each roughly $25 silicon chip, so each computer cost roughly $3.

CLIP's chip and total computer costs are rapidly decreasing in price because of higher yields from a redesigned CLIP4 chip; thus the present estimate is $4 per chip, or 50 cents per computer [4]. Therefore, a system with 250,000 or so computers could probably be built in the next two or three years for less than $1 million, possibly substantially less. This would mean that each picture element in a 512×512 TV array could be stored in a different computer's memory, and all could be processed as fast as a conventional computer could process one.

The arrays that have actually been built are, however, a good bit smaller than 512×512 (and their costs are a good bit higher than $3 per computer—the 64×64 DAP costs over $1,000,000 as an add-on to a $3,000,000 ICL mainframe; the MPP cost over $5,000,000, including design costs). As shown in Figure 23.4, *picture arrays larger than the array of computers can easily be handled*, in either of two ways:

1. A subarray of the picture array is stored in each computer's memory (e.g., an 8×8 subarray of a 1024×1024 array might be stored in each cell of a 128×128 computer array).
2. A subarray of the picture array that is the same size as the computer array is input to the computer array, storing one cell of the picture array in each computer's memory; this process is repeated until the entire picture array has been processed.

Each of these alternatives has its problems:

1. Both, of course, take much longer, since they must serially iterate the smaller computer array through the larger picture array.
2. The first also needs additional time (or/and hardware) to index through the

1a	1b		2a	2b		3a	3b		4a	4b
1c	1d		2c	2d		3c	3d		4c	4d
5a	5b		6a	6b		7a	7b		8a	8b
5c	5d		6c	6d		7c	7d		8c	8d
9a	9b		10a	10b		11a	11b		12a	12b
9c	9d		10c	10d		11c	11d		12c	12d
13a	13b		14a	14b		15a	15b		16a	16b
13c	13d		14c	14d		15c	15d		16c	16d

(a)

```
1 1 1 1        2 2 2 2
1 1 1 1        2 2 2 2
1 1 1 1        2 2 2 2
1 1 1 1        2 2 2 2
3 3 3 3        4 4 4 4
3 3 3 3        4 4 4 4
3 3 3 3        4 4 4 4
3 3 3 3        4 4 4 4
```

(b)

Figure 23.4 Processing image arrays larger than the computer array. (a) Subarrays of the image are stored in each computer's memory. Each cell with the same number is stored in the same computer's memory. The example pictured in this figure indicates how a 16 × 16 picture array is mapped and stored into a 4 × 4 array of 16 computers (numbered 1 through 16). Each computer's memory has a 2 × 2 subarray of the image (numbered a through d). (b) Subarrays of the image array, each the same size as the array of computers, are processed in turn. The example pictured in this figure decomposes a 16 × 16 picture array into four 4 × 4 subimages that are input into and processed by a 4 × 4 array of computers one at a time.

subarray, and additional memory to store the entire subimage and all resulting transformations.

3. The second must handle windows at the interior borders, where cells adjacent to but outside the subarray being processed must be examined as part of the window. This means that either:
   a. The system must constantly roll in needed border rows and columns.
   b. Overlapping subarrays must be processed, with commensurate additional slow-downs.
   c. Additional hardware must be added to make the borders accessible, by storing overlapping columns of the image in memories that can be treated as either the left border or the right border, and overlapping rows that can be treated as either the top border or the bottom border.

These very large arrays are feasible only because each computer has been made as simple as possible, and all execute the same instruction (which means that only one controller is needed). Today's arrays use processors that compute one bit at a time. This is appropriate and efficient for logical operations. *Logic operations* include a very important class of operations: the picture-processing operations, where several features are examined together and combined, and the resulting characteristic (often called a *label*) is represented as being either "present" or "absent" at each cell. Arithmetic and string-matching operations that must be executed on pieces of data longer than 1 bit are quite straightforwardly carried out, but only serially, 1 bit at a time, in what is called a "bit-serial" mode. This means that, for example, when 3-bit, 21-bit, 33-bit, 47-bit, 64-bit, or 89-bit arithmetic or comparison operations must be executed they will take $3k$, $21k$, $33k$, $47k$, $64k$, or $89k$ times as long as the 1-bit operation that must be iterated (k is a small constant, roughly on the order of 2 to 5, for the overheads due to the iterations effected by the software). Thus, at the price of occasional 3-fold, 64-fold, or even greater degradations when large numbers and strings must be handled serially, today's actual running systems gain 4000- to 16,000-fold increases in speed when large arrays of numbers can be handled in parallel. Tomorrow's systems, each with hundreds of thousands or millions of computers, will achieve five, six, or even more orders-of-magnitude increases in speed.

CLIP4's basic instructions operate on the entire 3×3 window (plus a tenth number got as a second operand from the cell's memory only). This gives a further 10-fold speed-up—when and if the instruction looks at and uses all 10 pieces of information.

Probably the most important feature of the array, a feature that will become increasingly compelling over the next 10 to 20 years, is the fact that it can easily be expanded. A 512×512 array is simply a 4×4 array of 128×128 arrays. There are also economies of scale, since mass-production savings will be achieved as larger numbers of identical chips are fabricated, and as larger numbers of identical boards and other modules are tied together. With future VLSI technologies these parallel arrays will become *increasingly attractive*, because of their simple micromodular and highly replicated design.

In stark contrast, very little more can be done to make today's serial computers more powerful. And they are rapidly reaching the point where they cannot possibly be made faster (except by introducing special-purpose parallel processors into the single CPU) because they will have reached the ultimate "speed-of-light" barrier.

23.5 THE ENORMOUS VARIETY OF MORE GENERAL NETWORK STRUCTURES

Because of the limits on arrays (which, ultimately, are imposed by considerations of costs)—their 1-bit processors, the single controller, and the near-neighbor connectivity between processors—many people would prefer to see other kinds of networks built. A larger number of *networks* have been designed, including ring [6], n-cube [25], lattice [38], star [26], snowflake [7], lens [8], tree [14], x-tree [2], pyramid, and a variety of other graph structures (see, e.g., [31,35]). (For our purposes, a graph is simply a set of computer nodes linked together into a single multicomputer; sometimes the nodes are processors or memories rather than the entire processor-memory computer.)

But very few such networks have actually been built, and only two with more than 50 processors: *Cm** [26] and the *Genoa machine* [15]. The possibilities are potentially infinite, since they include all conceivable interconnections among (as technology improves and grows cheaper) a continually growing number of processor components—that is, all possible graphs. But the problems are enormous. Today, people are just beginning to attack the very difficult problems of designing such multicomputers, developing parallel algorithms, programming and mapping these algorithms properly into the network, and coordinating the many different processors that are executing the same program to work with reasonably efficiency (see Uhr [35,36,37] for an examination of these interrelated aspects of the problem).

23.6 ALGORITHM-STRUCTURED MULTICOMPUTER NETWORKS

Among the most attractive possibilities are networks whose structures mirror the structures of the algorithms that they execute. Among the most interesting examples of such structures are trees, especially the augmented *x*-trees [2] and hyper-trees [9], and the PASM reconfigurable array of conventional computers [20]. *Trees* have good structure for a variety of tasks where information is sorted, compared, or in other ways reduced and reorganized; and also where information is stored, accessed, or broadcast. Arrays have good structure for passing information locally, and the addition of reconfiguring switches [19] follows directly from the flow of data when computing fast Fourier transforms [24], together with a variety of other remappings of information.

The 1-bit arrays examined above are also good examples of such algorithm-structured architectures. The pyramid multicomputers to be examined next appear to be among the very best, since they very efficiently handle not only parallel local operations but also global passing and combining of information.

23.7 THE ARCHITECTURE OF HARDWARE MULTICOMPUTER PYRAMIDS OF ARRAYS

Arrays can come in several different flavors, depending on the details of links between computers. But let us begin with a concrete example. One of the simplest and most attractive schemes has each computer linked to its four square neighbors ("4-connected"), so that it can fetch one word of information from any one of their memories (or from a fifth memory, its own).

Now one of the simplest ways to start building a *pyramid* is to take such an array and, starting in any corner (say the upper left), linking each 2×2 subarray of its computers to a new computer, at the same time linking this collection of new computers together as a 4-connected array. This gives a second array with one-fourth the number of computers.

Now continue linking 2×2 subarrays to a single node in the next-higher array, until the "apex" array, with only one single node, is reached.

This gives a pyramid of arrays, with $\log (N) + 1$ arrays, where the base array is

$N \times N$ and each array one-fourth the size of its next-lower array. The diameter (i.e., the shortest distance between the two most distant nodes) of such a pyramid is $\log_4 (N)$.

23.8 LINKING SUCCESSIVELY SMALLER ARRAYS TO BUILD A PYRAMID

A pyramid can be built by stacking successively smaller arrays, as shown in Figure 23.5.

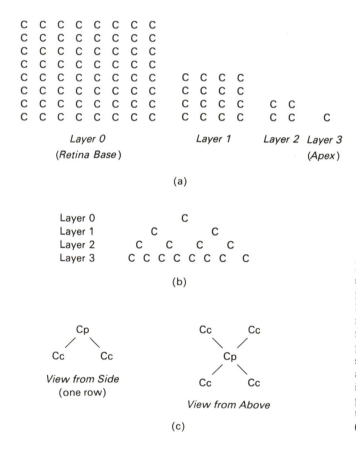

Figure 23.5 shows how an array can be linked to a lower-level children array that is twice the length and twice the width of the parent array, and therefore has four times as many computers.

Figure 23.5 Building a pyramid from successively smaller arrays: a four-layer pyramid with an 8×8 retinal base and 2×2 convergence from each layer to the next. (a) Four arrays of computers are first constructed, forming the layers of the pyramid. (b) These array/layers are stacked, each successively smaller layer above the last (side views are given, showing only one row of each layer). (c) Computers are linked from layer to layer so that each parent (Cp) has enough children (Cc) so all children have parents.

23.9 CONSTRUCTING PYRAMIDS BY LINKING EACH PLY OF A TREE INTO AN ARRAY

An illuminating alternative construction procedure starts with a tree of degree 4 (i.e., with the tree's "root" node linked to four "children" nodes, each of which is in turn linked to four children nodes, until the tree's "bud" nodes are reached). Thus each new "ply" (i.e., layer of children) has four times as many nodes as the previous ply. Now augment this tree with lateral links between "sibling" nodes (at the same ply) to turn each set of nodes at the same ply into an array. The root array is a 1×1 array of one single node. The array at the next layer of the tree is a 2×2 array, with the root's four children now augmented with the additional array links between them—that is, sibling links. Each node's set of four children is similarly arranged geometrically in a 2×2 subarray that is placed appropriately next to the subarrays of its siblings and the other nearby nodes in its layer. Thus the two-dimensional geometric constraints of the array are superimposed on the topology of the tree.

The term *quad tree* has been used for a variant on the above design—for a tree whose largest ply, that is, its buds—is augmented into an array, or receives as input the image of an array. But the intermediate plies do not have any lateral links to form intermediate arrays. Quad trees are very useful structures for compressing, storing, describing, and reconstructing large two-dimensional images. An "oct tree" has eight children, arranged in a $2 \times 2 \times 2$ three-dimensional solid; therefore, it can be used as a four-dimensional tree that looks at and processes a three-dimensional image (or other set of information) input to its base array of buds.

More generally, we can define a *C*-converging, *D*-dimension C^D tree. To give some examples:

A 2^2 tree is a quad tree.

A 2^3 tree is an oct tree.

A 3^2 tree is a degree 9 tree where each node has nine children arranged in a 3×3 subarray.

And, for example, a 3^5 tree is a degree 243 tree where each node has 243 children arranged in a five-dimensional solid each of whose sides is three nodes long.

We can extend this definition to include the number of *P*, plies, in the tree, to C^D, *P* trees.

The number of plies is $\log_{C^D} B$, where *B* is the number of buds in the tree. And, of course, $B = N \times N$, where *N* is the length and width of the $N \times N$ base array that augments the buds.

Thus a $2^2, \log_{22} (256)$ tree is a four-ply quad tree whose base array has 256^2 buds arranged in a 256×256 array.

A C^D,P,S pyramid is exactly like a C^D,P tree, but with every ply augmented laterally with *S* sibling links to form an array.

23.10 PYRAMIDS CAN BE VARIED IN MANY DIFFERENT WAYS

A number of variations on this basic pyramid theme are possible. The following are among the most interesting.

1. Convergence (C) can straightforwardly be by any small integer factor. (It is also possible to converge fractionally by interpolating, and this may be an attractive way to deepen the sequence of transformations through a pyramid.)

2. There can be different amounts of convergence between different pairs of layers.

3. There can be no convergence between particular pairs of layers or between all pairs (in which case the system becomes a three-dimensional rectangular solid array); that is, convergence can be set equal to 1.

4. Row convergence and column convergence might differ (this can be useful in handling rectangular as opposed to square images).

5. Parents can be placed step distance (S) away from one another. Then when $C = S$ the pyramid has no *overlap*; when $C > S$ each parent shares children with its adjacent parents, giving overlap. An attractive possibility is $C = 3$, $S = 2$—with each parent linked to a 3×3 window of children, and parents placed above every other child (giving 2×2 convergence).

6. Rather than converge basically by the degree of the tree/pyramid, giving $O(\log N)$-diameter pyramids, a very simple alternative scheme will converge in N steps. This is effected simply by building arrays with links between diagonal as well as square neighbors, and eliminating all nodes outside the pyramid, as shown (for two dimensions) in Figure 23.6.

   These can be thought of as $O(N)$ pyramids, in contrast to the $O(\log N)$ pyramids previously defined and described. They may well be attractive for a variety of applications. But they have not to my knowledge been explored at all. And when built large enough to handle real-world images, that is, with base arrays on the order of 512×512 or even larger, their $N/2$ distances for transforming the image and for message passing, and their $O(N^3)$ [rather than $O(N^2)$] nodes, appear to be too great.

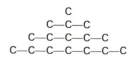

Figure 23.6 $O(N)$ pyramid, in two dimensions. Each processor is linked diagonally (not shown) to its two neighbors in its two adjacent layers (or only one neighbor in the smaller, higher layer if the processor is at the edge). Thus each layer has two fewer processors than its lower neighbor layer, and the whole pyramid has $N/2$ layers (N = the length of one side of the pyramid).

7. Links to siblings and children can be varied, for example, nine links to siblings, four to children; or vice versa.

8. The pyramid can be "truncated," with the very small highest arrays eliminated, or replaced by different kinds of more powerful computers.

23.11 IMPLEMENTATIONS OF PYRAMIDS

A large and growing number of people have been examining pyramids for their properties and potential uses (see, e.g., Uhr [32,33], Hanson and Riseman [10], Levine [13], Tanimoto and Klinger [29], and Rosenfeld [18]); and a number of pyramid algorithms have been devised for basic image-processing operations. But only three or four researchers have seriously investigated the architectural issues involved and reached the point of proposing multicomputer pyramid architectures (Dyer [5], Tanimoto and Pfeiffer [30], Tanimoto [27,28], and Uhr [34–37]).

Only one of these has to date resulted in actual construction. This is the joint Boeing Aerospace–University of Washington pyramid image processor (PIP) [21] being built along the lines of Steven Tanimoto's design. The PIP is planned to have a 128×128 array of computers at its base. It converges at the steady rate of 2×2, without overlap, and it is a complete pyramid. So the total pyramid contains the eight layers: 128×128, 64×64, 32×32, 16×16, 8×8, 4×4, 2×2, and the 1×1 apex.

Each computer's processor links to its four offspring in the lower layer (this is what gives its 2×2 convergence), to its eight neighbors (and also to its own memory) in its own layer, and to its single parent. Thus a single basic instruction will fetch a word of information from any subset of these 14 computers' memories (including its own) and compute some function of that information. This means that information can be passed laterally, within the processor's array, or/and upward, or/and downward.

A chip is presently being constructed that will handle 64 processors (linked to off-chip RAMs) in either of the following two ways (depending on how well the separate modules now being checked out can be packed together): four processors will serially iterate through 16 pixels each, in 2 μs in toto, or eight processors will serially iterate through eight pixels each, in 1 μs in toto.

23.12 PROBLEM-STRUCTURED ARCHITECTURES THAT TRANSFORM DATA FLOWING THROUGH

A program is designed to execute an algorithm, which in its turn is chosen to execute a set of transformations on a set of information (data). Both transformations and information have an inherent structure, and a good algorithm will share in that structure. Such a *problem-structured algorithm* is best mapped into the multicomputer network so that operations are set up as in a complex assembly line, and information is flowed through, again much as in an assembly line. This gives a pipeline-like flow, but of a two-dimensional image, and this flow takes place through a three-dimensional structure that is much more complex than today's one-dimensional pipelines. This kind of structure suggests something rather analogous to the living brain, through which data input via the sensory organs flow.

Such a well-structured network can actually contain among its subnetworks a variety of different types of structures, including arrays. It can also contain switches that, under program control, can actually be thrown and therefore be used to reconfigure the network, to better fit the structure of different programs, and even of different processes within a single program.

A large two-dimensional image maps nicely into a large array, and a large-array multicomputer can very effectively apply sequences of image-processing operations to the successively transformed image.

A pyramid in addition allows the programmer to converge and reduce the information stored in a transformed image when and as those data are reduced. For example, when a 3×3 window is used to discover a gradient the image can be passed to the next layer of a 3×3 converging pyramid. When several strokes or other features that are contained in a 2×2, 3×4, or whatever $N \times M$ window, are combined into a single higher-level feature, the array into which their output is stored can be commensurately smaller. A pyramid has both the local near-neighbor links appropriate for processing information whose interactions are local, and also the converging structure appropriate for efficient data reduction and message passing. Pyramids appear to offer enormous potential power for handling, in real time, streams of images (e.g., from a television camera that inputs an image frame every 30 ms), by pipelining these two-dimensional images through the complex sequences of successively more global operations effected at the different layers of the three-dimensional pyramid.

23.13 THE APPROPRIATENESS OF ARRAYS AND PYRAMIDS FOR VLSI

The processors used in arrays and in pyramids of arrays are routinely kept extremely simple. In almost all cases these have been 1-bit processors with only 100 to 800 gates. This is because to achieve four or more orders-of-magnitude increases in speed and power by using increasingly large numbers of processors in parallel, their architects have opted to use the simplest possible 1-bit processor, executing K-bit-serial operations to process K-bit numbers or strings.

The amount of memory each such processor needs appears to be a function of the total amount of memory needed to handle the images or other large sets of data given the system. Therefore, each processor appears to need relatively small amounts of memory (present implementations have 32 to 4096 bits of memory per processor).

Computers are linked together in a highly micromodular grid fashion, giving one of the very best candidates for dense packing into a VLSI design.

Today four, eight, or even more of these 1-bit computers are fabricated on a single VLSI chip. Because of the highly iterated micromodular design of such systems, it should soon be possible to fabricate hundreds, or even thousands, of processors, each with its own memory, on a single chip. This is in sharp contrast to VLSI realizations of conventional single-CPU computers, which, although one or several entire CPUs might be packed on a single chip, will continue to need several chips to handle each computer's several million bytes of high-speed memory.

Multicomputer arrays and pyramids of arrays with 1024^2 or more computers will be buildable from only a 16×16 array of 256 chips, each with a 64×64 array of 4096 computers, each with a 400-device processor and a 512-bit memory. Such an array, or a pyramid whose base was such an array (which will have at most one-third as many total computers as its base array alone), will be realizable with a small enough handful of 10^7-device chips to be packed onto a single wafer of chips. The highly replicated grid-connected micromodular array or pyramid is an especially attractive candidate for fault-tolerant wafer-scale integration.

23.14 CONCLUDING REMARKS

This chapter first describes and examines parallel arrays, pipelines, and a variety of other multicomputer topologies, because they offer enormous potential increases in speed and power. Indeed, we now have the opportunity to embody any possible graph into an actual multicomputer, where the nodes of the graph indicate the individual computers, and the edges of the graph indicate the direct links between pairs of individual computers.

It then concentrates on a relatively new topology that appears to be especially appropriate for image processing, pattern recognition, and computer vision. This is one that combines successively smaller arrays into an overall pyramid structure. Each layer of a pyramid can achieve the same potentially enormous parallel speed-up as a comparably sized array—since each layer is itself an array. In addition, all layers of the pyramid can be executing simultaneously.

Of great importance, the underlying tree topology of a pyramid means that information can be converged together as the image is successively transformed. Parts of the same object that originally were distant from one another move successively closer (and at the same time they are successively transformed into larger, more global subobjects). Thus a tree structure of a large number of each simple and local feature detectors can serve to recognize large and complex objects. From the point of view of message passing, the pyramid's $O(\log N)$ diameter is a major improvement over the array's $O(N)$ diameter. Indeed, when processing images in the range of 256×256 to 1024×1024, this is a crucial difference from the practical point of view.

Arrays, pipelines, and in particular pyramids offer potentially orders-of-magnitude increases in speed and power over traditional single-CPU computers. And they appear to be especially appropriate for image processing, pattern recognition, and computer vision. They also map especially well into VLSI technology constraints, because of their iterated micromodular design.

REFERENCES

[1] K. E. Batcher, "Design of a Massively Parallel Processor," *IEEE Trans. Comput.*, *29*:836–840 (1980).

[2] A. M. Despain and D. A. Patterson, "X-Tree: a Tree Structured Multi-processor Computer Architecture," *Proc. 5th Annu. Symp. Comput. Arch.*, Apr. 1978, pp. 144–151.

[3] M. J. B. Duff, CLIP4: "A Large Scale Integrated Circuit Array Parallel Processor," *Proc. IJCPR-3, 4*:728–733 (1976).

[4] M. J. B. Duff, personal communication, 1982.

[5] C. R. Dyer, "Pyramid algorithms and machines," in: K. Preston, Jr., and L. Uhr, eds., *Multicomputers and Image Processing*, Academic Press, New York, 1982, pp. 409–420.

[6] D. J. Farber and K. C. Larson, "The System Architecture of the Distributed Computer System—The Communications System," *Symp. Comput. Networks*, Polytechnic Institute of Brooklyn, Apr. 1972.

[7] R. A. Finkel and M. H. Solomon, "Processor Interconnection Strategies," *Comput. Sci. Dept. Tech. Rep. 301*, University of Wisconsin, 1977.

[8] R. A. Finkel and M. H. Solomon, "The Lens Interconnection Strategy," *Comput. Sci. Dept. Tech. Rep. 387*, University of Wisconsin, 1980.

[9] J. R. Goodman and A. M. Despain, "A Study of the Interconnection of Multiple Processors in a Data Base Environment," *Proc. 1980 Int. Conf. Parallel Process.*, 1980, pp. 269–278.

[10] A. R. Hanson and E. M. Riseman, "A Progress Report on VISIONS," *COINS Tech. Rep. 76-9*, University of Massachusetts, 1976.

[11] E. W. Kozdrowicki and D. J. Thies, "Second Generation of Vector Supercomputers," *Computers, 13*:71–83 (Nov. 1980).

[12] B. Kruse, "The PICAP Picture Processing Laboratory," *Proc. IJCPR-3, 4*:875–881 (1976).

[13] M. D. Levine, "A Knowledge-Based Computer Vision System," in A. Hanson and E. Riseman, eds., *Computer Vision Systems*, Academic Press, New York, 1978, pp. 335–352.

[14] G. A. Mago, "A Cellular Computer Architecture for Functional Programming," *Proc. COMPCON Spring 1980*, 1980, pp. 179–187.

[15] R. Manara and L. Stringa, "The EMMA System: An Industrial Experience on a Multiprocessor," in M. J. B. Duff and S. Lerialdi, eds., *Languages and Architectures for Image Processing*, Academic Press, London, 1981.

[16] S. F. Reddaway, "DAP—A Flexible Number Cruncher," *Proc. 1978 LASL Workshop Vector Parallel Process.*, Los Alamos, 1978, pp. 233–234.

[17] S. F. Reddaway, personal communication, 1982.

[18] A. Rosenfeld, ed., *Multi-resolution Systems for Image Processing*, North-Holland, Amsterdam, 1983, in press.

[19] H. J. Siegel, "A Model of SIMD Machines and a Comparison of Various Interconnection Networks," *IEEE Trans. Comput.*, *28*:907–917 (1979).

[20] H. J. Siegel, "PASM: A Reconfigurable Multimicrocomputer System for Image Processing," in M. J. B. Duff and S. Levialdi, eds., *Languages and Architectures for Image Processing*, Academic Press, London, 1981.

[21] W. Snapp, personal communication, 1982.

[22] S. R. Sternberg, "Cytocomputer Real-Time Pattern Recognition," *8th Pattern Recognition Symp.*, National Bureau of Standards, 1978.

[23] S. R. Sternberg, personal communication, 1982.

[24] H. S. Stone, "Parallel Processing with the Perfect Shuffle," *IEEE Trans. Comput.*, *20*:153–161 (1971).

[25] H. Sullivan, T. Bashkov, and D. Klappholz, "A Large Scale, Homogeneous, Fully Distributed Parallel Machine," in *Proc. 4th Annu. Symp. Comput. Arch.*, 1977, pp. 105–124.

[26] R. J. Swan, S. H. Fuller, and D. P. Siewiorek, "Cm*—A Modular, Multi-microprocessor," *Proc. AFIPS NCC*, 1977, pp. 637–663.

[27] S. L. Tanimoto, "Towards Hierarchical Cellular Logic: Design Considerations for Pyramid Machines," Computer Science Dept. Tech. Rep. 81-02-01, University of Washington, 1981.

[28] S. L. Tanimoto, "Programming Techniques for Hierarchical Parallel Image Processors," in K. Preston, Jr., and L. Uhr, eds., *Multi-computers and Image Processing*, Academic Press, New York, 1982, pp. 421–429.

[29] S. L. Tanimoto and A. Klinger, eds., *Structured Computer Vision*, Academic Press, New York, 1980.

[30] S. L. Tanimoto and J. J. Pfeiffer, Jr., "An Image Processor Based on an Array of Pipelines," *Proc. Workshop on Computer Architecture for Pattern Analysis and Image Data Base Management*, IEEE Computer Society Press, 1981, pp. 201–208.

[31] K. J. Thurber, *Large Scale Computer Architecture*, Hayden, Rochelle Park, N.J., 1976.

[32] L. Uhr, "Layered 'Recognition Cone' Networks That Preprocess, Classify and Describe," *IEEE Trans. Comput.*, *21*:758–768 (1972).

[33] L. Uhr, "'Recognition Cones' and Some Test Results," in A. Hanson and E. Riseman, eds., *Computer Vision Systems*, Academic Press, New York, 1978, pp. 363–372.

[34] L. Uhr, "Converging Pyramids of Arrays," *Proc. Workshop on Computer Architecture for Pattern Analysis and Image Data Base Management*, IEEE Computer Society Press, 1981, pp. 31–34.

[35] L. Uhr, *Algorithm-Structured Computer Arrays and Networks: Parallel Architectures for Perception and Modelling*, Academic Press, New York, 1984.

[36] L. Uhr, "Pyramid Multi-computer Structures, and Augmented Pyramids," in M. Duff, ed., *Computing Structures for Image Processing*, Academic Press, London, 1983, in press.

[37] L. Uhr, "Pyramid Multi-computers, and Extensions and Augmentations," in D. Gannon, H. J. Siegel, L. Siegel, and L. Snyder, eds., *Algorithmically Specialized Computer Organizations*, Academic Press, New York, 1984, in press.

[38] L. D. Wittie, "MICRONET: A Reconfigurable Microcomputer Network for Distributed Systems Research," *Simulation*, *31*:145–153 (1978).

24

Parallel Algorithms for Image Analysis

AZRIEL ROSENFELD

University of Maryland
College Park, Maryland

24.1 INTRODUCTION

Image processing and analysis (IPA) systems employ a wide variety of techniques for image encoding, transformation, segmentation, and property measurement. Many of these techniques are suitable for efficient parallel implementation. This chapter defines some general classes of IPA algorithms, and indicates how such algorithms can be implemented in parallel using various types of "cellular" multiprocessor architectures.

IPA has a large and rapidly growing literature. There are at least a dozen textbooks [1–12] covering major parts of the subject, aside from books on specific topics and collections of papers. An annual bibliography (the most recent is [13]), covering primarily the non-application-oriented U.S. literature, currently includes about 1000 references per year. References on specific (classes of) algorithms will not be given in this paper.

It was proposed about 25 years ago [14] that many IPA algorithms could be implemented in parallel using a "cellular array" machine (Figure 24.1): a two-dimensional array of processors ("cells"), operating synchronously, each of which can communicate with its neighbors in the array. Several machines of this type, with array sizes of up to 128×128, have actually been constructed. Numerous IPA algorithms suitable for implementation on a cellular array have been developed; for general discussions of this subject, see [15,16]. Recently, there has been some interest in using "pyramids" of cellular arrays, of sizes $2^n \times 2^n$, $2^{n-1} \times 2^{n-1}$, ..., 2×2, 1×1, where each cell can communicate not only with its neighbors ("brothers") on its own level, but also with its four "sons" on the level below and with its "father" on the level above [15,17] (see Figure 24.2).

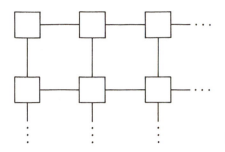

Figure 24.1 Cellular array.

Cellular arrays or pyramids are suitable for many types of IPA operations at the pixel level. On the other hand, some image analysis operations, involving regions in an image, do not make use of pixel arrays, but rather use other types of data structures to represent regions and their relationships. For such operations, a more general class of graph-structured cellular machines would be appropriate, in which the cells correspond to the nodes of a graph, and can communicate with their neighbors as defined by the arcs of the graph. On such "cellular graph" machines, see [18,19]; on architectures corresponding to more specific types of data structures, see [20–22].

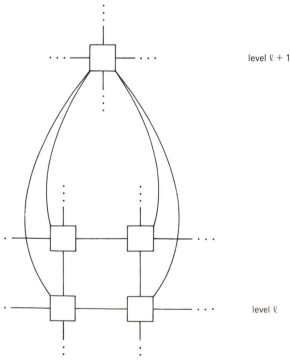

level $\ell + 1$

level ℓ

Figure 24.2 Cellular pyramid.

24.2 PIXEL-LEVEL OPERATIONS

A *digital image* is a rectangular array of pixels ("picture elements"). A pixel is usually integer-valued (most commonly, 8-bit integers are used), but it can also be real- or complex-valued, or vector-valued (in the case of color or multispectral imagery). In this section we describe IPA algorithms that operate on pixel arrays.

24.2.1 Point and Local Operations

Most of the operations commonly performed in IPA take images into images, where the value of a pixel in the output image depends only on the value(s) of the corresponding pixel, and possibly of some of its neighbors, in the input image (or images). Some examples of such operations are:

1. Contrast enhancement by gray-scale transformation: the new value of a pixel depends only on its old value, as defined by a given mapping.
2. Sharpening by, for example, Laplacian filtering, involving the difference between the pixel and the average of its neighbors.
3. Smoothing by local averaging [taking an average, possibly weighted, of the pixel and (some of) its neighbors], by median filtering (using the median of the pixel and its neighbors), by averaging of multiple images in register, and so on.
4. Segmentation by thresholding (the new value of a pixel is 1 or 0 depending on whether or not the old value exceeds a threshold), or more generally, by classification of the pixels based on a set of property values (color components, local property values, etc.).
5. Edge (or other local feature) detection, based on computing differences between neighboring pixels.
6. Expanding or shrinking: the new value of a pixel is the max or min of the values of a set of its neighbors [in some cases (e.g., that of thinning operations) additional conditions must also be satisfied before the value of a pixel is changed].

Note that some of these operations are linear (and hence are convolutions), but most of them are not.

Operations of these types can be performed very efficiently on a cellular array machine in which (ideally) there is a processor associated with each pixel. Each processor collects the values of its neighbors, if necessary, and then computes the required function of these values. The time required to do this depends on the neighborhood size and the complexity of the operation, but not on the size of the image. If there are not enough processors, we can process the image blockwise, using enough overlap between blocks to avoid border effects.

24.2.2 *Transforms*

Various types of integral transforms (or their discrete versions) are often performed on images; the Fourier transform is the most common example. Here again the output image is an array of the same size as the input image, but there is no longer a correspondence between their pixels. The transforms are usually separable, so that they can be performed first row-wise, then column-wise; the value of a pixel in the transform is then a linear combination of the values of the pixels in that row (or column) of the image.

When transforms are done on a conventional computer, one can use efficient algorithms (e.g., the fast Fourier transform) which require $O(n \log n)$ operations, rather than $O(n^2)$, on each row (or column); the total computational cost for an $n \times n$ image is thus $O(n^2 \log n)$. (We ignore here the problem of accessing the image from peripheral storage, and the possible need to transpose the image in order to access it efficiently column-wise as well as row-wise.) On a cellular array machine, the rows (or columns) can be transformed in parallel, and the time required for each row is $O(n)$ (each pixel must be multiplied by n coefficients, and the results must be grouped and summed); thus the overall time is also $O(n)$.

Many useful types of operations (e.g., convolution operations) can be performed on an image by taking its Fourier transform, multiplying the transform pointwise by an appropriate weighting function (or multiplying the transforms of two images pointwise), and then taking the inverse Fourier transform of the result to obtain the processed image. This also requires only $O(n)$ time on a cellular array machine. For convolutions involving large numbers of weights, this may be more efficient than performing the convolution directly by parallel collection of information from the neighbors of each pixel.

24.2.3 *Geometric Operations*

Another class of image-to-image operations involves geometric transformations of an image [e.g., rescaling, rotation, or arbitrary "warping" (to correct geometric distortions, or to achieve registration with another image)]. Here the output pixels do correspond to the input pixels, but not in a simple one-to-one fashion (even a transformation such as rotation, when performed digitally, is not one to one). To perform such a transformation on an image, one must compute, for each pixel in the output array, the corresponding positions in the input array (which will not, in general, coincide with the position of an input pixel). One must then assign a value to that output position by interpolation on the nearby input values.

The basic method of performing a geometric transformation on a cellular array machine is to assign an output pixel to each processor, and scan the input image over the array in such a way that each processor can collect the input values that it needs to compute its output value. To illustrate how this can be done in $O(n)$ steps, let S_{ij} denote the set of input pixels needed to compute the output value at pixel (i, j); let R_i denote the ith row, and C_j the jth column. Let us assume that, for all i and j, the set $|(\cup_i S_{ij}) \cap R_i|$ is of bounded size—in other words, the number of input pixels in a given row R_i whose values are needed by all the output pixels in a given column C_j is bounded. (This is not true for all possible transformations; for example, it is false for the operation of transposing. However, for any of the commonly encountered geometric transformations, there exists a pair of directions—not necessarily the "row" and "column" directions—for which the analogous property holds.) Then we proceed as follows: cyclically shift the input data along the rows, and let each cell (i, j) collect all the values in $|(\cup_i S_{ij}) \cap R_i|$ as they pass it. Then cyclically shift the collected data along the columns, and let each cell (i, j) select the values it needs (i.e., those in S_{ij}).

24.2.4 Property Measurement

We now consider operations that map an input image into a (set of) property value(s) rather than into an output image. Examples of such operations include:

1. Determining the presence or absence of a particular pixel value in the image, or computing statistics of the values (min, max, median, range, mean, standard deviation, etc.).
2. Counting the number of occurrences of a particular value: for example, the number of 1's in a two-valued image gives the area of the set of 1's; the numbers of pixels having each possible value define the gray-level *histogram* of the image; the numbers of pairs of pixels in a given relative position that have each possible pair of values define a gray-level "cooccurrence matrix" of the image, which is useful in describing its texture. Note that the last two examples involve sets of k or k^2 properties, where k is the number of gray levels.
3. Counting the number of connected components of pixels having a particular value (this is the standard method of counting objects in a segmented image).

On a cellular array machine, such operations require $O(n)$ time, since the pixel values must be brought together in one place in order to count them or compute their statistics, and this requires a number of communication steps proportional to the array diameter. Counting connected components requires a preliminary step in which each component is reduced to a single pixel, but this too can be done in time $O(n)$ using a special type of shrinking process.

Statistics computation and counting can be done in time $O(\log n)$ on a cellular pyramid machine (see Section 24.1). Each pixel passes its value to its "father" on the level above it, which counts or consolidates the values received from its sons and passes on the results to its own father; thus after $\log n$ steps (the number of levels), the cell at the apex of the pyramid has the final desired value. Note that this process makes use only of the vertical connections (between levels) in the pyramid, but not of the connections within a given level; thus it requires only a cellular *tree* machine having the pixels at its leaves. Connected component counting does require horizontal connections in the base of the pyramid in order to carry out the shrinking step, which still takes $O(n)$ time; thus a pyramid provides no great advantage in the case of component counting, and it is also of no great benefit in image-to-image operations.

24.3 REGION-LEVEL OPERATIONS

24.3.1 Region Representations

A region in an image, or in fact any subset of an image, can be represented by a two-valued "overlay" image in which the pixels belonging to the subset have value 1, and all other pixels have value 0. This representation has the advantage of being in registration with the original image, but it has the disadvantage of requiring n^2 bits of memory no matter how simple the given region may be. Regions can be represented in other ways which require less memory for simple regions. Moreover, we can compute properties of regions, and derive new regions from given ones, by operating directly on the representations. Such operations can also be implemented in parallel, but an array of processors is no longer the appropriate architecture, since the representation is no longer array-like. In this section we discuss some standard region representations and the possibility of operating on them in parallel using appropriate multiprocessor architectures.

A region can be defined by specifying its borders (there is more than one border if the region has holes); for each border, this requires the coordinates of a starting point, together with a sequence of codes defining the succession of moves from pixel to pixel around the border [3 bits per move, since successive pixels are neighbors, and a pixel has only eight immediate neighbors; see Figure 24.3(a) and (b)]. A natural architecture for parallel processing of border codes [20] consists of processors connected in ring structures, with each ring representing a border (one code per processor). Computing properties of the regions represented in this way takes time $O(m)$, where m is the border length, as it would on a sequential machine; but certain tasks that take time $O(m^2)$ when done sequentially can be done in $O(m)$ on a ring machine (e.g., computing the border codes of the union or intersection of two regions, given the codes of the regions).

Another way to represent a region is to regard each row of the image that meets the region as a succession of runs (= maximal sequences) of 0's alternating with runs of 1's. Each row is determined by specifying the starting value (1 or 0) and the sequence of run lengths [Figure 24.3(c)]. Region properties and run-length codes of derived regions can be computed directly from the code(s) of the given region(s). A simple architecture for processing run-length codes in parallel might consist of strings of processors, where each string contains the run lengths for a given row. Greater efficiency could be achieved by allowing direct connections between strings representing adjacent rows, with the processor representing a given run connected directly to the processors representing runs on the adjacent rows that overlap the given run. This approach to parallel region processing does not seem to have been systematically investigated (but see [21]).

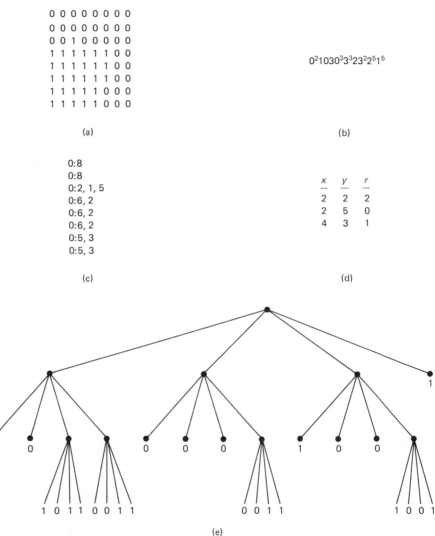

0 0 0 0 0 0 0 0
0 0 0 0 0 0 0 0
0 0 1 0 0 0 0 0
1 1 1 1 1 1 0 0
1 1 1 1 1 1 0 0
1 1 1 1 1 1 0 0
1 1 1 1 1 0 0 0
1 1 1 1 1 0 0 0

(a)

$0^2 1 0 3 0^3 3^3 2 3^2 2^5 1^5$

(b)

0:8
0:8
0:2, 1, 5
0:6, 2
0:6, 2
0:6, 2
0:5, 3
0:5, 3

(c)

x	y	r
2	2	2
2	5	0
4	3	1

(d)

1 0 1 1 0 0 1 1   0 0 1 1   1 0 0 1

(e)

Figure 24.3 Region representations. (a) Region (1's). (b) Border code (clockwise from upper left); $i = 90i°$, and i^k denotes k repetitions of i. (c) Run-length codes: value of first pixel, followed by list of run lengths, for each row. (d) Medial axis transformation: centers (x, y) and radii (r) of maximal (upright square) blocks; lower left corner of image is (0, 0). (e) Quad tree; order of sons of each node is NW, NE, SE, SW. Leaf nodes are labeled with their values.

Runs are maximal horizontal "strips" of constant-value pixels; a more compact way of representing a region is to use maximal two-dimensional blocks of constant-value pixels. Each such block is defined by specifying its center and radius, and the region is then the union of the blocks [Figure 24.3(d)]. A representation of this type, known as the medial axis transformation, was introduced about 20 years ago; but it has not been used extensively for region processing, because it is difficult to compute region properties or to derive new regions from it directly, due to the fact that the blocks overlap one another and are not organized in a systematic way. In some cases it may be possible to represent a region as a union of "generalized ribbons," where each ribbon is a union of maximal blocks whose centers all lie on a curve. Such a representation would be much more manageable, and could be processed in parallel by assigning the code of each curve (i.e., the sequence of moves and the corresponding radii) to a string of processors; this possibility has not been investigated.

Another type of maximal-block region representation can be constructed by recursively subdividing the given two-valued image into quadrants, subquadrants, ... until blocks of constant value are reached. The resulting block structure can be represented by a tree of degree 4 (a *quadtree*) in which the root corresponds to the entire image, and the sons of a node correspond to its quadrants [Figure 24.3(e)]. For an $n \times n$ image, where n is a power of 2, the height of the tree is at most $\log_2 n$, and each leaf of the tree represents a block of the image consisting entirely of 0's or 1's. Region properties, and quadtree representations of derived regions, can be computed from the quadtree(s) of the given region(s) sequentially by traversing the tree(s). Quadtree-connected sets of processors can be used to perform many of these operations in parallel very efficiently [22]. A generalization of this approach can be used to represent a multivalued image as a union of homogeneous blocks (e.g., blocks of constant value, or blocks in which the standard deviation of pixel values is low), where we divide a block into quadrants iff it is nonhomogeneous.

24.3.2 Region Properties and Relations

The region representations described above are especially useful in manipulating data bases of regions (e.g., in digital cartography). In image analysis, such representations are used for measuring region properties and for deriving new regions from given ones. In this section we consider a more abstract level of processing in which regions are not completely specified, but are represented by lists of their properties. An image segmentation can be represented, at this level, by a graph structure in which the nodes correspond to regions, labeled with lists of property values; and the arcs correspond to related pairs of regions (e.g., adjacent), labeled with relation values (e.g., length of common border). For a toy example, see Figure 24.4.

A segmentation can be modified by merging pairs of regions based on information provided by the graph representation, without any need to refer to the original image; and the graph of the new segmentation can be constructed directly from that of the given segmentation. For example, suppose that the graph contains information about the area, perimeter, and average pixel value of each region, and the length of common border of each adjacent pair of regions. The following are some possible criteria for merging a pair of adjacent regions: their average values

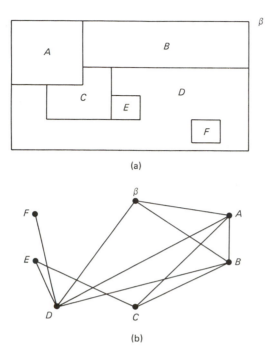

(a)

(b)

Figure 24.4   Region adjacency graph: (a) regions; (b) graph.

are very similar; their areas are very different; their length of common border is a large fraction of (one of) their perimeters. Their criteria can be checked directly from the graph. Moreover, if we decide to merge two regions, we can construct the new graph directly from the old one by replacing the two old nodes with a single node connected to all of the old nodes' other neighbors. The properties of the new node can be computed as follows. Its area is the sum of the old nodes' areas; its average pixel value is the weighted average of the old nodes' averages, weighted by their areas; its perimeter is the sum of the old nodes' perimeters minus the length of their common border. Finally, the lengths of common border between the new node and its neighbors can be computed immediately from these lengths for the old nodes; they remain the same except in the case of a neighbor common to both of the old nodes, where the two lengths must be added.

Region-merging processes such as that just described can be carried out in parallel using a network of processors in which a processor is assigned to each region, and the processors corresponding to adjacent regions can communicate directly—in other words, the processor network is isomorphic to the region adjacency graph. Thus each processor can examine the information stored at its neighbors (and at their arcs) and decide whether merging is possible. It should be pointed out that when merging is done in parallel, the decision to merge a pair of regions must be agreed to by both of them; if a region were allowed to make such a decision on its own, we might find that region *A* merges with region *B* and at the same time *B* merges with *C*, leading to an inconsistency (there are new nodes representing

$A + B$ and $B + C$, but no node for $A + B + C$, which in any case may not be an appropriate merge). To avoid this, only disjoint pairs should be allowed to merge. For further discussion of parallel region-level processing, see [18].

24.4 CONCLUDING REMARKS

This chapter has reviewed some of the basic types of operations used in image processing and analysis, at both the pixel and region levels, and has described idealized multiprocessor configurations suitable for carrying out such operations in parallel.

We have seen that for pixel-level operations taking images into images, a cellular array architecture, with processors connected in a regular grid, is very natural. For image property measurement, on the other hand, greater efficiency can be achieved by using tree-structured connections, with the processors at the leaves of the tree. For parallel processing of regions defined by border codes, ring-connected processors are appropriate. Other connection schemes are suitable if the regions are defined by maximal blocks (e.g., by run-length codes or by quadtrees). At a more abstract level, when regions are represented by lists of properties, region merging can be carried out in parallel using a network of processors connected in the same way as the region adjacency graph.

Parallel region-level processing generally requires a much smaller number of processors than does parallel processing at the pixel level. A cellular array machine for parallel processing of a 512×512-pixel image, one processor per pixel, requires $\frac{1}{4}$ million processors, which is not yet practical; but a region-level processor might require only a few hundred processors per region (depending on their complexity), or even fewer processors to handle a region adjacency graph (depending on the complexity of the segmentation). These numbers of processors are quite manageable, but their interconnections pose a problem. For pixel-level processing, the images to be processed will all be of the same size, and the neighbor interconnections are the same for every image, so that a cellular array machine can be hard-wired once and for all. For processing at the region level, on the other hand, the interconnections vary from image to image, since the shapes of the regions cannot be predicted in advance. Worse yet, we may even want the interconnections to vary in the course of a computation, as new regions are defined or old regions merged. This calls for some type of reconfigurable multiprocessor architecture [19], where ideally the reconfiguration itself should take place in parallel. For some types of representations (e.g., border codes, for which linked rings of processors can be used), such reconfiguration may be relatively easy; but for other representations, requiring tree of graph interconnections, parallel reconfiguration may not be easy to realize in such a way as to avoid serious interprocessor communication bottlenecks. As advances in hardware technology make it possible to build large multiprocessor networks, the problems involved in designing efficient systems for parallel image processing and analysis, both at the pixel and region levels, will have to be addressed.

ACKNOWLEDGMENTS

The support of the U.S. Air Force Office of Scientific Research under Grant AFOSR-77-3271 is gratefully acknowledged, as is the help of Janet Salzman in preparing this paper.

REFERENCES

[1] H. C. Andrews, *Computer Techniques in Image Processing*, Academic Press, New York, 1970.

[2] D. Ballard and C. Brown, *Computer Vision*, Prentice-Hall, Englewood Cliffs, N.J., 1982.

[3] K. R. Castleman, *Digital Image Processing*, Prentice-Hall, Englewood Cliffs, N.J., 1979.

[4] R. O. Duda and P. E. Hart, *Pattern Classification and Scene Analysis*, Wiley, New York, 1973.

[5] R. C. Gonzalez and P. Wintz, *Digital Image Processing*, Addison-Wesley, Reading, Mass., 1977.

[6] E. L. Hall, *Computer Image Processing and Recognition*, Academic Press, New York, 1979.

[7] T. Pavlidis, *Structural Pattern Recognition*, Springer-Verlag, New York, 1977.

[8] T. Pavlidis, *Algorithms for Graphics and Image Processing*, Computer Science Press, Rockville, Md., 1982.

[9] W. K. Pratt, *Digital Image Processing*, Wiley, New York, 1978.

[10] A. Rosenfeld, *Picture Processing by Computer*, Academic Press, New York, 1969.

[11] A. Rosenfeld and A. C. Kak, *Digital Picture Processing*, Academic Press, New York, 1976; 2nd ed. (2 vols.), 1982.

[12] J. Serra, *Image Analysis and Mathematical Morphology*, Academic Press, New York, 1982.

[13] A. Rosenfeld, "Picture Processing: 1981," *Comput. Graphics Image Process., 19*:35–75 (1982).

[14] S. H. Unger, "A Computer Oriented toward Spatial Problems," *Proc. IRE, 46*:1744–1750 (1958).

[15] A. Rosenfeld, *Picture Languages*, Academic Press, New York, 1979.

[16] C. R. Dyer and A. Rosenfeld, "Image Processing by Memory-Augmented Cellular Automata," *IEEE Trans., PAMI-3*:29–41 (1981).

[17] C. R. Dyer and A. Rosenfeld, "Triangle Cellular Automata," *Inf. Control, 48*:54–69 (1981).

[18] A. Rosenfeld and A. Wu, "Parallel Computers for Region-Level Image Analysis," *Pattern Recog., 15*:41–50 (1982).

[19] A. Rosenfeld and A. Wu, "Reconfigurable Cellular Computers," *Inf. Control, 50*:64–84 (1982).

[20] T. Dubitzki, A. Wu, and A. Rosenfeld, "Parallel Computation of Contour Properties," *IEEE Trans., PAMI-3*:331–337 (1981).

[21] J. S. Todhunter and C. C. Li, "A New Model for Parallel Processing of Serial Images," *Proc. 5th Int. Conf. Pattern Recog.*, 1980, pp. 493–496.

[22] T. Dubitzki, A. Wu, and A. Rosenfeld, "Region Property Computation by Active Quadtree Networks," *IEEE Trans., PAMI-3*:626–633 (1981).

25

VLSI Architectures for Pattern Analysis and Image Database Management

KING-SUN FU, KAI HWANG, AND BENJAMIN W. WAH

Purdue University
West Lafayette, Indiana

25.1 INTRODUCTION

In this chapter VLSI computing structures are introduced for the analysis and management of imagery data. Information scientists have long recognized the fact that *one picture is worth a thousand words*. Machine intelligence would be greatly enhanced if a new generation of computers could be designed to process multidimensional imagery data above the string processing of alphanumerical information by present computers. Over the last decade, extensive research and development has enabled us to achieve the capabilities of pattern analysis and image understanding by computers. Practical applications of such computers include the analysis of biomedical images, the recognition of characters, fingerprints, and moving objects, remote sensing, industrial inspection, robotic vision, military intelligence, and data compression for communications [11,12,24].

Image analysis refers to the use of digital computers for *pattern recognition and image processsing* (PRIP). On-line imagery data need to be restored on disks and quickly retrieved for PRIP applications. A VLSI-based image analysis machine should integrate both pattern-analysis and image-database-management capabilities into a unified system design [1,10].

25.2 VLSI AND COMPUTER IMAGING

We are in the era of *very large scale integration* (VLSI). Integrated circuits have penetrated all areas of human civilization. IC wafer size has increased from 1 in. to 6 in. in 20 years. Bell Laboratories have produced some 8-in. wafers and 1 megabit memory chips. Fujitsu in Japan has announced a new nonsilicon device, *high-electron mobility transistor* (HEMT), which has a 17-ps switching time, 30 times faster than the fastest silicon counterparts. The *very high speed integrated circuit* (VHSIC) project was challenged to produce 4-ps devices with some success. In the research community, *wafer-scale integration* (WSI) has been vigorously considered in implementing algorithmically specialized computer structures.

Advances in VLSI technology have triggered the thought of implementing many signal/image processing algorithms directly on specialized hardware chips. To promote image understanding, a back-end image database machine will be highly desirable in future computers. The new concept of data-driven computations can be adopted for artificial intelligence applications. New concepts on VLSI computing architectures and asynchronous data-flow multiprocessors should be seriously explored for potential use in image processing and pattern recognition. Extending control flow computers from two-dimensional arrays to three-dimensional pyramid structures is also a viable approach. Multiple-pipeline computers are also good candidates for parallel image processing if task scheduling problems can be efficiently solved.

The principal deficiency of today's computers lies in their input/output mechanisms. Computers still cannot communicate efficiently with human beings in natural forms, such as spoken or written languages, pictures or images, documents, and

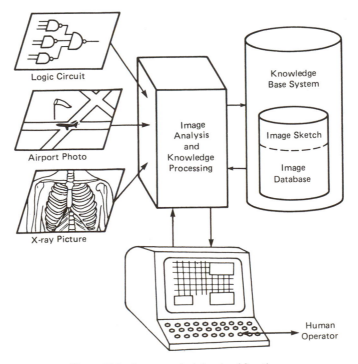

Figure 25.1   Image analysis/retrieval functions.

illustrations. Existing computers are far from satisfactory in their slowness in I/O and lack of speech, vision, translation, and real-time responses. To develop a "human-oriented" interactive computer will require, first, upgrading their capability to understand "natural" information representations and to respond to them intelligently and perhaps more reliably than human beings. Natural language and speech processing are beyond the scope of this presentation. We focus below on developing intelligent computers with image analysis/retrieval functions.

To establish these functions, one needs to develop subsystems for *input*, *output*, and *analysis* of imagery data retrieved from a large *image database system*. Pictorial functions of such an intelligent image analysis computer are conceptually illustrated in Figure 25.1. The input forms may be line drawings or gray-scaled images such as logic circuit diagrams, chest x-ray pictures, and airphoto images. The corresponding outputs may be the classification and/or interpretation of the input data, such as the layout of a VLSI circuit design, a precise description of the abnormality in the lung area, or a combat map generated in real time. The image database machine may be a part of a large-knowledge-base system. Many image-analysis functions need to be built into the middle processing section for image enhancement and segmentation, feature extraction, pattern classification, structural analysis, image description, and interpretation, as listed in Table 25.1. Some of these functions can be implemented directly by VLSI hardware and some by special software packages. Advances in image I/O devices, pictorial query processing, and image database management techniques are demanded to construct an integrated image analysis/retrieval system.

TABLE 25.1   CANDIDATE IMAGE ALGORITHMS FOR VLSI

Image processing	Enhancement, filtering, fhining, edge detection, segmentation, registration, restoration, clustering, texture analysis, convolution, Fourier analysis, etc.
Pattern recognition	Feature extraction, template matching, statistical classification, graph algorithms, syntax analysis, change detection, language recognition, scene analysis and synthesis, etc.
Image query processing	Query decomposition, query optimization, attribute manipulation, picture reconstruction, search/sorting algorithms, query-by-picture-example implementation, etc.
Image database processing	Relational operators (JOIN, UNION, INTERSECTION, PROJECTION, COMPLEMENT), image-sketch-relation conversion, similarity retrieval, data structures, priority queues, dynamic programming, spatial operators, etc.

25.3 VLSI IMAGE PROCESSORS

Recently, many attempts have been made in developing VLSI devices for signal/ image processing and pattern analysis. The statistical approaches to PRIP often involve large-scale matrix computations. The structural approaches require the performance of syntax analysis and parsing operations. We present below example designs of pipelined VLSI architecture, for statistical feature extraction and classification. The pipeline stages are constructed with modular arithmetic devices as functionally specified in Figure 25.2. These VLSI arithmetic modules will be used

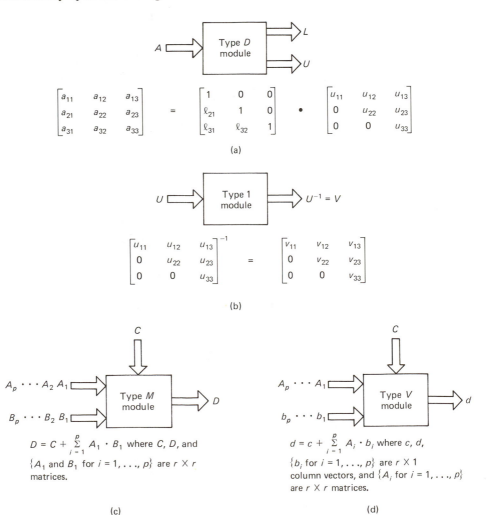

$$
\begin{bmatrix} a_{11} & a_{12} & a_{13} \\ a_{21} & a_{22} & a_{23} \\ a_{31} & a_{32} & a_{33} \end{bmatrix} = \begin{bmatrix} 1 & 0 & 0 \\ \ell_{21} & 1 & 0 \\ \ell_{31} & \ell_{32} & 1 \end{bmatrix} \cdot \begin{bmatrix} u_{11} & u_{12} & u_{13} \\ 0 & u_{22} & u_{23} \\ 0 & 0 & u_{33} \end{bmatrix}
$$

(a)

$$
\begin{bmatrix} u_{11} & u_{12} & u_{13} \\ 0 & u_{22} & u_{23} \\ 0 & 0 & u_{33} \end{bmatrix}^{-1} = \begin{bmatrix} v_{11} & v_{12} & v_{13} \\ 0 & v_{22} & v_{23} \\ 0 & 0 & v_{33} \end{bmatrix}
$$

(b)

$D = C + \sum_{i=1}^{p} A_1 \cdot B_1$ where C, D, and $\{A_1 \text{ and } B_1 \text{ for } i = 1, \ldots, p\}$ are $r \times r$ matrices.

(c)

$d = c + \sum_{i=1}^{p} A_i \cdot b_i$ where c, d, $\{b_i \text{ for } i = 1, \ldots, p\}$ are $r \times 1$ column vectors, and $\{A_i \text{ for } i = 1, \ldots, p\}$ are $r \times r$ matrices.

(d)

Figure 25.2   Primitive VLSI matrix arithmetic modules: (a) submatrix decomposition module; (b) submatrix inverter; (c) matrix multiplier; (d) matrix-vector multiplier.

iteratively in submatrix computations. This pipeline architecture is based on the "partitioned" matrix algorithms developed in [9] for *L-U decomposition, matrix multiplication, inversion of triangular matrices,* and *solving triangular systems of equations.*

Figure 25.3 shows the pipelined structure for implementing the "partitioned" *matrix inversion* algorithm being outlined in four steps. Each VLSI module performs an $r \times r$ submatrix computation where n is the order of the input matrix U. In practice, $n = kr$ and $n \gg r$ are assumed. The case of $k = n/r = 4$ is shown in Figure

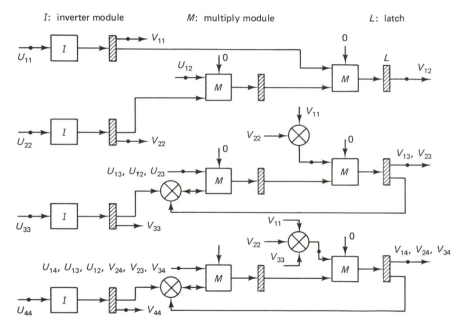

I: inverter module   *M*: multiply module   *L*: latch

Note: All U_{ij}, V_{ij} are $m \times m$ submatrices

$$
U^{-1} = \begin{bmatrix} U_{11} & U_{12} & U_{13} & U_{14} \\ 0 & U_{22} & U_{23} & U_{23} \\ 0 & 0 & U_{33} & U_{34} \\ 0 & 0 & 0 & U_{41} \end{bmatrix}^{-1} = \begin{bmatrix} V_{11} & V_{12} & V_{13} & V_{14} \\ 0 & V_{22} & V_{23} & V_{21} \\ 0 & 0 & V_{33} & V_{34} \\ 0 & 0 & 0 & V_{44} \end{bmatrix} = V
$$

Step 1: $V_{11} = U_{11}^{-1}$; $V_{22} = U_{22}^{-1}$; $V_{33} = U_{33}^{-1}$; $V_{44} = U_{44}^{-1}$ (*I* modules)

Step 2: $V_{12} = -V_{11} \cdot (U_{12} \cdot V_{22})$; $V_{23} = -V_{22} \cdot (U_{23} \cdot V_{33})$; $V_{34} = -V_{33} \cdot (U_{34} \cdot V_{44})$ (*M* modules)

Step 3: $V_{13} = -V_{11} \cdot (U_{12} \cdot V_{23} + U_{13} \cdot V_{33})$
$\qquad V_{24} = -V_{22} \cdot (U_{23} \cdot V_{34} + U_{24} \cdot V_{44})$ (*M* modules)

Step 4: $V_{14} = -V_{11} \cdot (U_{12} \cdot V_{24} + U_{13} \cdot V_{31} + U_{14} \cdot V_{41})$ (*M* modules)

Figure 25.3 VLSI matrix inversion pipeline based on Hwang/Cheng's partitioned algorithm.

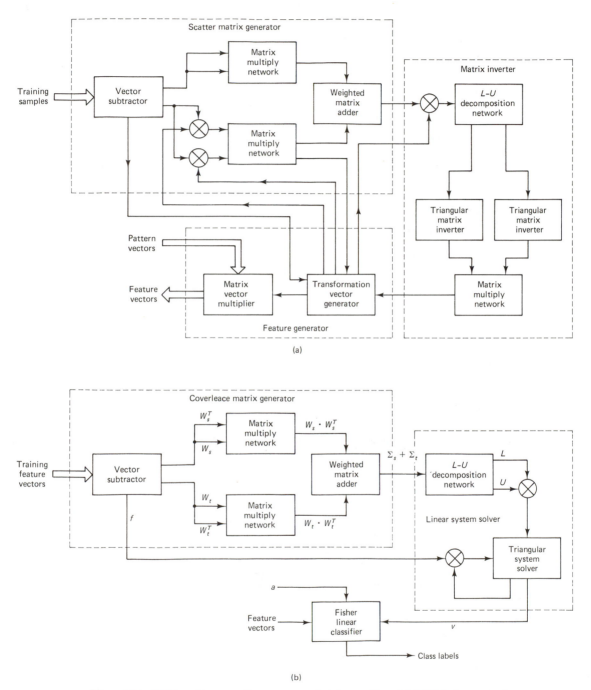

Figure 25.4 VLSI architectures for pattern analysis/recognition: (a) VLSI feature extractor; (b) VLSI pattern classifier.

25.3. For large k, this pipeline requires $O(k) = O(n/r)$ VLSI modules to implement. The total time delay to generate $V = U^{-1}$ will be $O(n^2/r)$. Similar arithmetic pipelines can be constructed for *matrix multiply*, *L-U decomposition*, and *solving triangular systems*. Details can be found in [8].

VLSI feature extraction.   Figure 25.4(a) shows the functional design of a VLSI feature extractor. This extractor is constructed with three subsystems: *scatter matrix generator*, *matrix inverter*, and *feature generator*, as shown by dashed-line boxes. The vector subtractor is implemented with modified V modules for generating the sample offset matrices and the mean difference. Two matrix multiply networks are used to perform orthogonal matrix multiplications. Each network contains n/r M modules. The weighted matrix adder can be implemented by n/r M modules with some special constant inputs. The inversion of the scatter matrix is done by employing an L-U decomposition network, two triangular matrix inverters, and one multiply network to yield the computation $A^{-1} = (L \cdot U)^{-1} = U^{-1} \cdot L^{-1}$. The feature generator can be implemented by V modules with modified constant inputs. Finally, the matrix-vector multiplier is also implemented with V modules.

VLSI pattern classification.   The functional design of a VLSI pattern classifier is sketched in Figure 25.4(b). The schematic design of the *covariance matrix generator* is similar to the scatter matrix generator in Figure 25.4(a). The *linear system solver* is composed of an L-U decomposition network and a triangular system solver. This matrix solver is needed to triangularize a dense system. The Fisher classifier is implemented by some combinational logic circuits and modified V modules.

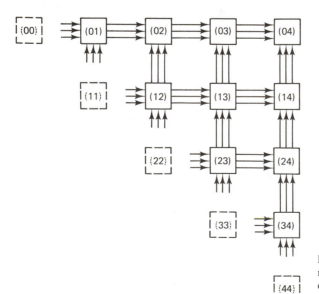

Figure 25.5 VLSI architecture for fast recognition of context-free languages (the case of $n = 4$ is shown). Solid square: cell; dashed square: boundary location.

Context-free language recognition.   In syntactic image analysis, an image pattern is often represented by a string in a context-free language. Recognition of an image pattern is accomplished through a parsing of the string with respect to a given pattern grammar [14]. A VLSI systolic array for high-speed recognition of context-free languages is shown in Figure 25.5. The recognition process is based on the Cocke–Kasami–Younger algorithm. This pipelined triangular array, constructed of $n(n + 1)/2$ processing cells, can be applied in syntactic pattern recognition. Each cell has two unidirectional data channels and one control line along each direction. Data appear as strings of symbols flowing through the recognition matrix from left to right and bottom to top as shown in Figure 25.5. This two-dimensional array can recognize any input string of length n in $2n$ time units. This context-free language recognizer and its extension to recognize finite-state languages are described in more detail in [3]. A VLSI architecture for high-speed recognition of context-free languages using Earley's algorithm has recently been proposed [21].

VLSI seismic classification.   A special-purpose VLSI processor is presented below for fast classification of seismic waveforms [20]. This special-purpose processor which contains three systolic arrays can be attached to a host computer.

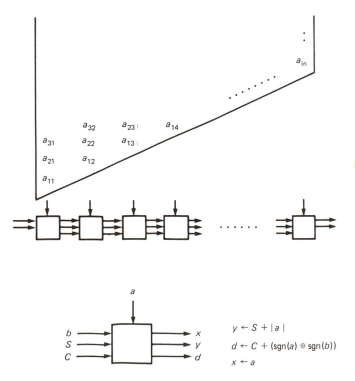

Figure 25.6 Processor array, data movement and operations of each processor for feature extraction.

Each systolic array has time complexity $O(1)$ provided that input data can be properly supplied.

The systolic array for feature extraction contains linearly connected processing elements, as shown in Figure 25.6. The input data, which are the digitized and quantized seismic waveform coded in binary form, are stored in separate memory modules in a skewed format. Two features, zero-crossing count and sum of absolute magnitudes, are computed. Zero crossing is detected by checking the signs of every two consecutive points. An exclusive-OR circuit is used for the detection of sign change. All the n processing elements (PEs) compute the two features simultaneously and pass the partial results to the next PEs.

For primitive recognition, we compute the distance between an unknown feature vector and each reference vector, for example, mean vector, of each cluster (primitive), and then assign the unknown feature vector to the cluster of the minimum distance. This procedure can be divided into two steps: compute the distances between the unknown feature vector and the reference vectors, and then select the

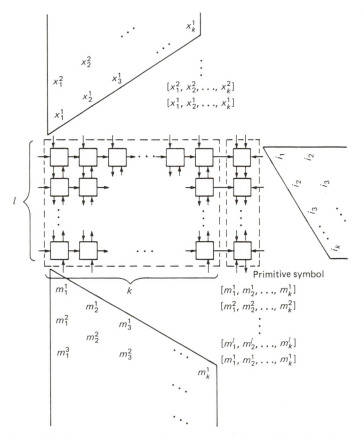

Figure 25.7   Processor arrays and data movement for primitive recognition.

one associated with the smallest distance. We use a processor array which contains *compute* processors for distance computation, and a processor array which contains *compare* processors for distance comparison. Suppose that there are l primitives; each primitive i has a reference feature vector $[m_1^i, m_2^i, \ldots, m_k^i]$, where k is the total number of features. A processor array of l by k which performs the distance computation is shown in Figure 25.7. The reference vectors of the primitives enter from the bottom and move up while the unknown feature vectors enter from the top and move down.

It is well known that the Levenshtein distance between two strings can be computed by a dynamic programming procedure. Therefore, it can be implemented by parallel processing on VLSI architectures. For Levenshtein distance, each insertion, deletion, and substitution is counted as one error transformation. We have developed a processor array for this string-matching computation in Figure 25.8. The proposed string matcher can be used for any problem where the Levenshtein distance computation is required. It can be used for string matching in our seismic recognition, for character string matching in information retrieval or for pattern

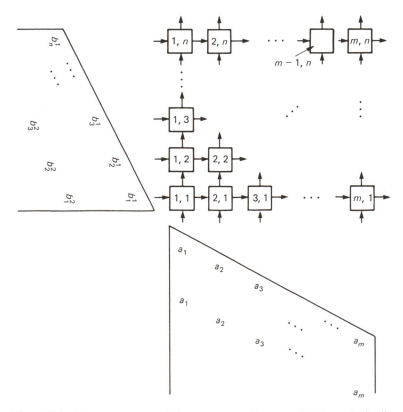

Figure 25.8 Processor array and data movement for computing Levenshtein distance.

matching in shape analysis if the object can be represented by a string, for example, using chain codes. The primitive recognizer can also be applied to any minimum-distance recognition problem and vector pattern matching.

Simulations have been performed for the three systolic arrays: feature extraction array, primitive recognition array, and string matching array. The design of the systolic arrays has been shown to be correct and the operations are as expected. Details of the systolic arrays and simulation results above were reported in Liu and Fu [20].

25.4 IMAGE DATABASE MACHINES

An image database system provides a large collection of structured imagery data (digitized pictures) for easy access by a large number of users. It provides both high-level quay support and low-level image access. Most image database systems are implemented with specially developed software packages upon dedicated pattern-analysis systems. It is highly desirable to develop a dedicated back-end database machine for image database management. So far, several hardware attempts were suggested [4,16,22]. But none of them has been actually implemented for image database management.

Image database management functions and peripheral supports are depicted in Figure 25.9. First, we need faster and intelligent image input devices. The image features and structures (shape, texture, and spatial relationships) extracted by the host image processor should be converted into symbolic image sketches stored in a *logical* image database. For those unconverted raw images, the system must convert them into efficient codes stored in the *physical* image database. Flexible image manipulation and retrieval functions must be established using high-level image manipulation languages and image description languages. The logical database is

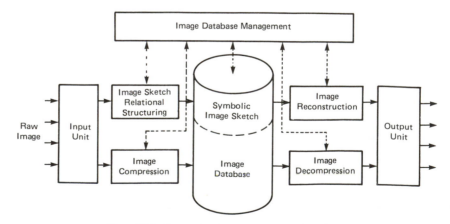

Figure 25.9   Image database management functions.

used for image reconstruction from relational sketches. The compressed raw images must be decompressed for high-resolution console display. The output unit is responsible for extracting results to be sent to the host computer. The image database management functions above should be supported with specially designed hardware units that constitute an image database machine.

A database machine for image processing can be identified to have the following functional features. High-level database functions such as selection, projection, and join are implemented. These operations are useful for manipulating the image database. On the other hand, low-level image processing operations such as histogramming and edge detection are also implemented. An image database machine is, therefore, a conventional database machine enhanced with low-level image-processing hardware.

A general assumption about VLSI chips is that they are inexpensive. For complex operations, this is not really true, due to the fact that external control, timing, memory, and software must be provided. Furthermore, as the types of VLSI chips increase and the degree of replication is large, the system becomes expensive. A solution to this problem is to use a resource-sharing interconnection network so that a pool of common resources can be used. This concept is illustrated in Figure 25.10. VLSI chips are distributed into each storage module. They can be used to real-time off-the-track processing. A pool of common resources are also shared

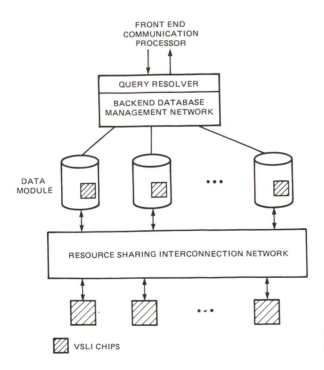

Figure 25.10 Conceptual view of an image database machine.

among the storage modules. The resource-sharing interconnection network connects these resources to the storage medium. The shared database operators will be used for data filtering, projection, join, or other operations if a relational image database is established. Some VLSI database operators are listed in Table 25.1.

25.5 DISTRIBUTED SCHEDULING OF RESOURCES

In general, an interconnection network routes requests from a set of source points to a set of destination points (they may coincide with each other). In a *resource-sharing mode*, the destination points are identical (or sets of identical) resources such as special-purpose VLSI chips for which requests or tasks can be delegated. In this respect, jobs initiated at source processors can be sent to any one of the free resources of a given type at the destination. This is the important point that differentiates resource sharing from address mapping.

Since the system operates continuously, requests from source processors can be initiated at random times. At any time, a set of processors may be making requests and a set of resources are free. It is the function of a scheduler to route the requests in order to connect the maximum number of resources to the processors, that is, to have the maximum resource utilization.

The earliest study of networks for resource sharing has been realized with centralized control. A unibus is used in a time-shared fashion for connecting peripheral I/O devices to the CPU. Multiple time-shared buses have been used in the PLURIBUS minicomputer multiprocessor. A cross-bar switch has been used in C.mmp, although the network is mostly used in the address mapping mode. The single- or multiple-bus approach is a source of bottleneck, and is the least expensive design. The cross-bar switch is the most expensive network but has the least degree of blocking. A compromise is to use a less expensive network than the cross-bar switch and which has less blocking probability than single-bus systems. This has been studied with respect to the Banyan network. A tree network is proposed to aid the scheduler in choosing a resource to allocate. The tree network has a delay of $O(\log_2 n)$ in selecting a free resource (n is the total number of resources).

The scheduling algorithms studied earlier are centralized and use address mapping interconnection networks. For mapping n requesting processors to n resources, the scheduling algorithm has a worst-case complexity of $O(n \log_2 n)$. This complexity depends on the number of requesting processors. This is practical when n is small or when requests are not very frequent. A solution that avoids the sequential scheduling of requests is to allow requests to be sent without any destination tags, and it is the responsibility of the network to route the maximum number of requests to the free resources. In this way, the scheduling intelligence is distributed in the interconnection network. This approach permits multiple requests to be routed simultaneously. We termed this network a *resource-sharing interconnection network*.

The Omega and generalized cube networks belong to a class of networks with the property that the delay from a source to any reachable destination is proportional to the logarithm of the number of source points. The basic element in these networks is a two-input two-output four-function interchange box which allows a straight, exchange, upper broadcast, or lower broadcast connection. For a network connecting N inputs to N outputs (N is a power of 2), there are $\log_2 N$ stages and $(N/2) \log_2 N$ interchange boxes. The delay in the networks is, therefore, $O(\log_2 N)$. The $O(N \log_2 N)$ hardware complexity is much better than that of the cross-bar switches, $O(N^2)$.

Since the networks have nonzero blocking probability, some of the feasible mappings from sources to destinations do not lead to maximal resource allocation. A centralized scheduler has to examine all the different possible ordered mappings in order to allocate the maximum number of resources. Suppose that x processors are making requests and y resources are free. The scheduler has to try a maximum of $\binom{x}{y} y!$ (for $x \geq y$) or $\binom{y}{x} x!$ (for $y > x$) mappings in order to find the best one. Suboptimal heuristics can be used [19], but will only be practical when x and y are small.

On the other hand, a distributed scheduling algorithm allows all the requests to be scheduled in parallel. The resource scheduling overhead is, therefore, proportional to the delay time in the network, $O(\log_2 N)$, and independent of the number of requesting processors.

The distributed algorithm is implemented by distributing the routing intelligence into the interconnection network so that there is no centralized control. Each exchange box can resolve conflicts and route requests to the appropriate destinations. If a request is blocked, it will be sent back to the originating exchange box in the previous stage. Request routing is thus dynamic, and all the exchange boxes operate independently.

The distributed algorithm is illustrated in Figure 25.11 on an 8×8 Omega network. Suppose that resources R_0, R_1, R_4, and R_5 are available and status information is passed to the processors. The numbers on the output/input ports represent the status information received/sent. Assuming that P_0, P_3, P_4, and P_5 are requesting one resource each, the requests are sent simultaneously to the network after new status information arrives. In stage 0, no conflict is encountered. $B_{1,1}$ in stage 1 receives two requests. Since only one output terminal leads to free resources, the request originating from $B_{0,3}$ is rejected. This request, subsequently, finds another route via $B_{1,3}$ and $B_{2,2}$ to R_5. In this example, each request has to pass through 3.5 exchange boxes on the average before it finds a free resource. For clarity, status changes due to new requests are not indicated in the figure.

The resource-sharing network discussed here is a generalization of address-mapping interconnection networks with routing tags. An address-mapping network is a resource-sharing network connecting processors and multiple types of resources with one resource in each type. In a resource-sharing mode, multiple resources are allowed in each type. This resource-sharing network has a uniform structure suitable for VLSI implementation.

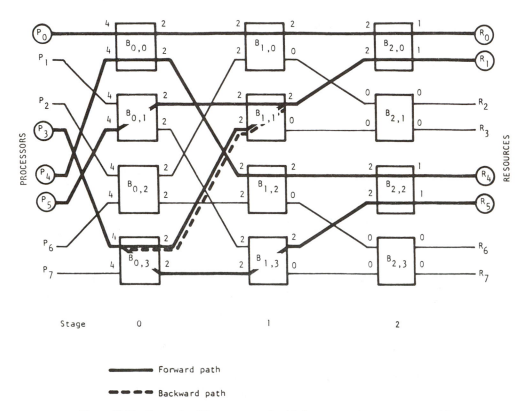

Figure 25.11 Example of Omega network with four requesting processors and four free resources (25% of requests are blocked and backtracked; 100% resource allocation; average delay = 3.50 units).

25.6 CONCLUDING REMARKS

Feature extraction and pattern classification are initial candidates for possible VLSI implementation. The Foley–Sammon feature extraction method [5] and the Fisher linear classifier have been proposed for VLSI implementation [8]. Other methods, such as the eigenvector approaches to feature selection and Bayes's quadratic discriminant functions, should be also realizable with VLSI hardware. It is highly desirable to develop VLSI computing structures for smoothing, image registration, edge detection, image segmentation, texture analysis, multistage feature selection, syntactic pattern recognition, pictorial query processing, image database management, and so on. The potential merit lies not only in speed gains, but also in reliability and cost-effectiveness.

Image analysis and image database management are two functions that cannot be separated in an efficient pictorial information system. The integrated system approach is supported by the merging VLSI technology and by various parallel processing techniques. Cost-effectiveness is the key issue in developing special-purpose machines for image processing, recognition, and database management. Toward the eventual VLSI realization of an integrated image analysis/retrieval computer, we suggest below a number of important research projects.

1. Develop systematic design methodology for mapping PRIP algorithms into VLSI hardware architectures [3,6–9,15,20,21].

2. Develop VLSI devices for image description, image manipulation, and pictorial query processing [2,11,13,17].

3. Develop back-end image database machines, including both image database structures and management policies [4,16,22].

4. Develop the resource arbitration networks for a multiprocessor system with a shared VLSI resource pool [1,18,19].

5. Investigate the possible use of the data flow concept in designing VLSI systems for PRIP and artificial intelligence computations [1,7,23].

6. Integrate VLSI architectures for image analysis with those for natural language and speech processing [10,23,24].

ACKNOWLEDGMENT

This work was supported by National Science Foundation Grant ECS 80-16580.

REFERENCES

[1] F. A. Briggs, K. S. Fu, K. Hwang, and B. W. Wah, "PUMPS Architecture for Pattern Analysis and Image Database Management," *IEEE Trans. Comput.*, Oct. 1982, pp. 969–982.

[2] S. K. Chang, J. Reuss, and B. H. McCormick, "Design Considerations of a Pictorial Database System," *Intl. J. Policy Anal. Inf. Syst.*, *1*(2):49–70 (Jan. 1978).

[3] K. H. Chu and K. S. Fu, "VLSI Architectures for High-Speed Recognition of General Context-Free Languages and Finite-State Languages," *Proc. 9th Intl. Symp. Comput. Arch.*, Austin, Tex., Apr. 1982, pp. 43–49.

[4] T. DeWitt, "DIRECT: A Multiprocessor Database Machine," *IEEE Trans. Comput.*, 1979, pp. 395–406.

[5] D. H. Foley and J. W. Sammon, "An Optimal Set of Discriminant Vectors," *IEEE Trans. Comput.*, Mar. 1975, pp. 281–289.

[6] M. J. Foster and H. T. Kung, "The Design of Special-Purpose VLSI Chips," *Comput. Mag.*, Jan. 1980, pp. 26–40.

[7] K. Hwang and F. A. Briggs, *Computer Architectures for Parallel Processing*, McGraw-Hill, New York (in press to appear).

[8] K. Hwang and S. P. Su, "VLSI Architectures for Feature Extraction and Pattern Classification," *J. Comput. Graphics Image Process.*, (accepted to appear in 1983).

[9] K. Hwang and Y. H. Cheng, "Partitioned Matrix Algorithms for VLSI Arithmetic Systems," *IEEE Trans. Comput.*, Dec. 1982, pp. 1215–1224.

[10] K. Hwang and K. S. Fu, "Integrated Computer Architectures for Pattern Analysis and Image Database Management," *Computer*, Jan. 1983, pp. 51–61.

[11] M. Onoe, K. Preston, and A. Rosenfeld, eds., *Real-Time/Parallel Computing: Image Analysis*, Plenum Press, New York, 1981.

[12] K. Preston, Jr., and L. Uhr, eds., *Multicomputers and Image Processing*, Academic Press, New York, 1982.

[13] N. S. Chang and K. S. Fu, "Picture Query Languages for Pictorial Database Systems," *Computer, 14*, Nov. 1981.

[14] K. S. Fu, *Syntactic Pattern Recognition and Applications*, Prentice-Hall, Englewood Cliffs, N.J., 1982.

[15] E. E. Swartzlander, "VLSI Architecture," in D. F. Barbe, ed., *Very Large Scale Integration (VLSI): Fundamentals and Applications*, Springer-Verlag, New York, 1980.

[16] B. W. Wah and S. B. Yao, "DIALOG—A Distributed Processor Organization for Database Machines," *AFIPS Conf. Proc.*, Vol. 49, 1980, NCC, pp. 243–253.

[17] M. Yamamura, N. Kamibayashi, and T. Ichikawa, "Organization of an Image Database Manipulation System," *Proc. Workshop Comput. Arch. for PAIDM*, Hot Springs, Va., 1981, pp. 236–241.

[18] D. W. L. Yen and A. V. Kulkami, "The ESL Systolic Processor for Signal and Image Processing," *Proc. Workshop Comput. Arch. for PAIDM*, Hot Springs, Va., 1981, pp. 265–272.

[19] B. W. Wah, "A Comparative Study of Resource Sharing on Multiprocessors," *IEEE Trans. Comput.*, Aug. 1984.

[20] H. H. Liu and K. S. Fu, "VLSI Systolic Processor for Fast Seismic Classification," *Proc. 1983 Intl. Symp. VLSI Tech. Syst. Appl.*, Taipei, Taiwan, Mar. 31, 1983.

[21] Y. T. Chiang and K. S. Fu, "A VLSI Architecture for Fast Context-Free Language Recognition (Earley's Algorithm)," *Proc. 3rd Intl. Conf. Distributed Comput. Syst.*, Oct. 1982.

[22] K. Yamaguchi and T. L. Kunii, "PICCOLO Logic for a Picture Database Computer and Its Implementation," *IEEE Trans. Comput.*, Oct. 1982, pp. 983–996.

[23] K. Hwang, "Computer Architectures for Image Processing," (Guest Editor's Introduction), *Computer*, Jan. 1983, pp. 10–13.

[24] K. S. Fu, ed., *Applications of Pattern Recognition*, CRC Press, Boca Raton, Fla., 1982.

26

Signal Processing in High-Data-Rate Environments: Design Trade-offs in the Exploitation of Parallel Architectures and Fast System Clock Rates

B. K. Gilbert, T. M. Kinter, D. J. Schwab, B. A. Naused,
L. M. Krueger, and W. Van Nurden

Mayo Foundation
Rochester, Minnesota

R. Zucca

Rockwell International Microelectronics
Research and Development Center
Thousand Oaks, California

26.1 INTRODUCTION

This book explores the methods by which the emerging very large scale integration (VLSI) technology (i.e., the ability to place more than 10,000 logic gates on a single integrated circuit) can be exploited for the solution of difficult signal processing problems. The following discussion will concentrate on a highly specialized subset of the total signal processing environment (i.e., that small minority of such problems in which a single unprocessed data stream appears at the input of a digital processor in real time and at very high data bandwidths). These high-volume data streams must be processed, at least by the "front end" of the signal processor, at clock rates equal to or greater than the rates at which they are delivered; in later stages of processing, it may be possible to partition the single high-speed data stream into a series of lower-speed substreams, and to institute parallel processing on the substreams. Since the early 1970s this laboratory has investigated potential solutions to these high-data-rate problems, and has compared these problems with the capabilities of silicon VLSI, as well as other technologies, with which they may be addressed.

The evolution of digital integrated circuit technology for commercial application, and even to some extent in the military world, has been directed to the fabrication of ever-smaller physical structures and device geometries on mass-produced integrated circuits. This desire for high device count and high packing density has resulted in an emphasis on structures such as NMOS, CMOS on bulk silicon, and CMOS on sapphire (CMOS/SOS), whose device characteristics permit gate propagation delays in the range 2 to 20 ns (Figure 26.1); the most advanced of these structures can support system clock rates in the range 30 to 50 MHz. However, several classes of problems (e.g., electronic warfare, radar signal processing, wideband spread-spectrum military communications, and even certain computation intensive biomedical tasks) may require unprocessed input data bandwidths as great as 0.1 to 2×10^9 bytes/s (e.g., [1]), necessitating system clock rates as high as 2×10^9 Hz. High-density, low-clock-rate VLSI technology is at present incapable of solving these types of problems; alternative methods must be identified. Several examples of these difficult computational problems derived from the world of biomedicine will be discussed in the following section.

26.2 EXAMPLES OF COMPUTATION— INTENSIVE PROBLEMS FROM BIOMEDICINE

The impact of computers upon biomedicine has perhaps been best publicized in the discipline of x-ray computed tomography (CT). The x-ray computed-tomography scanner melds an x-ray source and detector technology, which collects x-ray penetration data in a nonimage format, with a computer, which reorganizes the information content of these data to create an image of the body from an aspect not achievable with conventional radiographic equipment [2,3].

The developmental trends of these machines have been apparent since their announcement in 1973; enthusiasm for CT imagery has created a demand for ever-increasing spatial and gray-scale resolution, the ability to image ever-thinner slices, and ever-shorter durations between patient exposure and the availability of the processed results. Although these enhanced capabilities are being achieved, an increasing burden is being placed on both the sophistication of the sensors and on the computer technology. The image-reconstruction algorithms employed by modern CT scanners are extremely computation intensive; 1000 to 10,000 arithmetic operations are required to reconstruct the gray-scale value of each pixel in a CT image. In 1973 an 80×80 pixel image could be reconstructed on a computer executing 2×10^5 operations/s within 5 to 6 minutes. By the late 1970s, the resolution of the CT images (up to 512×512 pixels) and the speed with which they were required to be generated had raised the minimum computational capabilities incorporated into the CT scanners to 10 million arithmetic instructions per second (usually consisting of a large minicomputer and a commercially available array processor); several hundred million separate steps are now required to produce a typical x-ray CT image. Nevertheless, when multiple cross sections are required for a given examina-

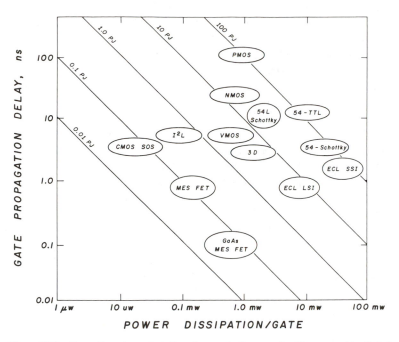

Figure 26.1 Operating characteristics of several silicon and gallium arsenide digital logic families.

tion, the physician may be compelled to wait for processed results for 1 to 2 hours after completion of the scan sequence.

A second mode of medical diagnostic imaging which has been widely investigated in the past few years relies on ultrasound as the probing energy. Ultrasonic transceivers can be operated in the so-called echo mode: in a manner analogous to submarine sonar, in which images are generated on the basis of backscatter of acoustic energy from reflecting substructures within the volume of tissue under study. However, it must be noted that these images are qualitative in nature, depending in large measure for their usefulness on the peculiarities of each machine design and on the skill of the operator.

The limitations of ultrasound echo scans have motivated investigations of more quantitative ultrasound imaging techniques relying on extensions of x-ray computed tomography. In x-ray CT, the interaction between the impinging energy and the tissue through which it passes is limited to an absorption phenomenon that varies with beam energy. Ultrasound energy interacts with tissue in a more complex manner described by so-called wave equations, which predict the interaction of the tissue and the ultrasound waves. The tissue is assumed to be an inhomogeneous medium possessing values of compressibility, density, attenuation, scattering cross section, and so on, which vary with location and orientation, and with the center frequency and spectral content of the impinging energy source.

Generalized wave equations written for such complex (i.e., typical) tissues cannot at present be evaluated in closed form with sufficient ease to allow the computation of the ultrasonic material properties for each element of tissue being imaged, as would be necessary to create a material property-dependent ultrasonic CT "image." Hence investigators have applied various simplifying assumptions and approximations to the general wave equations, thereby creating the basis of a transmission ultrasound CT imaging technique [4,5]. Ultrasound CT, although not exhibiting the spatial resolution of x-ray CT, nonetheless has provided quantitative images whose characteristics are independent of the equipment or the operator. The various interaction modalities of ultrasound with tissue can all be exploited to create distinct sets of images related to the various material properties of tissue, such as attenuation and acoustic speed, which yield independent but supportive sets of results [6].

Ultrasound CT images have not yet attained the spatial and gray-scale resolution of, and are not as artifact-free as, x-ray CT images. The image blurring and artifacts result in large measure from reconstruction errors traceable to the above-described simplifications of the generalized wave equations, which render their numerical solution on a computer less time consuming and hence less costly. In principle, reliance on more accurate representations of the wave equations which correctly account for the inhomogenieties in the tissue material properties would result in reconstructed images exhibiting improved definition. The generalized wave equations are highly nonlinear, but nonetheless appear to be soluble by a variety of numerical techniques [7]. Presently, the high cost of implementing such sophisticated numerical methods on large computers has precluded aggressive attacks on these problems. The magnitude of the computational task for true three-dimensional ultrasound imaging through exploitation of the generalized wave equations has been estimated to be from one to several orders of magnitude greater than for the various x-ray CT examples described earlier (i.e., in the range of 10^8 to 10^{11} arithmetic operations/s).

The capabilities of powerful digital computers and computer-driven calligraphic color displays are beginning to alter the tools and methodology of biochemical research as well. With the aid of specialized computer hardware and software, it is becoming feasible to study the detailed structure of complex macromolecules and their chemically active binding sites, and to investigate both the geometric interactions of receptors and the chemically active sites on potentially useful drugs and the reaction kinetics of these molecular interactions.

Formerly, most of the methods employed to study these facets of macromolecular structure and function could only indirectly probe the molecules under investigation and, in general, only x-ray crystallography could delineate the atomic substructure of biological molecules such as proteins. Based on the x-ray crystallographic data, the investigator was then compelled to build laboriously a physical three-dimensional stick-and-wire model of the molecule.

It is now feasible, however, to enter the crystallographic data for any macromolecule into a computer, then to display the molecule in color and, via rotation of the image, to create progressively a three-dimensional representation of the structures or any part thereof, from any desired angle of view [8]. Appropriate computer programs allow one to "peel away" the outer portions of a complex molecule to reveal its inner structure, which may contain important voids or invaginations, and even to present a view of the molecule from its inside looking out.

It has been pointed out that because macromolecules are flexible and can distort in one another's presence, sole reliance on crystallographic data to assess possible chemically active sites on these structures may be misleading. Hence computer programs have been developed to identify single covalent bonds in the vicinity of a potential active site, and to rotate, stretch, and bend these bonds while examining the resulting three-dimensional conformations of the active sites [9]. In addition, these flexion-torsion studies may be combined with calculations of the total bond energy in the structure which includes the active site, to assess the "most probable" shapes of the active regions [10].

In addition, specialized computer programs can substitute one substructural fragment for another in the "drug" under study; using a computer-controlled color graphics screen, the investigator can identify a small subsection of the entire complex molecule, interactively replace, for example, an existing structural subunit with another fragment of somewhat different three-dimensional conformation, and retest the "drug" for chemical activity. This computer-based approach not only allows a very rapid and thorough analysis of existing structures, but in a very real sense, allows true computer-assisted design of drugs.

The computer power of presently available drug design systems, which rely on a large minicomputer and an associated commercially available array processor, limits the exhaustiveness of computer searches of potentially matching drug-tissue receptor conformations to the rotation and bending of not more than four interatomic bonds, whereas it has been pointed out that the ability to test 8 to 10 bonds, or occasionally even more, would be a powerful extension of present capabilities [10]. Executed as a lengthy batch-processing task, the four-bond problem requires about 10 to 12 million arithmetic operations per second, whereas the 10-bond problem may require 40 to 50 million arithmetic operations per second. If performed in real time, or nearly so, as desired by the biochemists, these calculations would require execution at rates of 5 to 50×10^8 arithmetic operations per second.

Finally, there is continuing interest in the use of computers to evaluate mathematical models of biological processes, including the transport of gasses and nutrients across biological membranes, and the electrical conduction properties of nerve and muscle fibers as a function of local concentrations of ions in the bathing medium [11]. Analysis of the fluid mechanics of blood flow through the vasculature, through the cardiac chambers, and around prosthetic heart valves by the two-dimensional solution of simplified versions of the Navier–Stokes equations for non-Newtonian fluids (i.e., the blood) has been performed for the past decade by several investigators [12,13]. The stress and strain patterns in the walls of the cardiac chambers have also been investigated using finite-element analysis techniques originally developed to quantitate these forces in complex man-made structures such as airplane wings [14]. Many of these applications, especially the molecular modeling and blood flow problems, are of such large magnitude that they can easily consume the most powerful computational resources currently available or envisioned for the near future.

In summary of the comments above, several interesting and potentially useful biomedical research and clinical diagnostic projects currently under way are employing and even stressing the state of the computer art, requiring throughput levels which require imaginative computational approaches and advanced processor designs. The best approaches for achieving these high throughputs are the subject of the remainder of this discussion.

26.3 ARCHITECTURAL SOLUTIONS TO THE ENHANCEMENT OF PROCESSOR THROUGHPUT

The architectural approaches that have been proposed over the past 15 years for enhancing computational throughput usually assume that a very large number of individual logic gates can be configured to achieve very high throughput by means of computational parallelism [15]. Considerable effort has been expended to identify generic methods for "parallelizing" signal and data processing algorithms, which in turn can be executed on a variety of parallel architectures, or networks of cooperating processors, as well as special designs that straddle the boundaries between these classical structures. As will be delineated below, these efforts to "parallelize" processor architectures have considerable application to the majority of signal processing problems, particularly those implemented in VLSI. However, they may not be equally applicable to that subclass of signal processing applications characterized by wide input data bandwidths, in combination with a requirement for algorithm execution in real time.

Single-instruction, multiple-data (SIMD) structures are defined as those designs that implement a set of identical coprocessors operating simultaneously on parallel data streams, executing the same instruction in each coprocessor during each clock cycle; the individual instructions may be transmitted simultaneously to all coprocessors from a single control unit, or may be stored in advance in the local memory of each coprocessor [16]. These structures are moderately easy to design and build since all coprocessors are identical; the composite system is reasonably straightforward to program and also to test once fabricated. However, these architectures assume the existence of parallel streams of data; if the coprocessors are to be implemented in a VLSI technology, the individual data streams must be constrained to exhibit only a moderate bandwidth. If the input data bandwidth is high, it becomes necessary to design a "front-end" subsystem fast enough to partition the single high-speed data stream into multiple lower-speed substreams before presentation to the VLSI coprocessors of the SIMD computer. Of course, doubling, trebling, and so on, the throughput of an SIMD processor requires *at least* a doubling, trebling, and so on, of the total system gate count, which may become a limitation in very complex systems even for VLSI components. In addition, the SIMD structure is not applicable to those computational problems in which all data streams cannot be processed identically.

MIMD designs are those in which a group of cooperating, communicating, but largely autonomous processors execute different instruction streams on several data streams which may also be quite different from one another [17,18]. The individual subprograms are usually stored within the local memories of the independent processors. MIMD structures have proven in practice to be extremely resistant to elegant physical hardware design, and are difficult to program. Perhaps more important, but frequently overlooked, is the difficulty of establishing adequate control strategies for independently operating coprocessors. Either the separate coprocessors must be autonomous but communicating, which requires extensive communications hardware and software [18], or they must be orchestrated by a central control unit [16,17]; in the latter case, development of the global control software has proven to be a monumental task. A few MIMD structures have been reported which appear to overcome these difficulties by brute-force techniques, or to circumvent them by careful selection of the class of tasks to which the processor will be applied [19]; however, the MIMD design philosophy has not been uniformly successful. As the survivors of such development projects attest, their successful completion is a tour de force of major proportions [20].

Although there have been a few noteworthy exceptions, cooperating coprocessors rarely yield a performance enhancement *in a general data or signal processing environment* equal to the product of the capabilities of an individual coprocessor and the number of coprocessors recruited to any given task. Several major processor development projects have as a result avoided the use of very large numbers of coprocessors, whether in an SIMD or an MIMD configuration [17,21], and have relied on alternate approaches.

26.4 WHY A COMPLETE SOLUTION FOR A LARGE PROCESSING PROBLEM CANNOT ALWAYS BE ACHIEVED THROUGH ARCHITECTURAL APPROACHES: A CASE STUDY

A biomedical image/signal processing task requiring computational rates approaching several billion arithmetic operations per second has been under study in this laboratory since the early 1970s. A very high-performance image processor was required to support an advanced generation truly three-dimensional real-time x-ray computed-tomography capability, the dynamic spatial reconstructor (DSR). Unlike commercially available x-ray computed-tomography machines, which collect in 5 to 20 s sufficient projection data to reconstruct only a single cross-sectional image, the DSR collects within an 11-ms duration sufficient projection data to reconstruct up to 240 adjacent cross sections encompassing a cylindrical volume of tissue (e.g., the thoracic or abdominal contents) 22 cm in axial extent and 24 cm in diameter

[22,23]. Since the entire data collection procedure can be repeated 60 times each second, a scan lasting only 10 s produces sufficient projection data, *at an input data rate of 100 to 200 million bytes/second (megabytes/s)*, to require reconstruction of up to 150,000 cross sections. Reconstruction and appropriate three-dimensional display of these cross sections result in a time-varying, truly three-dimensional x-ray image of the structure under study, with potentially powerful biomedical research and clinical diagnostic possibilities [22,23].

To achieve maximum utility in a clinical diagnostic environment, the duration necessary to completely process and display the volumetric images must eventually be reduced to at most a few minutes for a few seconds of actual data collection. Numerous algorithmic and simulation studies were carried out to determine the optimum solution to this large computational task [7,22,24]. The best throughput times achieved to date using large minicomputers combined with presently available programmable array processors is approximately 2 to 10 s per cross section. It was determined, however, that several orders-of-magnitude speed improvement would be necessary in the near term to achieve the required throughput, equivalent to a computational rate of at least 4 billion arithmetic operations per second; usage projections indicated an eventual need for 90 billion operations per second by the late 1980s. Since such large throughputs did not appear supportable even by next-generation programmable array processors, it was recognized that a special-purpose processor executing a limited subclass of algorithms would have to be developed.

The most numerically efficient algorithm was selected and mapped into several feasible processor architectures; although implementational differences may be identified between these various architectures, a common design theme is that all are comprised of two different types of single-instruction-stream, multiple-data-stream (SIMD) systems. Following a brief description of the reconstruction algorithm, a generic version of these processor architectures will be described.

The CT reconstruction algorithm involves several stages. First, the digitized raw data vectors are each multiplied by a precalculated weighting vector of the same length to correct for nonuniformities in the intensity of the x-ray sources at the edges of their beams. Each of these "data vectors," which are employed to generate the reconstructed image, is then filtered by a direct convolution or Fourier transform filtration process to correct for aperture effects caused by the finite width of the x-ray beam. The filtered data vectors are then "back-projected" or "binned" into the image space to generate the gray-scale value of the cross-section image pixels. To generate the gray-scale value of a single pixel by back projection, one sample from each of the filtered data vectors is read from memory and multiplicatively weighted, followed by summation of all weighted samples. This procedure is then repeated to create all of the 16,000 to 250,000 pixels comprising a single cross-sectional image [3,14–16].

Figure 26.2 is a schematic diagram of a special-purpose parallel hardware processor designed to execute several closely related reconstruction algorithms [22,24]. The arithmetic unit depicted in the upper portion of the figure executes a linear filtration operation on individual projection data vectors each containing up

DIGITAL HIGH-SPEED PARALLEL PROCESSING FOR
DIVERGENT BEAM CONVOLUTION RECONSTRUCTION ALGORITHM
Generation of Projection Addresses (Ray Sum Indices)
& Back Projection Weights via Table Lookup

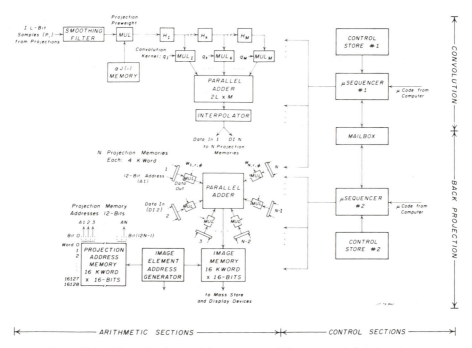

Figure 26.2 Schematic of a special-purpose parallel processor designed to execute several closely related biomedical image reconstruction (computed tomography) algorithms. Upper processor executes preweighting and filtration operations. Lower SIMD processor executes parallel binning operation. Control is via communicating microcontrollers. (*Reproduced with permission from Gilbert et al., IEEE Transactions on Nuclear Science, New York.*)

to 512 elements, and then transfers its intermediate results to the multiple subprocessors in the lower portion of the figure, which execute the "binning" portion of the algorithm.

In spite of several years of intensive work, major problems remain with the architecture described above. After considerable study of the fundamental CT image-reconstruction algorithm and the application of this algorithm to the specific project, it became evident that no reasonable mechanism exists for subdividing the computational process into more than at most 30 separate sections, which could be mapped onto no more than 30 separate coprocessors, with the entire system composed of approximately 225,000 logic gates (not including memory). Each coprocessor could, in turn, be conceptualized as a VLSI chip, especially since 28 of the coprocessors would be identical. If, then, this VLSI implementation was assumed to operate at a maximum clock rate of 20 MHz (a high clock rate for VLSI at present), it was estimated that the processor would be capable of *at most* 2 billion arithmetic operations per second. However, *the minimum* required computational throughput was estimated at 4 billion arithmetic operations per second, with a maximum eventual rate of 90 billion arithmetic operations per second, requiring system clock rates of 30 to 675 MHz. The same dilemma has appeared in a few biomedical computational tasks and an entire host of military situations; the problem can be only partially addressed through architectural solutions, algorithmic "tricks," parallelism, and pipelining.

This case study appears to be representative of a subclass of image and signal processing tasks which must satisfy a unique set of computational and economic constraints. In this environment, a processor is confronted with a signal stream of very high input data bandwidth which requires at the least that its front end, and in some cases its back end, be capable of operation at clock rates of 0.1 to 2 GHz. The central portions of such a computer may be able to operate at somewhat lower clock rates, although in many instances still greater than those attainable by current or near-future VLSI components. Note also that the microcycle clock rate of a processor must frequently be a multiple of the input data rate; for example, some algorithms for the preprocessing of radar return data require the execution of 10 to 20 microcycle steps within each intersample interval; that is, for a 100-megasample/s data rate, at least a portion of the processor executing such a task may require 1 to 2 GHz clock rates. In addition, in at least one other nontrivial subclass of such problems presently under study, the *entire processor* can be implemented in the most straightforward manner if it operates at the maximum system clock rate [25].

As a last, and perhaps the most fatal, constraint on the solution of these types of high-computational-demand problems, specialized components may be required which are optimized to the given task. However, as exemplified by the case study, for many signal processing applications requiring specially designed integrated circuits, the production runs of the components developed for a specific task may be so small that custom designs, especially custom VLSI designs, are simply unaffordable. The costs of development and layout of a large VLSI component are still formidable, even with marked improvements in computer-aided design during the past few years [26]. A fabrication run of even 50 to 100 processors, each containing custom components, is in general not nearly large enough to offset the costs of custom component design.

In summary, efforts to improve the throughput of processors for the widest-bandwidth, highest-clock-rate classical signal processing problems cannot always or entirely rely on architectural solutions intended for data processing problems, even with VLSI implementations thereof. A moderate number of system design projects, including several undertaken in this laboratory, when confronted by these constraints, have exploited enhancements in *processor speed* based on the exploitation of "faster" logic gates and the transistors from which they are assembled.

26.5 TECHNOLOGY-BASED METHODS
FOR ACHIEVING PROCESSOR SPEED

High transistor switching speed is a parameter which is difficult to exploit at the VLSI level, since the attainment of subnanosecond logic gate propagation delays in silicon integrated circuits requires the dissipation of large amounts of power (Figure 26.1). Power dissipation increases in this manner because large instantaneous current flows are required at a given logic voltage swing to charge the parasitic capacitances of the interconnecting lines between adjacent logical functions on the integrated circuit (or, for that matter, between integrated circuits). If each gate dissipates an average of 1 mW, a 10,000-gate VLSI component would dissipate 10 W, making it difficult to achieve adequate cooling of the integrated circuit in a straightforward manner. Silicon VLSI devices of 10,000-gate complexity should dissipate no more than 100 to 400 μW per gate; at this low power dissipation, switching currents are so small that it is difficult to achieve average gate propagation delays much less than 1 to 2 ns (Figure 26.1).

Although continued reductions in transistor and gate sizes could in theory result in a decrease in average gate propagation delays for the same power dissipation levels, this trend will not necessarily result in faster gate speeds on VLSI components. The reason for this is that designers will exploit the ability to fabricate smaller transistors to compact even more circuitry on a single integrated circuit than at present; for example, a 100,000-gate integrated circuit will have to constrain the power level of a single gate to 10 to 40 μW to maintain chip power dissipation in the 1 to 4W range. Figure 26.1 demonstrates that only one device technology, CMOS/SOS, approaches such a low level. Thus, for the foreseeable future, achievement of high system clock rates and high complexity levels will be incompatible design goals for silicon VLSI.

26.6 EXPLOITATION OF GALLIUM ARSENIDE DIGITAL
DEVICE TECHNOLOGY AT THE LSI LEVEL
TO OBTAIN PROCESSOR SPEED

How, then, can those signal processing problems be solved which are simultaneously constrained by high-input-data bandwidth and high system clock rate, and which require the design of special components in a low-volume-production, rapid-turnaround environment? One approach currently under intensive investigation in a number of research laboratories involves a synergistic interaction of new technologies which will allow system clock rates above 250 MHz, as well as rapid design and turnaround cycles both at the system and at the component level. It may be possible to develop such a capability by exploiting the physical properties of gallium arsenide (GaAs) for the integrated-circuit substrate rather than conventional bulk silicon or silicon-on-sapphire substrates. The potential advantages of GaAs when used to manufacture digital integrated circuits have been understood for many years. However, progress in the development of this technology has been retarded with respect to silicon as a result of several unique materials-related properties of GaAs which make it impossible to exploit fabrication technologies developed for silicon devices. The physical properties of the bulk crystal structure are much more complex than for silicon, and are presently under intensive investigation; equivalent levels of wafer technology and fabrication techniques evolved over a duration of 25 years for silicon are presently undergoing development and are simply not yet as finely tuned for gallium arsenide.

Counterbalancing these obvious disadvantages, however, are several major strengths of gallium arsenide integrated circuit technology. First, the mobility of the electrons in GaAs at a given voltage gradient is six to eight times greater than in silicon, which in turn results in faster electron transit across the channels of N-channel field-effect transistors (FETs) fabricated on GaAs substrates. Gate propagation delays can thus be more than three times faster than for similar structures in silicon, or gate power dissipation can be decreased for a given gate speed, or a combination of these two may be achieved (Figure 26.1). The effects of the high electron mobility of this material have already been demonstrated in functional integrated circuits at the MSI and LSI levels [27,28].

Gallium arsenide SSI and MSI integrated circuits in pilot-line, preproduction quantities (i.e., not laboratory-fabricated single integrated circuits) have been packaged and then installed on logic boards, and operated at gate propagation delays and clock rates unachievable with commercially available silicon subnanosecond emitter-coupled logic (ECL) integrated circuits. Figure 26.3 presents the performance of small multicomponent test structures operated over a range of system clock rates from 1 to 2.5 GHz. These data appear to be the first demonstration of the performance of multiple interconnected GaAs digital devices operating in a representative circuit-board environment. On-chip gate delays of 80 to 120 ps, and off-chip rise times of 250 to 350 ps, have been measured, with good system noise margins and waveform conformations, even though optimum packaging and interconnect technology were not employed in these early studies. The integrated circuits were designed with the most critical structural dimension, the FET gate length, set at 1 μm, a level of dimensional control which is quite feasible in gallium arsenide (but still very difficult in silicon) because of the simple structure of gallium arsenide FET transistors and the small number of requisite fabrication (mask) steps. In addition, there is now definite reason to believe that the advantages to be achieved from shrinkage of device geometries in silicon integrated circuits will apply at least as well, and perhaps even more, to gallium arsenide, including the onset of ballistic electron transport across the GaAs transistors at much larger dimensions than for silicon devices [29]. It appears feasible to reduce FET transistor gate lengths (which, in turn, determine transistor speeds and power dissipation for a given set of device substrate characteristics) at least to 0.7 μm, thereby decreasing on-chip gate propa-

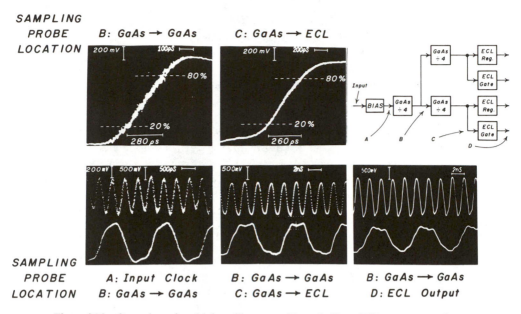

SAMPLING
PROBE
LOCATION

Figure 26.3 Operation of multiple gallium arsenide and silicon ECL components interconnected as depicted in diagram. Off-chip risetimes are consistently less than 300 ps; system clock rate is 2.07 GHz.

gation delays to the neighborhood of 50 to 80 ps, with the gain of additional performance benefits as well [29].

In addition, the energy bandgap characteristics of gallium arsenide and gallium aluminum arsenide allow the fabrication of certain types of structures which are not feasible in silicon at all. For example, "heterojunction" bipolar transistors comprised of a sandwich of aluminum gallium arsenide and gallium arsenide can be fabricated in several forms [28]; the f_τ of one such form of these heterojunction bipolar devices should by the mid-1980s support the design of gates with propagation delays of 10 to 20 ps at 300°K for logic gates of approximately equal size to those fabricated in silicon [30]. Conversely, a heterojunction transistor comprised of layers of GaAs and AlGaAs, but exploiting field-effect transistor structures, referred to as the high electron mobility transistor (HEMT), has also been fabricated and reported extensively (e.g., [31]). The *effective* electron mobility in the channel rises to a level at least 7 to 10 times greater than can be achieved for conventional GaAs FET transistors. These HEMT structures have already demonstrated gate propagation delays for transistors with 1-μm gate length in the range 30 to 40 ps at room temperature and as low as 12.8 ps at 77°K, as well as higher transistor gains and higher transconductances than available from the conventional GaAs FET transistors; in addition, HEMT-based GaAs integrated circuits are very likely to achieve VLSI density levels late in this decade.

26.7 PERFORMANCE COMPARISONS OF COMMUNICATIONS NODE FABRICATED IN SILICON AND GₐAₛ

A comparison of present and expected gallium arsenide transistor and gate performance characteristics, with similar values from high-speed silicon ECL structures, may be enlightening. For example, assuming a silicon-based subnanosecond ECL custom LSI design, estimates were carried out of the probable performance to be expected from a computer communications network node element with eight input lines and eight output lines. If this node element were fabricated using present state-of-the-art silicon ECL technology, a structure containing approximately 500 to 700 gates would be required and could be fabricated. The communications node would be able to transmit data from any one or more of the eight input ports to any one or more of the eight output ports at rates of up to 500 megabits/s per line, with a total on-chip power dissipation in the range of 4 to 8 W. An extrapolation to technology improvements likely to occur by 1985 to 1987 indicated that advances in lithography techniques would allow this silicon ECL structure to be fabricated with roughly the same number of gates, but exhibiting half the total power dissipation, or twice the throughput (i.e., to a rate of 1 gigabit/s per line), or some combination of decreased power dissipation and enhanced throughput.

The same communications node structure was then designed assuming currently available gallium arsenide integrated circuit device technology, using either of two possible designs to achieve the same network node performance. The results from the design exhibiting the lowest gate count are presented here. For single-component gate counts of approximately 500 gates, throughput data rates of 1.5 gigabits/s per line are feasible with 1982 technology at a power dissipation of 1.6 W (i.e., three times the performance of the 1982 silicon implementation for a power dissipation only 20 to 40% as high); in a second and more optimistic implementation of this same node, the throughput rates with present GaAs technology would be 3 gigabits/s per line at a power dissipation of 0.3 W. The performance achievable in a gallium arsenide implementation in 1985–1987 was also estimated, assuming transistors with 0.5- to 0.7-μm channel lengths; throughput rates would then increase to 4 gigabits/s per line at chip power dissipations of 0.2 W.

A second estimate was then made for a gallium arsenide implementation, also for the 1985–1987 time frame, but assuming the availability of HEMT transistor structures with channel lengths of 0.9 μm. Under this set of assumptions, and based on recently reported data, it may be feasible to achieve a 500-gate node structure which can be operated over the following range of data throughput and power dissipation levels: 18 gigabits/s per line at 0.2 W per chip; 14 gigabits/s per line at 0.058 W per chip; or 10 gigabits/s per line at 0.015 W per chip. Other workers have reported somewhat more conservative results, which extrapolate to throughput levels of 8 gigabits/s per line at 0.048 W per chip. These performance increases appear to be achievable in gallium arsenide with a reasonable input of research and development effort over the next 5 years.

26.8 GATE/CELL ARRAYS FABRICATED ON GaAs SUBSTRATES

The speed and power characteristics described above underscore several of the advantages of gallium arsenide technology whenever very short gate delays or fast system clock rates are a necessity. However, as alluded to earlier, an additional constraint faced by the practitioners of extremely high-speed signal processor design is the small production runs generally undertaken, whether or not the processor incorporates "off the shelf," custom, or semicustom integrated circuits. Recognizing this problem, several laboratories have begun the development of configurable gate arrays and/or configurable cell arrays based on gallium arsenide substrates. Mirroring recent results for first-generation silicon ECL cell arrays, GaAs gate/cell arrays in the 1000 to 4000 equivalent gate size are feasible [28], while even larger gate/cell arrays may be feasible. Trade-offs between the achievement of maximum packing density for custom circuits, and the rapid design cycle possible for the layout and fabrication of semicustom designs on a gate or cell array, favor the latter approach if high density is not required or appears cost-ineffective. Conversely, of course, the large percentage of the integrated circuit real estate devoted to routing channels removes the gate/cell arrays from the realm of VLSI densities for the foreseeable future. In fact, controversy has arisen regarding the usefulness of GaAs cell arrays, the primary negative argument being that the additional speed performance of the GaAs transistors will be dissipated by the long interconnects typical of these types of arrays.

26.9 CHARACTERISTICS OF SIGNAL PROCESSING ALGORITHMS WHICH ALLOW THEIR IMPLEMENTATION ON GaAs GATE/CELL ARRAYS

With GaAs gate/cell arrays effectively excluded from the attainment of VLSI densities for the present, is it reasonable to suggest the use of these structures as serious competitors to custom VLSI, whose major advantage is the ability to encompass portions of, or an entire, processor mainframe on a single integrated circuit? Supporting evidence must be presented that a semicustom device containing only 2000 to 10,000 gates will be acceptable or useful in the high-performance signal processing environment. The response to this issue has several ramifications. First, the characteristics of signal processing algorithms are quite different from those of data processing techniques, for the following reasons. Signal processing operations are generally performed on extremely long vectors of one, two, or more dimensions. The individual elements of these vectors are usually of relatively low precision, usually represented by fixed-precision operands 8 to 12 bits in length, occasionally as complex fixed-precision values in forms such as Real[12] + Imag[12], or even less frequently, as block-floating-point operands or conventional low-precision floating-point operands [21]. Full-precision floating-point computation, with its attendant hardware complexity, is rarely required.

Signal processing algorithms developed over the past three decades, and often employed in a variety of combinations, include autocorrelation, cross correlation, convolution, fast Fourier transformation, multiplication of a vector by a scalar, formation of dot and cross products, matrix operations, binning, and chirp-Z transformation. All of these algorithms are characterized (1) by the regularity of the primitive arithmetic operations (i.e., multiplications, additions, and subtractions, with which they are implemented); (2) by a negligible amount of decision branching; and (3) by the large ratio of computational steps to loop steps during their execution. Hence a large number of global operations are implemented in a very similar manner. In addition, the use of two's-complement fixed-precision arithmetic reduces addition and subtraction to a single operation. Generalized fixed-precision multiplication can in certain instances be implemented by combinations of binary shift operations, which are even easier to mechanize [3,32]. The basic signal processing operations are frequently chained together, and the internal structures of the resulting hardware embodiments are also very regular. As a result, signal processing algorithms tend to exhibit a "flow-through" behavior, with partially processed operands moving from one step to the next sequential step with almost no requirement for decision branching, looping, or operand feedforward or feedback.

When these low computational precision, low-complexity algorithmic structures are mapped onto integrated circuit substrates, the various subunits of the algorithm can often be positioned adjacent to one another, allowing very short on-chip propagation path lengths. The flow-through characteristics described above also allow the incorporation of pipeline registers at multiple sites throughout the hardware implementation, thereby doubling, tripling, or quadrupling the speed of the processor for a given system clock rate with a very low overhead (gate count increases of only 10 to 20% are not uncommon). Although the signal processing algorithms are in many cases inherently "parallelizable" [3,33], the hardware overhead is always considerably higher for a given throughput increase than is achieved by pipelining the algorithm.

It is also feasible for many signal processing problems to exploit so-called "bit slicing" of their structures; that is, an entire processor can be partitioned vertically so that one integrated circuit of only moderate gate count performs calculations on all least significant portions of data operands, while an adjacent, interconnected, and often identical component (or components) performs the computations on the most significant portions of the same operands. This technique is somewhat better suited to algorithms employing fixed-precision or block-floating-point arithmetic than to high-resolution floating-point data processing problems.

The foregoing comments only partially address the disparity between (1) large configurable gate or cell arrays in the complexity range of 1000 to 10,000 equivalent gates; and (2) custom VLSI components (albeit operating at much slower clock rates) in the range 10,000 to 50,000 gates. While large custom-design silicon VLSI components clearly will be very useful in the majority of signal processing tasks, recent studies conducted in this laboratory indicate that very powerful high-speed signal processors can be fabricated with configurable gate/cell arrays of only modest complexity. In these studies, a variety of classical signal processing elements, including arithmetic units, input/output ports, and intelligent memory management subprocessors, have been designed to the SSI/MSI macrocell level, and have then been subjected to partitioning onto hypothetical gate/cell arrays varying in complexity from 1000 to 16,000 equivalent gates, and with signal input/output pin counts from 100 to 200.

The typical signal processor substructures described above, including communications nodes, memory management structures, and arithmetic units, can be partitioned into gate/cell arrays containing not less than 2000 to 4000 gates; input/output signal pinout requirements range from approximately 125 to 180 pins. As the cell array complexity rises above 4000 equivalent gates, partitioning problems in some cases become very much simplified, and also in some cases the minimum required number of input/output pins decreases by 15 to 20%. At 8000- to 10,000-gate integration levels, low-precision floating-point arithmetic primitive operations, as well as fixed-precision complex arithmetic structures, can be accommodated. At this complexity level, almost all processor structures examined to the present can be mapped onto a single part type. Thus it should be possible to fabricate high-throughput signal processors using the gallium arsenide devices of only modest gate count which will be available in the near and intermediate term.

26.10 NEAR-TERM TECHNICAL FEASIBILITY OF GaAs-BASED PROCESSORS

Architectural considerations aside, it will also become technically feasible in the near and intermediate term to assemble and operate gallium arsenide processors, or mixed gallium arsenide/silicon processors, at clock rates of 100 MHz and above. First, more than 30 separate custom-designed GaAs integrated-circuit part types have been fabricated and tested at small-scale and medium-scale integrated-circuit density levels. Wafer yields have become sufficiently high that these MSI-level devices are now being included in brassboard versions of processors intended for eventual manufacturing and field installation [25]. Custom-designed large-scale integrated circuits consisting of more than 1000 equivalent gates have already been successfully tested, although device yields are still unacceptably low. However, both the yields and the device densities on these custom designs appear to be improving in a manner similar to the advances in silicon-LSI-device densities and yields during the middle and late 1970s.

Notwithstanding the skepticism regarding the practicality of GaAs gate/cell arrays mentioned earlier, progress is becoming apparent for these devices as well.

Gate and cell arrays using both I²L bipolar [28] and field-effect transistor technology (Figure 26.4) are currently in layout and will undergo testing in 1984. Cell arrays are inherently more attractive for high-performance devices, particularly on gallium arsenide substrates, since any gate structure may be located wherever desired on the integrated circuit to provide minimum spacing between a signal source and its intended destinations (Figure 26.4). Finally, such gate/cell arrays can also benefit from the exploitation of high-speed bipolar or HEMT structures to maximize transistor fan-out capacity and integrated circuit throughput.

This combination of technologies thus provides for the design of very high-clock-rate, high-throughput processors relying on "custom" GaAs integrated circuits rather than silicon VLSI, while preserving the option to perform these design and development projects for very small production runs.

BUFFERED FET LOGIC CONFIGURABLE CELL ARRAY
FABRICATED IN GALLIUM ARSENIDE

(4000 Equivalent gates, 200 psec Propagation Delays)

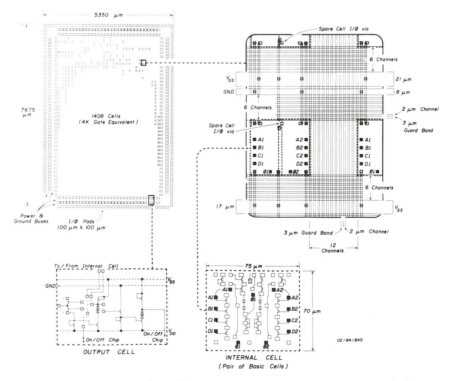

Figure 26.4 Structure of a gallium arsenide configurable cell array presently in development. Note arrangement of cell islands and routing channels between them, and the ability to tailor each cell to one of a variety of gate types.

26.11 COMPUTER AIDED DESIGN TOOLS REQUIRED TO SUPPORT THE LAYOUT OF GₐAₛ -BASED SIGNAL PROCESSORS

The tools and supporting capabilities necessary to achieve these high computational throughputs in the next generation of gallium arsenide processors must be identified and provided. First, the availability of suitable test equipment, as well as personnel experienced in high-clock-rate system design and checkout, are mandatory, as is the ability to design and fabricate appropriate integrated circuit encapsulation and multilayer printed-circuit boards. Because of the fast wavefronts generated by GaAs devices, interconnects between components on these circuit boards must behave as transmission lines. A review of conventional printed wiring board layout software packages has revealed that these programs do not actually perform the layout task in a manner consonant with transmission-line design rules, and must be modified to exploit properly the stripline and microstrip features inherent in multilayer printed-circuit boards [34,35].

Of even greater importance than the physical facilities for the design and fabrication of transmission-line logic boards and integrated-circuit encapsulation, the availability of a computer-aided-design (CAD) capability optimized for the layout of digital systems operating at microwave clock frequencies (L-band and S-band) is mandatory. Such a computer-aided-design package must, of course, support a hierarchical, top-down approach to system design, which begins at the coarsest level with large conceptual blocks, with increasing amounts of detail inserted for the lowest-level logical structures; this is the approach that is gaining support across the entire CAD industry. However, CAD to support the design of very high-clock-rate processors must also possess an elegant capability for bottom-up design in those portions of the processor architecture where the components must be taxed to their technological limits to perform a subtask properly. These critical high-technology portions must be hand designed with great care, and integrated to a sufficiently high level of functionality that their interfaces to the remainder of the system no longer represent a technological challenge. These high-risk functional blocks may then be employed as predesigned "macros" without concern that their technology-sensitive "cores" will exhibit the minor inefficiences typical of top-down design, nor subject the more conventional portions of the system to unnecessary design risk.

This hierarchical approach, also incorporating bottom-up design capabilities, is strongly recommended for the high system clock rates, short signal rise times, and wide bandwidths proposed for gallium arsenide signal processors. It is becoming increasingly difficult to subdivide such a system into individual regions of design responsibility carried out by different engineers, with integration of these separate blocks only when the processor is finally assembled. Logical partitioning, interchip communications protocols, chip carrier packaging, and board design and assembly all interact with one another at very high frequencies in a manner not observed in processors fabricated with conventional integrated circuits.

Computer aided design tools can more easily support an integrated, hierarchical design philosophy if the development of "custom" chips for specific portions of the system exploits the concepts of gate arrays, cell arrays, or so-called "standard cells." A gate/cell array is conceptually similar to a circuit board, with preestablished routing or interconnect channels between the "components," or logical building blocks (Figure 26.4). Just as integrated circuits are plugged into the sockets of a logic board, macros are assigned to cell regions of a gate/cell array. Hence, if the basic CAD package skeletal structure and database are designed from the outset to support both logic board and gate/cell array designs, the same software modules can be used for both structures in a reentrant fashion.

The high-frequency CAD package under development in the Mayo laboratory is divided into (1) an on-line, real-time design graphics input system based on a 1024×1280 resolution color raster graphics terminal with its own internal processor, and (2) a much larger portion of the CAD package which resides on a VAX host computer. The VAX-hosted software supports the partitioning into physical hardware of a logical design established on the graphics terminal, as well as timing verification and simulation, noise and crosstalk analysis, power budget calculations,

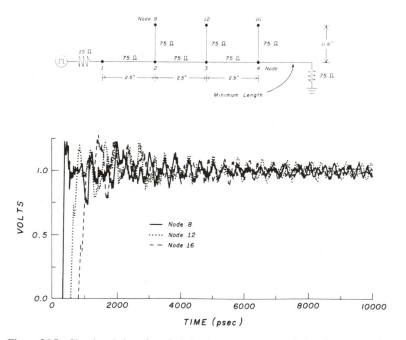

Figure 26.5 Simulated time-domain behavior of 75-Ω transmission line connecting one signal source to three destinations. The rising-edge duration of the step signal is 100 ps. Three lines are signals seen by the integrated circuits inside their packages (nodes 8, 12, 16). Note serious "ringing" of the signals, which does not decrease to acceptable levels for at least 4 ns. This interconnect would yield unacceptable performance on a GaAs circuit board, and would require redesign.

placement and transmission-line routing on printed wiring boards, and macro placement and signal routing analysis for gate/cell arrays. This software establishes and verifies interconnects and signal loading by direct time-domain simulation of the line behavior, with the results therefrom used to refine the layout of the transmission lines on the multilayer printed wiring boards (Figure 26.5). Output formats are prepared by the CAD package which are compatible with a computer-controlled plotter for the creation of logical schematics, board layouts, logic board and cell array placement maps, and output files for the preparation of artwork for the multilayer boards.

Translation routines generate hardware description language representations of a design, required by some component manufacturers to convert logical designs into integrated circuit masks, or control instructions for an electron-beam direct-write-on-wafer machine to scribe the interconnect lines onto gate/cell arrays. This same software package also supports an on-line interactive feature which allows a hardware technician to verify the correct operation of a new processor through the direct presentation of pertinent test data from the design database originally used to fabricate the system.

ACKNOWLEDGMENTS

The authors wish to thank Dr. F. S. Lee, formerly of Rockwell International, for the design of the GaAs components used in these studies; A. Firstenberg, Rockwell International, and Dr. R. C. Eden, Gigabit Logic, Inc., and Dr. F. G. Prendergast, Mayo Foundation, for technical discussions and encouragement; and Mrs. E. Doherty and Mr. S. Richardson, for preparation of text and figures.

This work was sponsored in part by Contracts F33615-79-C-1875 from the U.S. Air Force, MDA903-82-C-0175 from the Defense Advanced Research Projects Agency, N00014-81-C-2661 from the U.S. Navy, and a research grant from the Fannie E. Rippel Foundation.

REFERENCES

[1] B. K. Gilbert, "New Computer Technologies and Their Potential for Expanded Vistas in Biomedicine. The 26th Annual Bowditch Lecture," *Physiologist*, 25(1):2–18 (1982).

[2] R. A. Brooks and G. DiChiro, "Principles of Computer-Assisted Tomography (CAT) in Radiographic and Radioisotopic Imaging," *Phys. Med. Biol.*, 21:689–732 (Sept. 1976).

[3] B. K. Gilbert, S. K. Kenue, R. A. Robb, A. Chu, A. H. Lent, and E. E. Swartzlander, Jr., "Rapid Execution of Fan Beam Image Reconstruction Algorithms Using Efficient Computational Techniques and Special-Purpose Processors," *IEEE Trans. Biomed. Eng.*, *BME-28*(2):98–115 (Feb. 1981).

[4] J. F. Greenleaf and R. C. Bahn, "Clinical Imaging with Transmissive Ultrasonic Computerized Tomography," *IEEE Trans. Biomed. Eng.*, *BME-28*(2):177–185 (Feb. 1981).

[5] J. F. Greenleaf, "Computerized Transmission Tomography," *Methods Exp. Phys.*, *19*:563–589 (1981).

[6] J. F. Greenleaf, S. A. Johnson, and A. H. Lent, "Measurement of Spatial Distribution of Refractive Index in Tissues by Ultrasonic Computer-Assisted Tomography," *Ultrasound Med. Biol.*, *3*:327–339 (1978).

[7] S. A. Johnson, T. H. Yoon, and J. W. Ra, "Inverse Scattering Solutions of Scalar Helmholtz Wave Equation by a Multiple Source Moment Method," *Electron. Lett.*, *19*(4):130–132 (Feb. 17, 1983).

[8] R. Langridge, T. E. Ferrin, I. D. Kuntz, and M. L. Connolly, "Real-Time Color Graphics in Studies of Molecular Interactions," *Science*, *211*(4483):661–666 (Feb. 13, 1981).

[9] G. R. Marshall, C. D. Barry, H. E. Bosshard, R. A. Dammkoehler, and D. A. Dunn, "The Conformational Parameter in Drug Design: The Active Analog Approach," American Chemical Society, ACS Symp. Ser. 112, American Chemical Society, Washington, D.C., 1979, pp. 205–225.

[10] D. C. Weaver, C. D. Barry, M. L. McDaniel, G. R. Marshall, and P. E. Lacy, "Molecular Requirements for Recognition at a Glucoreceptor for Insulin Release," *Mol. Pharmacol.*, *16*:361–368 (1979).

[11] G. W. Beeler, Jr., and H. Reuter, "Reconstruction of the Action Potential of Ventricular Myocardial Fibres," *J. Physiol. (Lond.)*, *268*:177–210 (1977).

[12] M. Bercovier and M. Engelman, "A Finite Element for the Numerical Solution of Viscous Incompressible Flows," *J. Comp. Phys.*, *30*:181–201 (1979).

[13] M. S. Engleman, S. E. Moskowitz, and J. B. Borman, "Computer Simulation: A Diagnostic Method in Comparative Studies of Valve Prosthesis," *J. Cardiovasc. Surg.*, *79*(3):402–412 (Mar. 1980).

[14] Y. C. Pao, G. K. Nagendra, R. Padiyar, and E. L. Ritman, "Derivation of Myocardial Fiber Stiffness Equation Based on Theory of Laminated Composite," *J. Biomech. Eng.*, *102*:252–257 (Aug. 1980).

[15] H. J. Siegel, L. J. Siegel, F. C. Kemmerer, P. T. Mueller, H. E. Smalley, and S. D. Smith, "PASM: A Partitionable SIMD/MIMD System for Image Processing and Pattern Recognition," *IEEE Trans. Comput.*, *C-30*(12):934–947 (Dec. 1981).

[16] L. J. Siegel, H. J. Siegel, and A. E. Feather, "Parallel Processor Approaches to Image Correlation," *IEEE Trans. Comput.*, *C-31*(3):208–218 (Mar. 1982).

[17] D. Katsuki, E. S. Elsom, W. F. Mann, E. S. Roberts, J. G. Robinson, F. S. Skowronski, and E. W. Wolf, "Pluribus—An Operational Fault-Tolerant Multiprocessor," *Proc. IEEE*, *66*(10):1146–1159 (Oct. 1978).

[18] L. D. Wittie, "Communication Structures for Large Networks of Microcomputers," *IEEE Trans. Comput.*, *C-30*(4):264–273 (Apr. 1981).

[19] A. Jagodnik, Raytheon Corporation, private communication, Jan. 1983.

[20] N. Lincoln, "Technology and Design Tradeoffs in the Creation of a Modern Supercomputer," *IEEE Trans. Comput.*, *C-31*(5):349–362 (May 1982).

[21] S-1 Project Staff, *The S-1 Project*, Vols. II and III, Lawrence Livermore Laboratory Rep. UCID 18619, 1979.

[22] B. K. Gilbert, A. Chu, D. E. Atkins, E. E. Swartzlander, Jr., and E. L. Ritman, "Ultra High-Speed Transaxial Image Reconstruction of the Heart, Lungs and Circulation via

Numerical Approximation Methods and Optimized Processor Architecture," *Comput. Biomed. Res.*, *12*:17–38 (1979).

[23] E. E. Swartzlander and B. K. Gilbert, "Supersystems: Technology and Architecture," *IEEE Trans. Comput.*, *C-31*(5):399–409 (May 1982).

[24] B. K. Gilbert, T. M. Kinter, and L. M. Krueger, "Advances in Processor Architecture, Device Technology, and Computer-Aided Design for Biomedical Image Processing," in K. Preston and L. Uhr, eds., *Multicomputers and Image Processing: Algorithms and Programs*, Academic Press, New York, 1982, pp. 385–407.

[25] Rockwell International/Mayo Foundation/U.S. Navy Internal Report, Aug. 1982.

[26] W. W. Latten, J. A. Bayliss, D. L. Budde, J. R. Rattner, and W. S. Richardson, "A Methodology for VLSI Chip Design," *LAMBDA* (now VLSI Design), *2*(2):34–45 (Apr.–June 1981).

[27] D. Kimell, "A 320 Gate GaAs Logic Gate Array," *1982 GaAs Integrated Circuit Symp. Tech. Dig.*, IEEE 82CH1764-0, pp. 17–20.

[28] J. Yuan, "GaAs Bipolar Gate Array Technology," *1982 Integrated Circuit Symp. Tech. Dig.*, IEEE 82CH1764-0, pp. 100–103.

[29] S. Swierkowski, Lawrence Livermore Laboratories, private communication, Mar. 8, 1983 (unpublished data).

[30] P. Asbeck, Rockwell International, private communication, Feb. 21, 1983 (unpublished data).

[31] P. Tung, D. Delagebeaudeuf, P. Delescluse, M. Laviron, I. Chaplart, and N. Linh, "High-Speed Low Power Planar Enhancement Mode Two Dimensional Electron Gas FET," *1982 GaAs Integrated Circuit Symp. Tech. Dig.*, IEEE 82CH1764-0, pp. 10–12.

[32] S. K. Kenue, "High-Speed convolving kernels for CT having triangular spectra and/or binary values," *IEEE Trans. Nuclear Sci.*, *NS-26*(2), Part 2:2693–2696 (Apr. 1979).

[33] Special Issue on Parallel Processors and Processing, *IEEE Trans. Comput.*, *C-26*(2):97–169 (Feb. 1977).

[34] H. Howe, *Stripline Circuit Design*, Artech House Press, Dedham, Mass., 1982.

[35] T. C. Edwards, *Foundations for Microstrip Circuit Design*, WiRey, New York, 1981.

Index